Aerospace Propulsion

Aerospace Propulsion

Dennis G. Shepherd

Sibley School of Mechanical Engineering

Cornell University

American Elsevier Publishing Company, Inc.

New York · London · Amsterdam

AMERICAN ELSEVIER PUBLISHING COMPANY, INC.
52 Vanderbilt Avenue, New York, N.Y. 10017

ELSEVIER PUBLISHING COMPANY
335 Jan Van Galenstraat, P.O. Box 211
Amsterdam, The Netherlands

International Standard Book Number 0-444-00126-3

Library of Congress Card Number 71-190302

Manufactured in the United States of America

CONTENTS

Preface

This text grew out of a course in aerospace propulsion given at Cornell University in recent years to seniors and first-year graduate students. It presupposes the usual undergraduate training in mechanics, thermodynamics and fluid dynamics, plus some knowledge of gas dynamics to the extent of simple normal and two-dimensional shocks. The elementary notions of vectors are introduced to a limited extent. Concepts which are commonly less familiar are reviewed briefly for specific purposes in question. For example, in Chapter 10 there is a review of the relationships governing a charged particle in an electromagnetic field, that is, plasma behavior.

The emphasis is on performance over the whole field of aerospace propulsion rather than on any detailed exposition of components and machines. The object is to be able to understand and assess the place of various propulsive units for given missions and to gain understanding of the advantages and disadvantages of various general types of engine.

Thus the treatment is wide, covering the field from turbojets to ion rockets, and there is more attention to mission analysis than is perhaps usual, with the idea that it is necessary to understand the requirements of space vehicles before any proper study of engines can be made.

The text applies previous study of the engineering sciences to an area of technical interest to many and has been found suitable as a way of utilizing and reinforcing basic concepts in a context of interest in a contemporary technological area. Although the space propulsion field is in its infancy, the text deals mainly in principles of operation and generalized performance so that the material should not readily become obsolescent in spite of the rapid advances in technology. Engineers in general should be able to use the text to answer some questions relating to development of propulsive units.

Most of the chapters have problems appended, not many in number but generally fairly searching in character. Drill problems are out of place in a study of this nature and where necessary can be supplied by instructors from examples.

The material has been collected and prepared over a period of time for use in the course as given. Much thought was given to the question of units and the desirability of using the International System (SI). However, the SI units are not yet used to any degree in the United States and until concurrence is reached on the advisability of their introduction generally in engineering, then their use here would only be confusing. Again the idea of giving units in both FPS and SI values was considered, but it became apparent that this would be very cumbersome and also for consistency would have required many diagrams to be duplicated. Hence the foot-pound-second system has been used, differentiating between pound force and pound mass by the use of the dimensional constant $g_c = 32.174$ $(\text{lb}_m/\text{lb}_f)$ (ft/sec^2). Experience has shown that students need repeated emphasis on usage of force, mass and weight and the use of g_c provides an explicit check of dimensional homogeneity.

The author is indebted to many, both in general and in particular. For a text of this nature, much is due to workers and previous writers in the field. Specific individual contributions have been acknowledged where they occur and publications of a general nature which the author has found useful in his own studies are listed at the end of the text. The actual writing was begun during a leave of absence at the Technische Hogeschool, Delft, Netherlands, and I am grateful to the institution for their hospitality, in particular to Professor R. W. Stuart Mitchell of the Gas Turbine Laboratory. My thanks are due to Mrs. Ruth Brockway for her skill in transliterating my modern version of Linear B, and to Mrs. Deanna Bakko for her efficient transcription of the illustrations.

Ithaca, N.Y.
January 1972

D. G. Shepherd

CHAPTER 1

Introduction

1.1 The Field of Propulsion

The twentieth century has been characterized by man's effort to attain mobility in the air and now, in the second half, into space. A considerable degree of success has been achieved and in fact the accomplishment has been singular compared historically with the relatively slow progress in propulsion on land and sea. While the wheel cannot be called a means of propulsion, it was of course a major step forward in mobility and when it became liberated from human or animal power and was harnessed to mechanical energy, then propulsion in the engineering sense began. Nevertheless speeds and vehicle sizes and weights increased incrementally. On water, the time lapse between the first steamship of Robert Fulton and today's behemoth tankers has been about 150 years and while the difference in size is great, that in speed is relatively small. In general the speed of mechanically propelled vessels is only incrementally larger than those which are wind-driven. (Although lately there are indications that the development of the "Hovercraft" and hydroplane types of vessel may signal a considerable step forward.) There is a much greater differential in size from the few hundred tons of sailing ships to the quarter of a million tons of oil tankers. On land, the railroad locomotive may be dated from about 1800, with 1825 marking the year of the first public railroad, inaugurated with Stephenson's steam locomotive "Locomotion." Less than 75 years elapsed before the walking speed of the "Locomotion" had evolved into the 100 mph of express trains, a unit order of magnitude difference. Ironically railroad speed then suffered a decline and only now, three-quarters of a century later, are we trying to exceed 100 mph average speed for train travel. On the road, automobile travel in any real sense began with the advent of the internal combustion engine but in almost a century, in spite of its ubiquity, the automotive vehicle has changed relatively little from the initial concept and performance. (Both on land and sea of course there are records of speeds many times those of ordinary travel, but whatever their uses in development, such speeds remain special events of a sporting or record-breaking character only).

In the air, on the other hand, the progress in heavier-than-air craft has been startling and has taken place by quantum jumps rather than incrementally. Only 66 years elapsed between the Wright brothers' flight and the first landing of man on the moon. One was a journey of minutes over a few hundred yards at a speed measured in terms of miles per hour, the other a journey of more than half a million miles over a continuous period of several days at speeds several magnitudes greater than that of the first powered flight. This is not to elevate the ability of aeronautical and astronautical engineers above that of marine, railroad and automotive engineers, because the nature of the terrain is a very severe restraint on surface vehicles. Also, economic and social factors have been more favorable for the rapid development of space travel. The technological environment is advancing at an exponential rate and any new device has an ever-increasing scientific and engineering background on which to draw for its development, thus attaining maturity much more quickly than its predecessors.

Nevertheless, the progress of aerospace propulsion has been remarkable and continues to be a major influence in the ever-changing world picture as we experience it today. We may rather arbitrarily note certain of the quantum levels that illuminate aerospace propulsion history. Following the original Wright brothers' flight in 1903, there was the Blériot flight across the English Channel in 1909, the trans-Atlantic flight of Alcock and Brown in 1919, the 1931 Schneider cup race won with a new speed record of 340 mph, the introduction of the turbojet engine operationally in 1944, the German V-2 rocket in 1944, supersonic flight in the 1950's, the Sputnik I satellite in 1957, and finally the first moon landing by men in 1969. Whereas only recently, say 1960, the technology was predominantly that of aeronautics, it is now aeronautics and astronautics or aerospace engineering. Although there is a clear difference in propulsion, since one requires an atmosphere and the other does not, there is intimate intermingling of the theory and practice of both.

The field is so new that much of the technology is tentative. It is not certain what propulsive means is the optimum for particular missions or exactly how such missions might be accomplished. In the past, some new power sources and propulsive vehicles have had a way of showing promise for a short period, attaining considerable vogue and being given a great deal of attention and support before fading away, due either to being quickly superseded by a later development or being halted by a technical barrier. Thus, the airship has all but disappeared in spite of much early development, and the same is true of the flying boat. Engines having unsteady flow have not prospered, in spite of apparent simplicities and theoretical advantages, and neither has the free-piston engine. So with the current propulsive devices and power sources, we may expect that some which look promising today may be almost forgotten within ten years. This is particularly true of rocket engines for use in field-free space, where the possibilities seem limitless, but the practice meager. However, we shall try to categorize many of the existing and suggested possibilities in the following section, pretending neither to exhaustiveness nor prescience.

1.2 Categories of Engine

Figure 1.1 presents the various types of propulsion engine in diagrammatic form, showing the interrelationships where they exist. No claim is made for any particular rationality in this classification. It is neither a family tree showing chronological derivation nor a logical design chart of function. It seeks simply to provide a classification by certain obvious physical bases of operation, with subdivisions into types which have achieved sufficient distinctive features to be differentiated, sometimes by application, sometimes by energy source, sometimes simply by a particular process characteristic of the device.

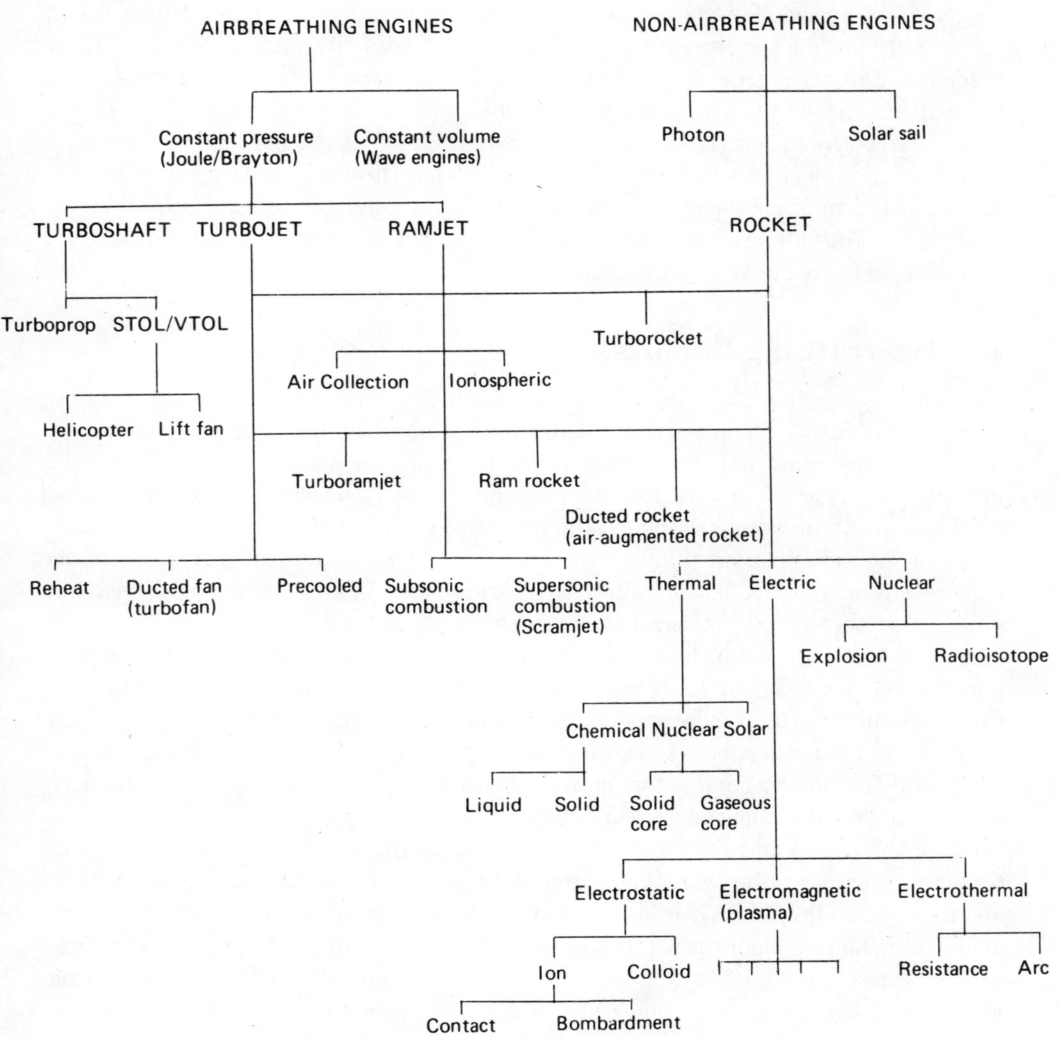

Fig. 1.1 Types of propulsive engines.

The major division is into airbreathing and nonairbreathing engines. The latter are sometimes generically termed "rockets," but if we take a common dictionary definition of rocket, there is an implication of oxidation of a liquid or solid propellant, thus implying a chemical reaction. In the categorization here the term rocket is used very widely but is restricted to the situation in which mass contained wholly in the vehicle is ejected rearwards for propulsion.

The fundamental principle of all aerospace propulsion is Newton's laws, namely that force is proportional to rate of change of momentum, and that action and reaction are equal and opposite. Thus the object is to eject rearwards a stream of material, the reaction to the force providing this change of momentum moving the vehicle forward. The stream of material is known as a "jet," hence turbojet and jet propulsion. For a given rate of flow of material, gaseous, liquid or solid, a higher velocity produces a greater reaction force or *thrust.* However, the real criterion is momentum rather than velocity and we shall discuss in detail this aspect of the propulsive performance. It should also be recognized that the propeller operates by exactly the same basic principle, but in this case the propulsive fluid is an unconfined stream of air separate from that which provides the working substance for the engine. Thus all propulsion systems are reaction machines, although this is often popularly attached only to jet engines.

1.3 Airbreathing Engines

Airbreathing engines are divided into constant-pressure combustion, steady-flow engines and into constant-volume, intermittent-combustion engines. The former are by far the most important and account for all operational types today. The intermittent type of engine has its proponents, as it does not necessarily need mechanical compression and hence the means to obtain it, thus allowing the possibility of high combustion temperatures, low weight and simplicity. A very simple type was used to considerable effect in World War I in the German V-I weapon, a pilotless "throwaway" aircraft launched from the coast of continental Europe to carry a bomb to the southern part of England. It was admirably suited to this purpose, as reliability or repeatable performance were not of first importance. In this particular form it consisted of spring-operated intake ports, a combustion chamber and a nozzle, shown diagrammatically in Fig. 1.2. Starting with a mixture of air and fuel in the combustor, ignition causes a sudden rise of pressure at more or less constant volume, the pressure closing the intake ports. Expansion then occurs through the nozzle and the gases are ejected rearwards for thrust. The discharge of the gases leaves a sub-atmospheric pressure behind the intake valves, which then open and mix air with fuel injected simultaneously. Ignition then starts the cycle over again. The frequency is comparatively low and depends on the relative volumes, as there is a resonance effect. The major difficulty is in sufficient life for the vibrating valves, which must be very light and quick-acting, but have great resistance to vibration fatigue. So far, no solution appears available, although from time to time, the idea is reformulated.

Other types of unsteady-flow engines are possible, based on the rapid closing of a port and consequent interruption of flow which sets up a compression wave that travels through the fluid and thereby increases the pressure level before combustion. Although such engines have considerable appeal for high performance on the theoretical level, so far no practical designs have been developed to the point of operational use.

Turning to the steady-flow type of engine on which all our current usage is built, these are based on the Joule or Brayton cycle. Air is taken in and compressed,

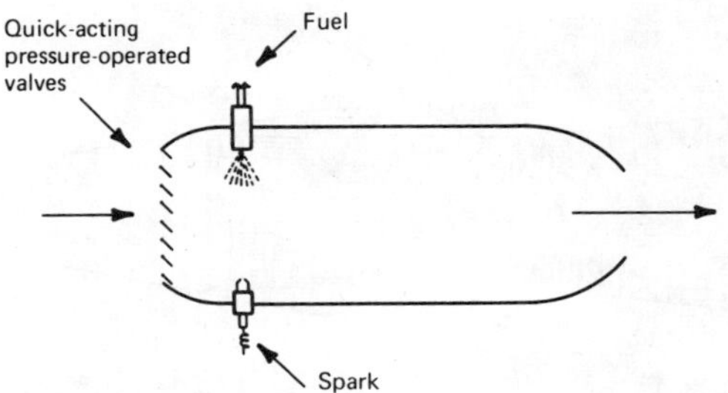

Fig. 1.2 Pulse-jet engine, diagrammatic.

fuel is injected and burned, and then the hot gases are expanded down to their original pressure level, the thermal energy being converted to directed kinetic energy to give thrust. In the turboengines (turboshaft, turbojet, turbofan), a large part of the compression is done mechanically by a turbocompressor, usually of the axial-flow type with several multibladed stages, although occasionally a radial-flow or centrifugal type is used. Some of the compression is obtained by ram action and this becomes very important at the higher speeds and is discussed in some detail in Chapter 3. After compression, which may be anywhere between 5 and 30 atmospheres, the temperature of the air is raised to some 2000°F by the combustion of a liquid hydrocarbon fuel. Then the gases pass through a turbine of one or two stages, the turbine being directly coupled to the compressor. This first expansion takes place only to the level where the compressor work is supplied, following which there is a second expansion, in another turbine if the engine is a *turboshaft* type or directly down to atmospheric level in a nozzle for the straight *turbojet*. In the case of the turboshaft there is still some residual jet energy as the shaft turbine expansion is never complete down to atmospheric level as this would require too many turbine stages. Both the turbojet and turboshaft engines may have two separate compressors each driven by its own turbine, known as a *twin-spool* arrangement. In this way, more flexibility of compressor performance is obtained at the higher pressure ratios. Also, the turboshaft engine may have the shaft output on a compressor-turbine shaft or may have a separate turbine. Some of these arrangements are shown in Fig. 1.3(a) and (b) for a turbojet and in Fig. 1.4(a), (b) and (c) for a turboprop, which is the use of a turboshaft engine to drive a propeller. Other varieties of the turboshaft engine are used for vertical or short takeoff and life (V/STOL) vehicles, of which the most prominent example at present is the *helicopter*. Helicopters are limited in forward speed and the V/STOL idea can be attained by a more or less normal type of winged aircraft which has a fan to provide direct lift at takeoff. In some designs, the fan has a separate engine or it may use the main engines with fan air being used for lift or thrust via a tiltable duct. In this way, high-speed aircraft can be made to have a very short takeoff (e.g., the British Hawker "Harrier").

The basic turbojet can be given improved performance by several means. The gases at turbine discharge are at a pressure above atmospheric and still contain a large fraction of oxygen, as the turbine stresses restrict the maximum temperature possible and the air-fuel ratio is still three or four times the stoichiometric value.

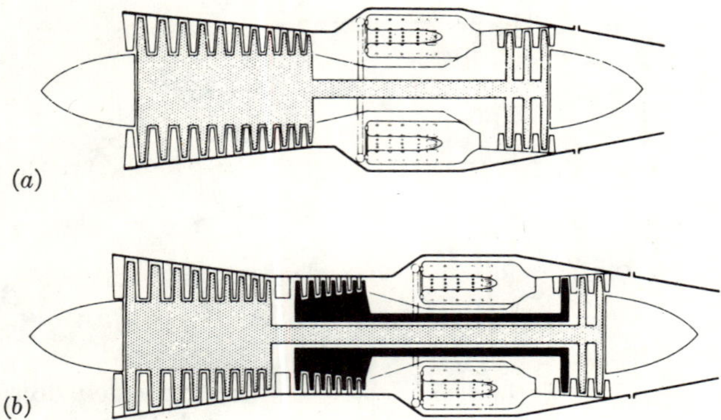

Fig. 1.3 (*a*) Turbojet engine—single shaft (courtesy Pratt & Whitney Aircraft).
(*b*) Turbojet engine—twin spool (courtesy Pratt & Whitney Aircraft).

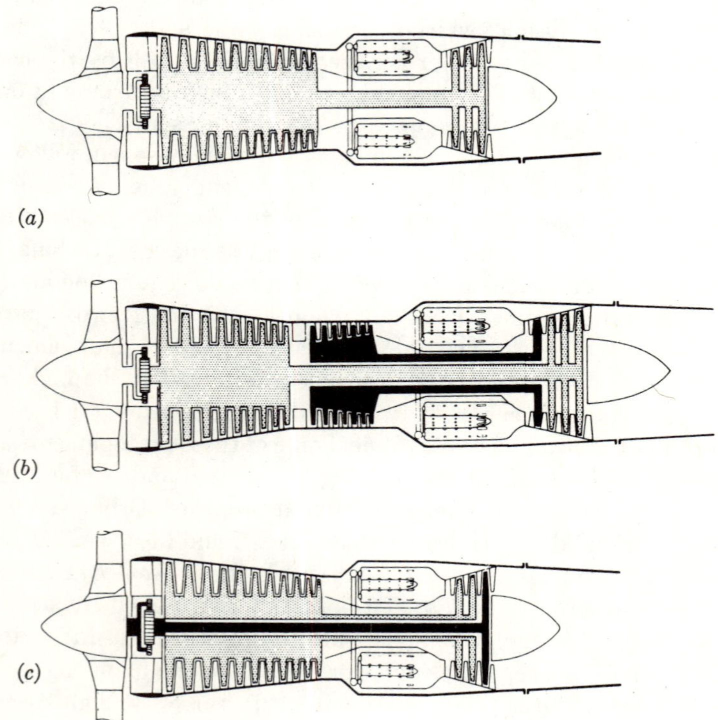

Fig. 1.4 (*a*) Turboprop engine—single shaft, direct drive (courtesy Pratt & Whitney Aircraft).
(*b*) Turboprop engine—twin spool, direct drive (courtesy Pratt & Whitney Aircraft).
(*c*) Turboprop engine—single shaft, separate propeller drive (courtesy Pratt & Whitney Aircraft).

Thus it is possible to burn more fuel in the gases before final expansion and in this case the temperature can be very high as no moving parts are involved. In this way, the exhaust can be given a higher velocity and hence the thrust is increased. This is known as *afterburning* or sometimes as *reheat,* and a diagrammatic afterburning turbojet is shown in Fig. 1.5.

It is also found advantageous for many aircraft and missions to add a second turbine and to use the shaft energy extracted by it to drive another fan that surrounds the main engine. This fan is analogous to a propeller but consists of smaller diameter, multibladed rotors of two or three stages enclosed in a duct and it provides a greater quantity of air at a lower velocity. This type of jet engine is known as a ducted fan or *turbofan,* and the fan may be at the front or the rear of the turboengine and its discharge may be mixed with the main jet or remain separate. Examples are shown in Fig. 1.6(*a*), (*b*) and (*c*). The turbofan is almost ubiquitous in present-day, large, high-speed subsonic transport aircraft.

Finally there is the *precooled* turbojet, in which liquid hydrogen is used as fuel and its cryogenic property is used to cool the incoming air. The lower the initial temperature of air before compression, the less work is required for a given final pressure and thus better performance is obtained by reducing compressor work.

The remaining branch of the airbreathing family is the *ramjet.* The basic cycle is the same as for the turboengines but the compression is obtained wholly from the stagnation or ram effect of forward speed. Thus no mechanical compressor or turbine is needed and so the combustion temperature can be high. Because the overall pressure ratio must be sufficiently high for good performance, the ramjet is unsatisfactory at low speeds and, of course, requires boosting to a minimum speed before it is self-driving. In its simple form, intake air is slowed to a low subsonic Mach number before combustion (Fig. 1.7), but an alternative form is available with supersonic combustion. This is known as the "scramjet." Because of its high-speed qualities, the ramjet in conjunction with a turbojet for the lower speeds appears as a viable engine for low supersonic flight speeds. It is known as the *turboramjet* (Fig. 1.8).

Two exotic forms of ramjet have been conceived but are not yet operational. The *air collection* unit is really an extension of the precooled turbojet, in which hydrogen as fuel is used to cool the intake air. In this case, the cooling is continued to the point where liquid air is formed and burned with the hydrogen in a rocket-type combustor in an environment beyond that suitable for an airbreathing engine. The process may be carried a stage further by separating the liquid oxygen (lox) from the nitrogen and storing it for a subsequent stage of the mission.

The other type of ramjet is known as the *ionospheric* or *recombination* ramjet and is intended for very high altitudes, above about 50 miles, where the air is dissociated into oxygen and nitrogen radicals (the ionosphere). A large intake scoop collects the "air," slows it down with consequent rise of pressure and passes it to a chamber where given time at the increased pressure, the radicals recombine, releasing energy which appears as increased temperature. Expansion takes place in a nozzle and thrust is produced. In this example, the energy is not added from an external source but is transformed from existing dissociation energy in the air to a useful form as enthalpy available for propulsion.

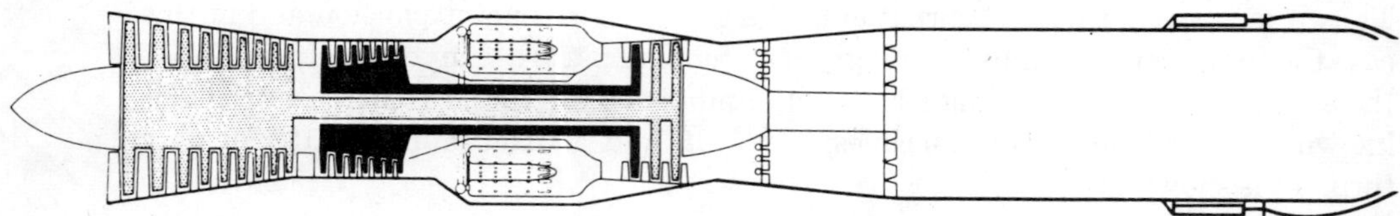

Fig. 1.5 Turbojet with afterburner (courtesy Pratt & Whitney Aircraft).

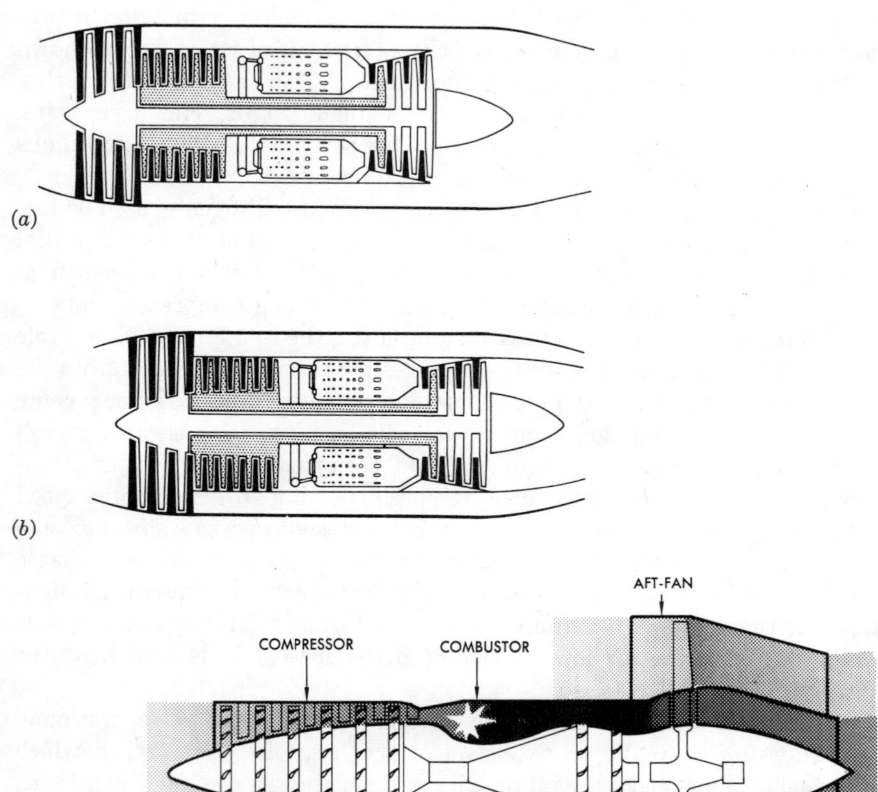

Fig. 1.6 (*a*) Turbofan engine with forward fan, mixed exhaust (courtesy Pratt & Whitney Aircraft).
(*b*) Turbofan engine with forward fan, unmixed exhaust (courtesy Pratt & Whitney Aircraft).
(*c*) Turbofan engine with rear fan, unmixed exhaust (courtesy of General Electric Co.).

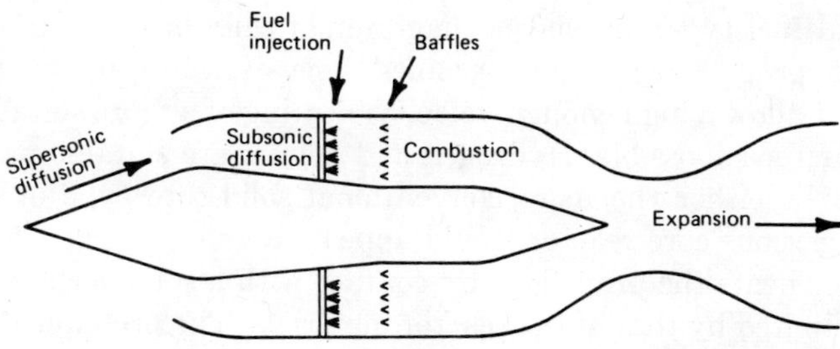

Fig. 1.7 Ramjet engine.

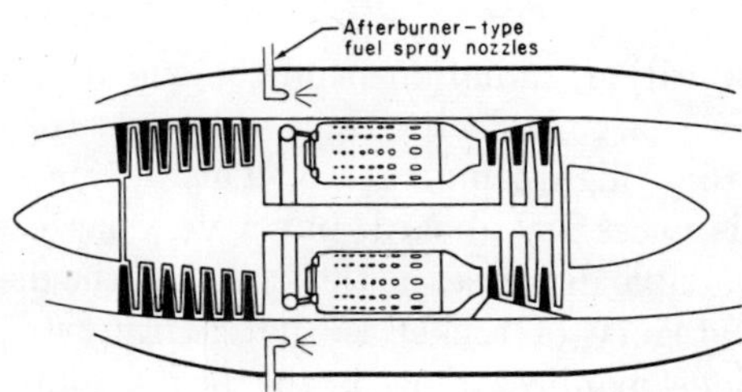

Fig. 1.8 Turboramjet engine (courtesy Pratt & Whitney Aircraft).

1.4 Non-airbreathing Engines

Although the practical engines realizable today in this category are rockets, we should mention two possibilities that may be developed someday. Both are based on the fact that radiation carries momentum and therefore can be used to provide a force. In the *photon* engine, as its name implies, photons of radiation, rather than gaseous matter or elementary particles, are ejected from the space vehicle. In the *solar sail*, radiation from the sun is used to "push" the vehicle, that is, use is made of the solar pressure. Neither of these means of propulsion is suited to the initial requirements of the space travel of today, but both present a possibility in situations where any great mass of propellant cannot be taken aboard.

Turning to the large class of *rockets*, we first have the major division into thermal, electric, and nuclear types. In *thermal* rockets, the jet is provided by heating a propellant under pressure, and there are many ways to do this. The method used to date is by chemical means, that is, by the combustion of a liquid fuel with a liquid oxidizer or by the combustion of a solid propellant containing both fuel and oxidizer materials. Familiar examples are the hydrocarbon-lox rockets of launch vehicles and the lox-hydrogen combination of many upper-stage rockets, or the military-missile type of solid rocket such as "Minuteman" and "Polaris." Although not yet used extensively, the hybrid chemical rocket with a liquid oxidizer and a solid fuel offers possibilities of less complication and better control. Chemical rocket engines are basically very simple, as there are no major moving components. Liquid and solid fuel types are shown diagrammatically in Fig. 1.9(*a*) and (*b*)

Where the propellant weight becomes excessive and where the chemical reaction does not allow a high enough velocity, heating of a more suitable propellant by a nuclear reactor is possible. Hydrogen and helium are suitable propellants and the reactor can be either the more conventional solid-core type or, when better developed, the gaseous-core reactor would appear to give better performance. The solid-core reactor heats the propellant by contact with a solid surface and thus the temperature is limited by that of the reactor materials. The principle of the gas-core reactor is that the energy is transferred to the propellant with both propellant and nuclear fuel in the gaseous state. The reactor must be so designed that the containing walls are protected from the very hot gas mixture resulting from the nuclear reaction and that fuel and propellant are separated before the latter discharges to the nozzle. There is still a way to go, however, before a nuclear reactor powers a space vehicle.

Finally in this classification, radiation from the sun may be used to heat the propellant. This requires a large collector that will focus the solar rays to allow heating the propellant to a high temperature. This scheme is attractive as no massive heating source is necessary aboard, but a very large collector area is required and the energy available decreases as the square of the distance from the sun.

The second main category of rocket is the *electric* type, so called because electrical energy in various manifestations is the prime source of the propulsive power. A power plant is needed to supply the electrical energy and at the present time it would appear that a nuclear reactor in conjunction with a liquid-metal Rankine-cycle or gaseous Brayton-cycle engine with an electric generator will be the first such plant. The energy can be used in various ways. One of the most promising for long-distance journeys, as to the outer planets, is the *electrostatic* engine, in which ionized particles are accelerated to very high velocities by the potential between electrodes. If atoms of a single element are used as propellant,

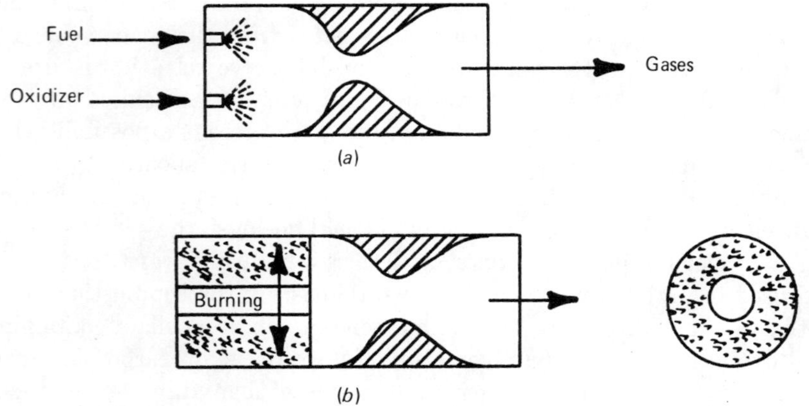

Fig. 1.9 (*a*) Rocket—liquid propellant.
(*b*) Rocket—solid propellant.

such as cesium or mercury, the propulsion unit is known as an *ion* engine, while, although the principle is the same, if very small particles consisting of aggregations of molecules of a substance are used, then it is called a *colloid* engine. Experimental ion engines have been operated successfully in space, but only in an extremely small size and without a long-term power plant for energy supply. Another method of accelerating particles in the form of a plasma (a mixture overall neutral but with a fraction of the propellant ionized) is by *electromagnetic* means, which offers possible advantages in size and in electrical supply. Such means are akin to the MHD (magnetohydrodynamic) generator which has been developed to some extent as a land-based plant, although the specific types and possible developments are too numerous to list individually. This appears to be an area where the situation is too confused by numbers of ideas and lack of adequate simple theory to be able to discern any real trends at this time.

The third subdivision of electric rocket is the *electrothermal* type, in which the power is used directly to heat the propellant. Two ways have been developed, both for small units but both reasonably successful. One is by simple *resistance heating* (the "resistojet"), the other by *arc heating*. Electrothermal rockets might be classified as a fourth division of thermal rockets, but it seems appropriate to include them in this category as they too require a complete power plant for operation.

Finally as a division of rocket engines, we have the true *nuclear* type. By this is meant that the actual propulsion is manifested by a nuclear reaction, as distinct from a nuclear reactor being used as a thermal energy source for direct propellant heating or for a power cycle. Under this heading, we have the *explosion* type in which it is proposed to lift extremely large loads against gravity by the action of a succession of atomic explosions. Although this appears to be technically feasible, it would also appear to be extraordinarily cumbersome and potentially dangerous. Another nuclear type of operation, however, is available and functions by the directed discharge of particles from a radioisotope. This is simple but is limited to extremely small thrusts, as the specific power is very low. However, there are possible uses where a very low thrust obtained by a simple, lightweight unit can be useful.

There are various hybrid, turboramjet-rocket engines. One is the *turborocket* (Fig. 1.10), which has compressor, turbine, and combustor similar to a turbojet, but the combustor is placed after the turbine, which is driven by the exhaust gases from a rocket. These gases, which are fuel-rich, mix and burn with the compressor air in the combustor, discharging directly to the nozzle for propulsion.

There are also some combinations of ramjet and rocket (Fig. 1.11). The *ducted rocket* consists of a rocket inside a duct with air intake and nozzle. The rocket exhaust acts as an ejector, drawing in air, and when the unit is in motion, air is forced through the duct by ram action. Air and gases mix to form a common exhaust and this combination is sometimes called the *air-augmented rocket*. A variation of this is the *ram rocket* in which there is further combustion of fuel-rich gases from the rocket with the augmenter air in the duct.

This delineates in skeleton form the range of propulsion vehicles in use or considered as having possible use in the future. The rest of this book attempts to analyze the performance characteristics of the main types in this family tree, so

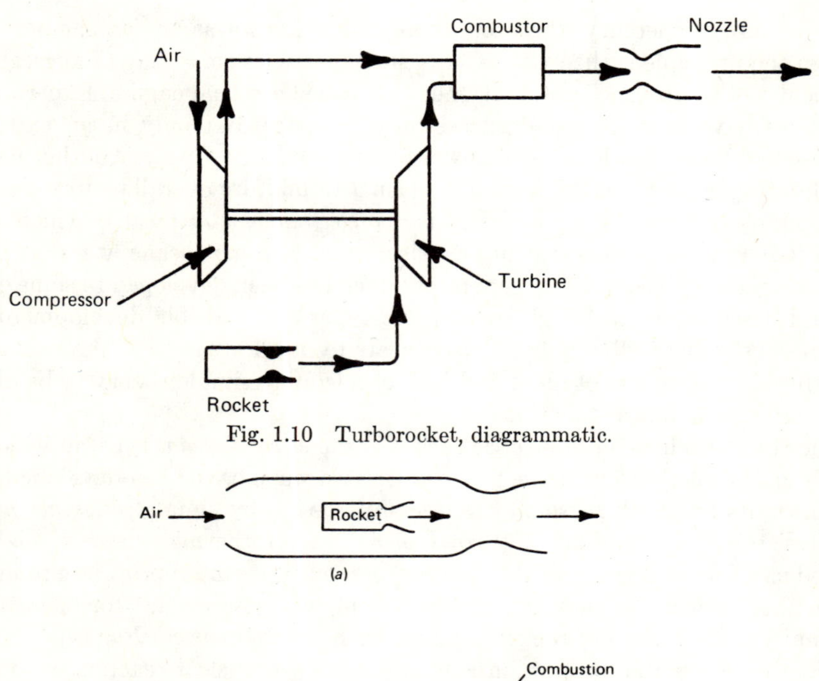

Fig. 1.10 Turborocket, diagrammatic.

Fig. 1.11 (*a*) Ducted rocket or air-augmented rocket, diagrammatic. (*b*) Ram rocket, diagrammatic.

that their place in the propulsion field can be assessed. It is not intended to deal in detail with each of the named types, but it should be possible at the end to be able to appreciate how they might fit in the general pattern. Little space will be given to the actual mechanical design and for this, recourse will have to be made to the abundant literature on the topic, mostly periodicals which often give in detail the design elements of current units, although there are a few books that can be helpful.[1]

1.5 The Atmosphere

The range of ambience of propulsion vehicles is almost infinite—from the surface of the earth to the far reaches of space. For us, the chief interest lies in the *atmosphere* surrounding the earth and in the cosmically small region of the solar system.

Near the surface of the earth we think of the environment as "weather" rather than atmosphere and we know only too well how variable it can be. However, there

[1] For example, G. P. Sutton, "Rocket Propulsion Elements," 3rd ed., Wiley, New York, 1963, and D. K. Huzel and D. H. Huang, "Design of Liquid Propellant Rocket Engines," 2nd ed., NASA SP-125, 1971.

is a general pattern and it is mandatory to standardize some figures for calculating and comparing performance of propulsion vehicles at various altitudes above the earth's surface. The domain of airbreathing vehicles extends at the present time up to almost 100,000 ft, with the highest, regularly used altitude being that for the supersonic transport, about 60–70,000 ft. Since aviation began, there have been various *standard* or *model* atmospheres used, such as the U.S. Standard Atmosphere, the International Civil Aviation Organization (ICAO) Atmosphere, and Air Research and Development Command (ARDC) Standard Atmosphere. There can also be various special models for latitude and season, for example, Tropical, North Atlantic, or Winter. There are only very small differences between the various models for a given region at the lower altitudes, but there are larger variations as altitude increases and in those that extend beyond 100,000 ft or so, there is considerable uncertainty so that figures are revised every few years. This is not surprising, as it is only recently with the advent of rockets as almost conventional pieces of apparatus, that direct observations have been possible. Even so there is still insufficient evidence to establish firm figures for the upper atmosphere and, indeed, there are data to indicate that there is continual variation, both diurnal from solar heat input and of larger period due to sunspot activity. It is interesting to note that considerable information on temperature in the outer atmosphere has been obtained from the decay in orbital period of artificial satellites, whose drag is a function of atmospheric density.[2]

The earth's atmosphere is generally regarded as divided by altitude into regions, although these are vaguely and variously defined. However, there seems to be a common definition of the *troposphere*, which extends up to a height of about 7

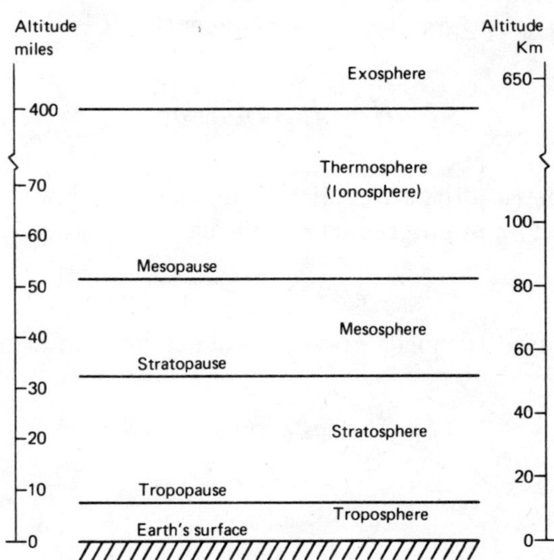

Fig. 1.12 Regions of the atmosphere and space.

[2] J. C. G. Walker and N. W. Spencer, "Temperatures of the Earth's Upper Atmosphere," *Science*, 162, 3861, Dec. 27, 1968).

miles and has specific temperature and pressure variations, terminating in the *tropopause.* Above this is the *stratosphere* and it is here that the definitions start their variation. We shall follow the usage of Craig,[3] which in turn is based on the recommendations of the International Union of Geology and Geodesics (IUGG 1960). In this system, shown diagrammatically in Fig. 1.12, the stratosphere extends from the tropopause to the *stratopause* at about 32 miles, comprising a region in which the temperature is either constant or increasing, hence the stratopause is at a local temperature maximum. Above the stratosphere is the *mesosphere* in which the temperature again decreases. At the upper limit of the mesosphere is the *mesopause* at an altitude of about 50 miles. Up to this point, the composition of the atmosphere is essentially the same as that at sea level, with oxygen and nitrogen predominating and having a molecular weight of 28.97. At this height, however, the molecular weight starts to decrease, with oxygen and nitrogen giving way to lighter constituents such as helium and, finally, atomic hydrogen. This region is known as the *thermosphere.* Also, although below this altitude the gravitational separation of light and heavy constituents is balanced by turbulent mixing, in the succeeding atmosphere a stable convective pattern exists because there is a continuous positive temperature gradient. At this altitude of about 300,000 ft, the pressure is of the order of 2×10^{-5} psia and upwards from this height the low density and absorption of very short ultraviolet radiation results in ionization. This region is sometimes called the *ionosphere.* This extends to about 400 miles, above which is the *exosphere,* in which the mean free path of particles is very great and molecular escape occurs. Even at 90 miles, the mean free path is about 100 ft.

A concept for vertical displacement used by geophysicists is that of *geopotential* altitude, defined as the increase in potential energy of unit mass raised from sea level to altitude Z against gravity. Thus geopotential H is then given by

$$H = \int_0^Z (g\,dZ/g_c) \tag{1.1}$$

where Z is the geometric altitude. Using the simple square law relationship for local acceleration of gravity g at any radius r from the center of the earth, we have

$$g/g_0 = R_0^2/r^2 \tag{1.2}$$

where g_0 is the standard (defined) gravitational acceleration at radius R_0. Thus

$$H = \int_0^Z (g_0/g_c)(R_0^2/r^2)\,dZ$$

and with $r = R_0 + Z$,

$$H = (g_0/g_c)R_0^2 \int_0^Z dZ/(R + Z)^2$$

[3] R. A. Craig, "The Upper Atmosphere," International Geophysics Series, Vol. 8, Academic Press, New York, 1965.

which upon integration yields

$$H = (g_0/g_c)[R_0 Z/(Z + R_0)] \quad \text{ft lbf/lbm} \tag{1.3}$$

With g_0 having the standard value of 32.174 ft/sec² numerically equal to g_c in (lbm)(ft)/(lbf)(sec²), and R_0 = 3959 miles = 2.09×10^7 ft, then

$$H = \frac{2.09 \times 10^7 Z}{Z + 2.09 \times 10^7} \quad \text{ft lbf/lbm} \tag{1.4}$$

The official definition of H is in units of energy/mass but if we disregard the differentiation between lbf and lbm, which is commonly done, then H is a geopotential height in feet, comparable to geometric height Z. The difference between Z and H is

$$\frac{Z - H}{Z} = 1 - \frac{H}{Z} = 1 - \frac{R_0}{Z + R_0} = \frac{Z}{Z + R_0} \tag{1.5}$$

At about 40 miles altitude, this difference is about 1% and at 100 miles, about $2\frac{1}{2}$%. Thus it can be neglected in first-order calculations but obviously not for precise trajectories.

The troposphere is a region from the earth's surface in which the atmospheric variation is defined by a constant *lapse rate* or decrease of temperature, denoted by λ. With a defined standard sea-level condition, the use of the hydrostatic equation together with λ then gives the accompanying pressure gradient. Thus

$$dp = -\rho(g/g_c)\, dZ \tag{1.6}$$

and with $\lambda = -dT/dZ$,

$$dp = \rho(g/g_c)(dT/\lambda) \tag{1.7}$$

Eliminating ρ by the perfect gas equation of state $\rho = p/RT$, then

$$\frac{dp}{p} = \frac{g}{g_c}\frac{1}{\lambda R}\frac{dT}{T} \tag{1.8}$$

Integrating between a datum state where $p = p_1$, and $T = T_1$ to any state p, T gives

$$p/p_1 = (T/T_1)^{g/g_c\lambda R} \tag{1.9}$$

Within the limits of altitude for which the lapse rate is valid, $g = g_c$ numerically and so $p/p_1 = (T_2/T_1)^{1/\lambda R}$. Integrating the lapse rate definition gives $T = T_1 - \lambda Z$, whence

$$p/p_1 = [1 - (\lambda Z/T_1)]^{g/g_c\lambda R} \tag{1.10}$$

Here we shall use the U.S. Standard Atmosphere (1962) as a model, which agrees with the ICAO model up to 20 geopotential kilometers for which the latter extends. For this model, the sea-level conditions are

$$T = 518.69°\text{R} \ (59°\text{F})$$

$$p = 2116.22 \text{ lb/ft}^2 \ (14.696 \text{ lb/ft}^3)$$

$$\rho = 0.076474 \text{ lb/ft}^3$$

The lapse rate is defined as $-0.0065°K$/geopotential meter, equivalent to 3.567°F/1000 ft geometric altitude, and using this allows the pressure at any altitude to be calculated.

The troposphere is assumed to extend to the defined *tropopause* at 11,000 geopotential meters, which translates to 36,152 ft geometrical altitude, better called 36,000 ft, as the accuracy is only in the conversion not in the physical state. At the tropopause, the temperature is 389.97°R (or 390°R = −70°F rounded off), and the pressure is reduced to about 3.3 psia. In the model atmosphere, the temperature then remains constant up to 20,000 geopotential meters or about 65,000 ft. This region of supposedly constant temperature has been a feature of the atmospheric pattern for some time, but as is typical of the state of the art in this area, it has recently been questioned and replaced by the condition of continuously increasing temperature. As an example of the lack of agreed standardization, the stratopause in the U.S. Standard Atmosphere is given at 20,000 m, while in the ARDC Model Atmosphere, it is at 25,000 m. The existence of this isothermal, or quasi-isothermal region, is a good reason for an altitude of about 35,000 ft being a common cruising altitude for long-range transport aircraft, as it has the lowest temperature for a feasible height and it will be seen later that low ambient temperature benefits turboengine performance. The pressure can again be calculated by using the hydrostatic equation together with the equation of state with constant temperature, thus

$$dp = -\rho(g/g_c)\, dZ = -(p/RT)(g/g_c)\, dZ$$

and

$$\int_{p_1}^{p} \frac{dp}{p} = -\frac{g}{g_c}\frac{1}{RT}\int_{Z_1}^{Z} dZ$$

whence with $g = g_c$,

$$p/p_1 = \exp\left[(Z_1 - Z)/RT\right] \tag{1.11}$$

At the stratopause, the pressure has decreased to about 1 psia.

Thence the temperature increases with a constant gradient to about 30°F at a height of 30 miles, the pressure decreasing to just under ½ in water. This is followed by another isothermal region, then a zone of decreasing temperature to about 55 miles, then increasing temperature again until at the limit of the model atmosphere, which is 700,000 m ≈ 400 miles, the kinetic temperature is about 3260°R. It has been found useful in eliminating another variable in atmospheric calculations to define a "molecular" temperature T_M, such that $T_M = T(M_0/M)$, where M_0 is the sealevel value of molecular weight and M is the local value. This is useful because temperature values at this altitude are derived from the velocity of sound or from densities, and thus are dependent on the molecular weight, which is poorly known. Hence the molecular temperature avoids the need of knowing molecular weights and does not have to be changed as knowledge of the composition changes. T and T_M do not differ until an altitude of nearly 60 miles is reached, when the molecular weight starts to decrease. At 400 miles, when $T \approx 3260°R$, T_M has increased to nearly 6000°R and the nominal molecular weight is 16.17. It should also be pointed out that in the high upper atmosphere, the "temperature" of the medium has several meanings—there is an electron temperature, an ion temperature, and a neutral-particle temperature. Thus the conventional meanings as related to a perfect gas are no longer valid. Although nominal temperature is very high, the effect on a body immersed in such an atmosphere is almost negligible, certainly small compared with a similar temperature on the earth. Thus it has been suggested that at these altitudes, the temperature should be considered as zero. The density, that is, the number of particles per unit volume, is so small that convective heating is negligible compared with radiation effects, either from the sun to the body or from the body to space.

For our immediate purposes, we only need to recognize, first, that within the present-day limits of airbreathing engines, the atmosphere is similar to that which we know on the earth's surface, but with prescribed varying temperature and pressure and, second, that within the satellite zone, the atmosphere is extremely thin but is not a vacuum, so that some very small resistance effect is present. It should also be emphasized again that model atmospheres are meant only for standardizing performance, as they do not yet have fixed values over a wide range and quoted values are only the best approximations of mean values known. Some of the foregoing atmospheric properties according to the U.S. Standard Model are summarized in Figs. 1.13 through 1.15, with Fig. 1.13 showing the temperature variation out to 350 miles and Figs. 1.14 and 1.15 showing temperature and pressure to a larger scale out to 120,000 ft.

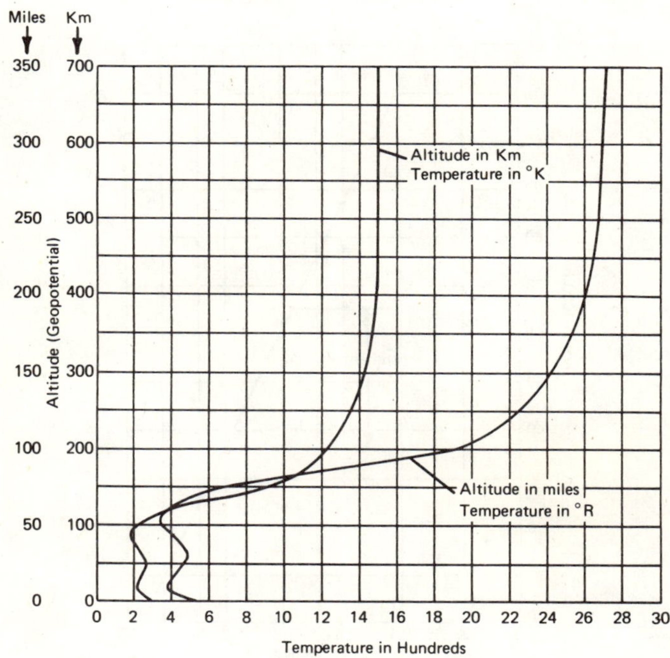

Fig. 1.13 Temperature vs altitude to 350 miles.

The outer atmosphere also has other attributes that may affect propulsion to a serious extent. One is the presence of solid matter as meteorites, which can be disastrous to propulsive vehicles. Considerable attention has been given to this problem because large meteorites can wreck a vehicle and a number of relatively

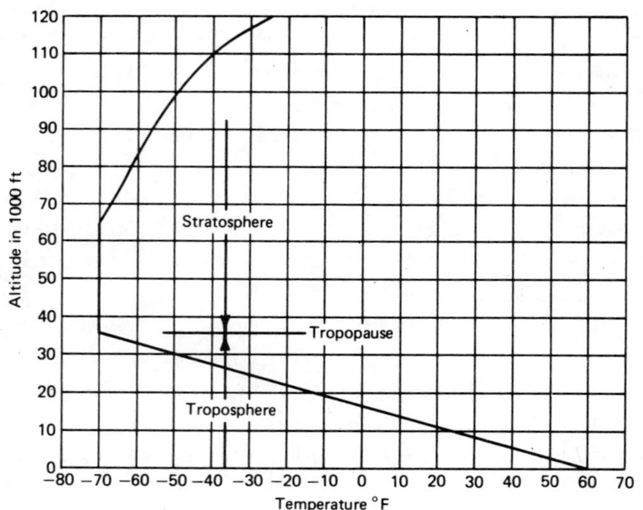

Fig. 1.14 Temperature vs altitude to 120,000 ft.

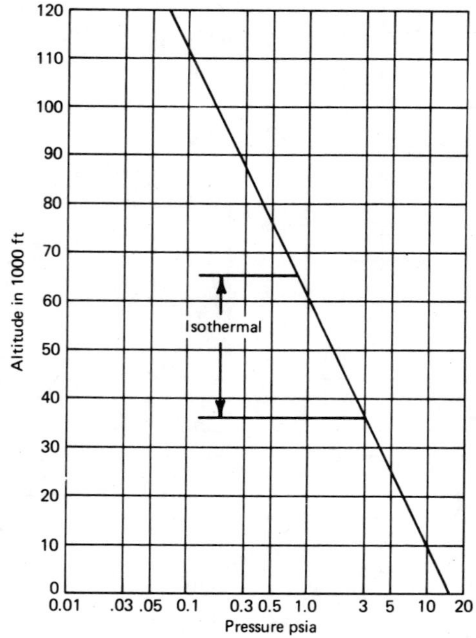

Fig. 1.15 Pressure vs altitude to 120,000 ft.

small collisions can alter the period of a flight lasting months. For example, power plants for interplanetary travel will have to have radiators to dissipate the rejected heat and because these radiators must be open to space to a considerable degree, shielding from meteorites cannot be total. Impact of meteorites can cause penetration of the containing material and lead to loss of the working substance. Data are being accumulated but only probabilities can be formulated. It would appear, however, that early estimates were probably somewhat pessimistic and although the improbable is bound to happen sometime, the picture is somewhat brighter than originally believed.

In addition to corporeal collisions, there arise the hazards of solar radiation in times of sunspots and flare activity and of the radiation from the well-known Van Allen belt. These effects are largely on human beings, although instruments are subject to deleterious effects and so are some materials. Here shielding is possible although the weight necessary may be inhibiting or even crippling.

Problems

1.1 Calculate the atmospheric pressure and temperature at an altitude of 50,000 ft using the U.S. Standard Atmosphere model.

1.2 It is sometimes useful, in drag analysis for example, to have available a simple expression for atmospheric density in the stratosphere. Devise such an expression in the form of a power function or transcendental function of density and altitude over a range of altitude starting at, say, 50,000 ft and going out as far as the expression is likely to be valid within a reasonable limit of accuracy. Give the likely maximum error over different parts of the range.

CHAPTER 2

Propulsion Performance Parameters

2.1 Thrust and Momentum Equations

The basic parameter for aerospace propulsion is the *thrust* force developed by the engine and it is necessary to have a clear idea of the component elements making up the thrust and also the conditions under which particular expressions are valid. The three conservation equations of mass, momentum, and energy are assumed to be known but the momentum equation will be rederived as it is a cardinal relationship in space propulsion and must be valid for a moving vehicle as well as a stationary one.

Newton's second law states that the force exerted on a body is equal to the time rate of change of momentum of the body, the familiar $F = Ma$ concept. However, this applies to a given quantity of matter and for an inertial reference frame. The problem for a space vehicle is one of the continuous flow of matter through a moving control volume, that is, we want to go from a consideration of particle properties to field properties. This can be rather complex, as for a general case with translation and rotation, the centrifugal, Coriolis and tangential accelerations must be included. Here we shall restrict ourselves to simple translational motion with a moving control volume, since the other terms are pertinent only for special maneuvers.

First, the momentum equation will be developed for a fixed control volume and then the moving condition will be taken into account by consideration of the vector sum of velocities. Looking at Fig. 2.1, consider fluid flowing through a fixed control volume or region R having a control surface S. At time $t = 0$, a certain mass of

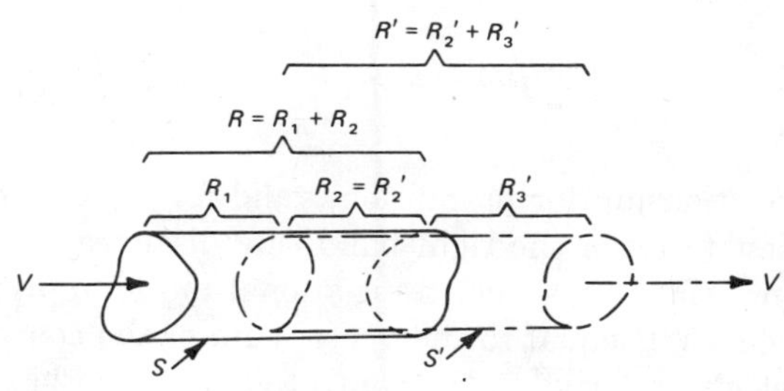

Fig. 2.1 Control volume for momentum analysis

fluid is contained in region R bounded by S. At time $t = \Delta t$, the same mass of fluid occupies region R' bounded by S', the prime indicating later time Δt. Region R can be divided into R_1 and R_2, region R' into R_2' and $R_3' \cdot R_2$ and R_2' are the same region but at different times. Thus at $t = 0$, the fluid occupies region $R = R_1 + R_2$ and at $t = \Delta t$ it occupies $R' = R_2' + R_3'$. Denoting momentum by $\mathbf{M}$, we have:

$$\text{At} \quad t = 0, \qquad \mathbf{M}_R = \mathbf{M}_{R_1} + \mathbf{M}_{R_2}$$

$$\text{At} \quad t = \Delta t, \qquad \mathbf{M}_{R'} = \mathbf{M}_{R'_2} + \mathbf{M}_{R'_3}$$

Hence the change of momentum $\Delta\mathbf{M}_R$ is

$$\Delta\mathbf{M}_R = \mathbf{M}_{R'} - \mathbf{M}_R = (\mathbf{M}_{R'_2} + \mathbf{M}_{R'_3}) - (\mathbf{M}_{R_1} + \mathbf{M}_{R_2})$$

$\mathbf{M}_{R'_3}$ is the momentum that has passed out of R during time Δt and $\mathbf{M}_{R_1}$ is the momentum that has entered during the interval. Rearranging this relationship and dividing through by Δt, we get

$$\frac{\Delta\mathbf{M}_R}{\Delta t} = \frac{\mathbf{M}_{R'_2} - \mathbf{M}_{R_2}}{\Delta t} + \frac{\mathbf{M}_{R'_3} - \mathbf{M}_{R_1}}{\Delta t}$$

In the limit as $\Delta t \to 0$, $R' \to R$, and $S' \to S$ and the time rate of change of momentum of the fluid becomes the sum of the time rate of change of momentum in the region R (or control volume) and the flux of momentum across the control surface S. These terms can then be written as

$$\lim_{\Delta t \to 0} \frac{\Delta\mathbf{M}_R}{\Delta t} = \frac{d\mathbf{M}}{dt} = \sum \mathbf{F}$$

the sum of the forces acting on the fluid within the control volume.

$$\lim_{\Delta t \to 0} (\mathbf{M}_{R'_2} - \mathbf{M}_{R_2})/\Delta t = (\partial/\partial t) \int_R \rho\mathbf{V}\, dR$$

and

$$\lim_{\Delta t \to 0} (\mathbf{M}_{R'_3} - \mathbf{M}_{R_1})/\Delta t = \int_S (\rho\mathbf{V})\mathbf{V}\cdot d\mathbf{S}$$

Hence

$$\sum \mathbf{F} = (\partial/\partial t) \int_R \rho\mathbf{V}\, dR + \int_S (\rho\mathbf{V})\mathbf{V}\cdot d\mathbf{S} \tag{2.1}$$

This is the basic relationship for propulsion, valid for a fixed control volume. For steady flow, the first term on the right-hand side disappears.

For the moving control volume, translational motion only, we have that the absolute fluid velocity $\mathbf{V}$ is equal to the vector sum of the control volume velocity $\mathbf{U}$ and the fluid velocity relative to the control volume $\mathbf{V}_r$, that is,

$$\mathbf{V} = \mathbf{U} + \mathbf{V}_r$$

Substituting for **V** in Eq. 2.1,

$$\sum \mathbf{F} = (\partial/\partial t) \int_R \rho_R(\mathbf{U} + \mathbf{V}_r)\, dR + \int_S \rho_s(\mathbf{U} + \mathbf{V}_r)\mathbf{V}_r \cdot d\mathbf{S}$$

Note that in the momentum flux term, only one **V** has been changed to $\mathbf{U} + \mathbf{V}_r$. The other becomes simply $\mathbf{V}_r$ because it is the velocity in the mass flow expression $\rho\mathbf{V}_r \cdot d\mathbf{S}$. In evaluating mass flow at any point, it is always the velocity with respect to the surrounding surface which is used, in this case $\mathbf{V}_r$.

Expanding the equation and dropping the vector notation, assuming we are concerned only with one-dimensional motion, then

$$\sum F = \underset{(1)}{(\partial/\partial t) \int \rho U\, dR} + \underset{(2)}{(\partial/\partial t) \int \rho V_r\, dR} + \underset{(3)}{\int \rho U V_r\, dS} + \underset{(4)}{\int \rho V_r V_r\, dS}$$

Let us examine these four terms.

Term 1: The vehicle velocity U is independent of the extent of the region so we can expand further to

$$(\partial/\partial t) \int \rho U\, dR = (\partial U/\partial t) \int \rho\, dR + U \int (\partial/\partial t)\rho\, dR$$

and because R is independent of time,

$$U \int (\partial/\partial t)\rho\, dR = U(\partial/\partial t) \int \rho\, dR$$

Thus term 1 becomes

$$(\partial U/\partial t) \int \rho\, dR + U(\partial/\partial t) \int \rho\, dR$$

$\partial U/\partial t$ is the control volume acceleration, $\int \rho\, dR$ the instantaneous mass and $\partial/\partial t \int \rho\, dR$ is the rate of increase of mass in the control volume.

Term 2: This is the rate of change of momentum within the control volume.

Term 3: U is independent of the control surface S, hence

$$\int \rho U V_r\, dS = U \int \rho V_r\, dS$$

and $\int \rho V_r\, dS$ is the flux of mass across the control surface or the mass flow rate.

Term 4: This is the flux of momentum across the control surface.

Adding all these expanded terms we have:

$$\sum F = (\partial U/\partial t) \int \rho\, dR + U(\partial/\partial t) \int \rho\, dR + (\partial/\partial t) \int \rho V_r\, dR + U \int \rho V_r\, dS + \int \rho V_r V_r\, dS$$

and rearranging,

$$\sum F = (\partial U/\partial t) \int \rho \, dR + (\partial/\partial t) \int \rho V_r \, dR + \int \rho V_r V_r \, dS$$
$$+ U \left[(\partial/\partial t) \int \rho \, dR + \int \rho V_r \, dS \right]$$

Now the last terms in parentheses represent the continuity principle, that is the sum of the rate of increase of mass in the region and the flux of mass across the surface and this is equal to zero. Thus

$$\sum F = (\partial U/\partial t) \int \rho \, dR + (\partial/\partial t) \int \rho V_r \, dR + \int \rho V_r V_r \, dS \qquad (2.2)$$

Comparing this with Eq. 2.1 for the fixed control volume, we see that there is an additional term representing the mass of the whole control volume and its acceleration, and that the other two terms are similar but with the fluid velocities relative to the control volume. For constant velocity of the control volume, the first term disappears and hence the expression is the same as for the fixed control volume but with relative fluid velocities. The first term disappears for a fixed control volume and as the relative and absolute fluid velocities are the same in this case, then Eq. 2.2 becomes Eq. 2.1.

Now $\sum F$ is the sum of the forces on the fluid and we must analyze these forces. In general there are pressure and viscous forces acting at the boundary of the region, that is, surface forces, and there are body forces acting on the mass of the region, for example, gravity and possibly electromagnetic forces. It is clearer to discuss these in terms of the propulsion units with which we are concerned and these will be divided into airbreathing (turbojet and ramjet) and non-airbreathing (rockets) units.

2.2 Thrust of Airbreathing Engines

Figure 2.2 shows a duct, schematically representative of turbojets (including turbofans) and ramjets. Performance of engines of this type is invariably given for either stationary or constant velocity conditions, hence the acceleration term disappears. Likewise we are most generally concerned with steady flow, hence the time rate term also disappears. As a further simplication usually justified in practice, the flow is considered uniform at entry and exit, with constant fluid properties. Thus the remaining integral term may be written as $\dot{m}_2 V_2 - \dot{m}_1 V_1$, where $\dot{m}_2$ and $\dot{m}_1$ are the mass flow rates and V_1 and V_2 are the inlet and discharge velocities. Using the more general case of the moving duct, V_1 becomes the vehicle velocity U, as this is the velocity of the air relative to the duct. V_2 will be called V_j, the *jet* velocity relative to the duct.
Hence

$$\sum F = -(1/g_c)(\dot{m}_2 V_j - \dot{m}_1 U) \qquad (2.3)$$

with the dimensional constant g_c introduced for the pound force, pound mass system of units.

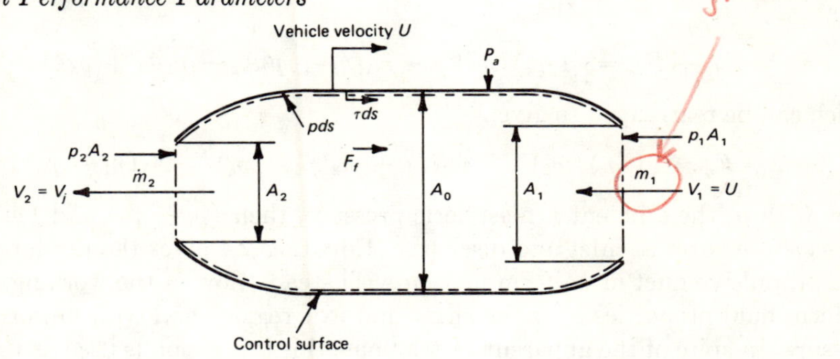

Fig. 2.2 Schematic duct arrangement for airbreathing engines.

Referring to Fig. 2.2, the surface forces on the fluid are those due to absolute pressure p_1 and p_2 at the inlet and discharge, p_1A_1 and p_2A_2, respectively, and pressure and viscous forces $\int p\, ds$ and $\int \tau\, ds$ at the internal interface of fluid and duct. Viscous forces on A_1 and A_2 are discounted, as the velocities are considered normal to the areas and hence no shear forces arise at inlet and discharge. However, internally there are shear forces along with the pressure forces but it is impossible to compute either of these in any real situation. Hence we replace their sum by an unknown force on the fluid F_f. Only level flight is considered so that no body force acts in the horizontal plane. Equation 2.2 with the simplifications and substitutions then becomes

$$F_f + p_aA_2 - p_1A_1 = -(1/g_c)(\dot{m}_2V_j - \dot{m}_1U)$$

The positive direction has been taken toward the right and F_f has been arbitrarily made positive. Then

$$F_f = -(1/g_c)(\dot{m}_2V_j - \dot{m}_1U) - p_2A_2 + p_1A_1$$

Now the internal force on the duct F_{d_i} must be equal and opposite to the internal force on the fluid F_f by Newton's third law, hence

$$F_{d_i} = -F_f = (1/g_c)(\dot{m}_2V_j - \dot{m}_1U) + p_2A_2 - p_1A_1$$

The total force on the duct is made up of this internal force due to passage of the fluid and an external force due to the ambient pressure and, if the duct is in motion, due to shear (drag). The pressure and shear drag can be complex, particularly in supersonic flight and it is difficult to determine what part of the total can be considered as due to the propulsive means and which to the general aerodynamic design. Hence, although the behavior of the propulsive means (the engine or motor) must be considered in relation to its drag-producing qualities, it is customary to debit the engine with only the external duct force due to pressure when stationary. This force is simply the ambient pressure p_a times the difference of the inlet and discharge areas (Fig. 2.3). The net positive external force is then

$$F_{d_e} = p_a(A_0 - A_2) - p_a(A_0 - A_1) = p_aA_1 - p_aA_2 = p_a(A_1 - A_2)$$

assuming that the ambient pressure p_a is uniform. The net force on the duct is then the sum of the internal and external forces, thus

$$F_n = F_{d_i} + F_{d_e} = (1/g_c)(\dot{m}_2 V_j - \dot{m}_1 U) + p_2 A_2 - p_1 A_1 + p_a A_1 - p_a A_2$$

which can be rearranged to give

$$F_n = (1/g_c)(\dot{m}_2 V_j - \dot{m}_1 U) + A_2(p_2 - p_a) - A_1(p_1 - p_a) \tag{2.4}$$

Now with p_a the ambient atmospheric pressure, then $(p_2 - p_a)$ and $(p_1 - p_a)$ are the *gage* pressures at inlet and discharge. Equation 2.4 gives the net force or *thrust* on a propulsive duct in uniform motion with steady flow of the working fluid, with uniform fluid properties over the entry and exit regions and with uniform ambient pressure. In spite of the apparent restrictions, this expression is used as the basis for the majority of performance analyses of airbreathing engines. It will be examined for some particular cases.

2.3 Thrust of Turbojets

For turbojets the difference between $\dot{m}_2$ and $\dot{m}_1$ is due to the fuel addition and this is usually small, about 2%. Furthermore it is very common to "bleed" a quantity of air from the compressor to act as a coolant for bearings, turbine blades and turbine disc, and this again is usually about 1–2% of the total air. Thus for all but a detailed analysis of a particular engine, $\dot{m}_1$ and $\dot{m}_2$ may be considered as equal and hence we write $\dot{m}_1 = \dot{m}_2 = \dot{m}$.

For subsonic flight, $p_1 = p_a$, and so this pressure term disappears. However if the discharge nozzle is convergent only, it is quite possible that it is choked under certain conditions, so that in general the exit-pressure term is retained. If however there is *complete expansion* of the gases down to p_a, then the term becomes zero.

For the stationary (test-bed or takeoff) case, the forward speed U is zero or negligible and hence $\dot{m}_1 U$ disappears.

These three cases may then be summarized as follows, using subscript j for the discharge (jet) station:

(1) Turbojet, incomplete expansion (choked nozzle), constant forward speed,

$$F_n = (\dot{m}/g_c)(V_j - U) + (p_j - p_a)A_j \tag{2.5a}$$

(2) Turbojet, complete expansion, constant forward speed,

$$F_n = (\dot{m}/g_c)(V_j - U) \tag{2.5b}$$

(3) Turbojet, complete expansion, stationary,

$$F_n = (\dot{m}/g_c)V_j \tag{2.5c}$$

Note that although Eq. 2.5a has the additive pressure term, the thrust actually will

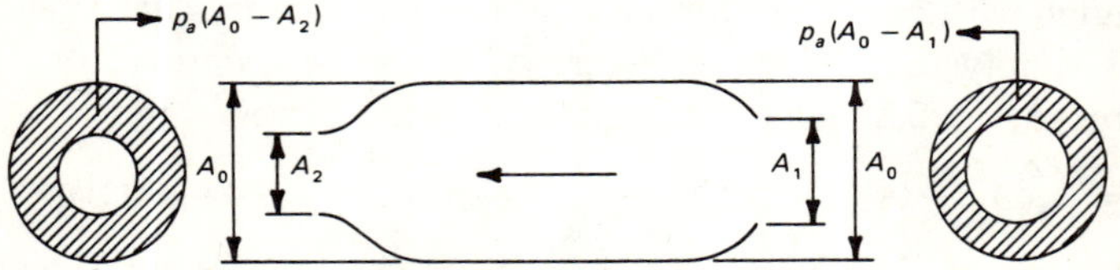

Fig. 2.3 External pressure forces on duct.

be less than for the equivalent fully expanded case as V_j is that for the nozzle throat, which is less than that at exit of the divergent part for full expansion.

We see also that the effect of forward speed is to decrease the thrust for a given mass flow and that the thrust goes to zero when the forward velocity equals the jet velocity. The "rating" of a turbojet in pounds of thrust must then be given at a particular forward speed. This is usually stationary at ground level so that a "20,000 pound thrust" engine usually implies this condition. Another general measure of performance might be the thrust at the aircraft cruising speed at the nominal cruising altitude.

2.4 Thrust of Ramjets

For ramjets, the fuel input is not negligible, being perhaps 5% of the air flow. Also there is no turbine to cool, hence little or no bleed from the compressed air. Because the entry momentum is a substantial fraction of the discharge momentum, that is, the net output is the difference of two large quantities, the additional discharge mass flow due to the fuel is significant.

This can be conveniently handled in terms of the fuel-air ratio f, so that $\dot{m}_2 = \dot{m}_1(1 + f)$. For constant forward speed, incomplete expansion, the thrust of the ramjet is then given by

$$F_n = (\dot{m}/g_c)[(1 + f)V_j - U] + (p_j - p_a)A_j \tag{2.6}$$

The stationary case is meaningless as the whole compression is achieved by the "ram" effect of stagnating the inlet air.

2.5 Thrust of Rockets

It is necessary to return to the basic equation (Eq. 2.2) for rockets, because as all the working fluid is contained within the vehicle itself, there is a continuously changing mass and hence varying momentum of the whole vehicle. The interest is in the flight performance of the vehicle as a whole, that is, the whole structure and not with drag attributed separately as for an airbreathing engine. We shall consider the general case of flight in a gravity field upward from the earth's surface for example, as shown in Fig. 2.4. The line of action of the thrust may be inclined to the direction of the gravity force at angle θ. The positive direction will be considered as the y direction (upward).

Considering Eq. 2.2, then the first term on the right-hand side can be written as $M\, dU/dt$, where M is the instantaneous mass. In the second term, $\int \rho\, dR$ is again the instantaneous mass inside the control volume, consisting of vehicle structure, stored propellants, payload, and so on, and this has no velocity V_r relative to the control volume. For steady flow, there is no time rate of change of velocity of the working substance with respect to the control volume. The second term therefore disappears. The third term is the momentum flux and there is no inflow in a rocket, so this term becomes simply $-\dot{m}V_j/g_c$. Equation 2.2 then becomes

$$\sum F = (M/g_c)(dU/dt) - (\dot{m}V_j/g_c)$$

There is a single pressure force across the unenclosed control surface, p_jA_j, and an unknown internal force on the fluid, F_f. Hence,

$$F_f = (M/g_c)(dU/dt) - (\dot{m}V_j/g_c) - p_jA_j = -F_{d_i}$$

the internal force on the duct. Note that M is the instantaneous mass within the control volume, which in this case is the whole vehicle.

Externally there is a force due to the ambient pressure, p_aA_j (considering the pressure to have a uniform value), a drag force D due to shear, and a body force due to gravity, that is, the weight, $Mg \cos\theta/g_c$. These forces, called F_{d_e}, are directed downwards or in the negative direction. We are considering the control volume as the whole vehicle and thus have accounted for all the forces for equilibrium. Hence $F_{d_e} + F_{d_i} = 0$ and

$$-p_aA_j - D - \frac{Mg\cos\theta}{g_c} - \frac{M}{g_c}\frac{dU}{dt} + \frac{\dot{m}V_j}{g_c} + p_jA_j = 0$$

Thus finally we may write

$$\frac{\dot{m}V_j}{g_c} + (p_j - p_a)A_j = \frac{M}{g_c}\frac{dU}{dt} + D + \frac{Mg\cos\theta}{g_c} \qquad (2.7)$$

It is seen that the left-hand side is the *thrust* of the rocket from the propellant and that the right-hand side is the mass × acceleration term plus the "resistance" forces of drag and gravity. We could have written directly that the net force, that is, thrust minus resistance forces, is equal to mass times acceleration.

2.6 Effective Jet Velocity

It is convenient for some analytical work to have a simple expression for thrust in terms of velocities only, without the pressure term. This implies complete expansion which is not always the case with turbojets and which seldom occurs with rockets. To help with this problem, the concept of the *effective* jet velocity, V_{je} is useful. This is defined as the jet velocity which, if the fluid were fully expanded, would give the same thrust as in the actual case, that is,

$$F_n = (\dot{m}/g_c)(V_j - U) + A_j(p_j - p_a) = (\dot{m}/g_c)(V_{je} - U)$$

Hence

$$V_{je} = V_j + (g_cA_j/\dot{m})(p_j - p_a) \qquad (2.8)$$

2.7 Efficiency of Propulsion Systems

It is usual in discussing or analyzing the performance of energy conversion systems to define efficiencies and to use these quite freely in a quantitative fashion. We do this in connection with the thermodynamic analysis of the component performance in particular types of propulsion engine, but seldom in the analysis of a complete propulsion system. Thus, we use compressor and turbine efficiencies in the analysis of turbojet performance as an engine but we do not use an overall efficiency of the turbojet as we would, for example, in the comparison of overall efficiencies of various types of plant in supplying electric power. This is inherent in the nature of propulsion, which implies a *mission* of a specified character rather than continuous production of a useful power. This is not to say we are not concerned in the efficiency of the operation in the economic sense because indeed we may be vitally interested, but the essence of propulsion is that some particular aspect of the mission is overriding or that several aspects combine to make the particular parameter of efficiency irrelevant. Thus weight is a major criterion in its own right, not only as a measure of cost, and time may be equally important. Hence although we are very much concerned with efficiency in the sense of expenditure of energy as fuel for a particular mission, we can seldom class alternative solutions in a neat packet of comparative efficiencies in the straightforward engineering sense. It is useful, however, to discuss various efficiencies which are important in specifying a propulsion system, as they enable us to compare methods in a qualitative if not precisely quantitative manner.

Figure 2.5 shows diagrammatically the disposition of the energy supplied to a propulsion system. This energy, denoted by E_{in} and given by the sum of its heating value and kinetic energy (if the vehicle is in motion), is supplied to and utilized by some sort of propulsion device. If the latter is a heat engine in the thermodynamic sense, part of this energy is unavailable by the second law of thermodynamics, even if the engine is an ideal, reversible one. So only part of E_{in} can be carried forward and this is denoted by E_{id}. Of this ideally available energy, some will be wasted by the inevitable losses of any real device. The remainder is denoted by E_{eng} and is equivalent to the shaft energy of an engine delivering shaft work. This energy E_{eng} may have to be changed to a form of kinetic energy suitable for propulsion, E_{kin},

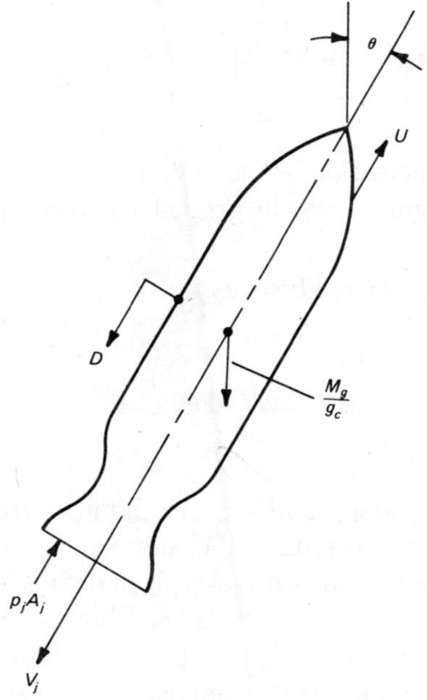

Fig. 2.4 Schematic duct arrangement for rockets.

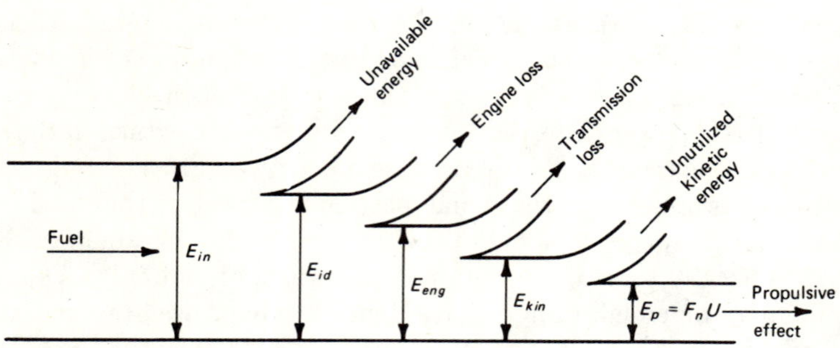

Fig. 2.5 Energy flow in propulsion engines.

and again some will be lost. An example here is a turboshaft engine, in which the shaft energy is given to an aircraft propeller or helicopter rotor. Of this kinetic energy E_k, not all will be used for producing a propulsion effect E_p, as some will be dissipated to the environment.

These energy quantities can be formed into efficiencies as follows:

$$\frac{E_{id}}{E_{in}} = \eta_{cyc} \qquad \text{(ideal cycle efficiency)}$$

$$\frac{E_{eng}}{E_{id}} = \eta_{eng} \qquad \text{(engine or internal efficiency)}$$

$$\frac{E_{eng}}{E_{in}} = \eta_{cyc} \times \eta_{eng} = \eta_{th} \qquad \text{(thermal efficiency)}$$

$$\frac{E_{kin}}{E_{eng}} = \eta_{trans}$$ (transmission efficiency, i.e., due to gearing and fluid dynamic losses in propeller or propulsion nozzle)

$$\frac{E_p}{E_{kin}} = \eta_p \qquad \text{(propulsion efficiency)}$$

$$\frac{E_p}{E_{in}} = \eta_p \eta_{trans} \eta_{eng} \eta_{cyc} = \eta_0 \qquad \text{(overall efficiency)}$$

The first two efficiencies, η_{cyc} and η_{eng}, are familiar from thermodynamics and their product E_{eng}/E_{in} is the thermal efficiency. For propulsive jets, there is no shaft power and the engine output is directly E_{kin}, defined as the increase in kinetic energy of the propellant, $(\dot{m}_2 V_j^2 - \dot{m}_1 U^2)/2_{g_c}$. Although there will usually be a loss of useful energy in the jet pipe and propelling nozzle of an airbreathing engine, this is usually charged to the engine itself, that is, the transmission efficiency η_{trans} is taken as unity or absorbed into η_{eng}.

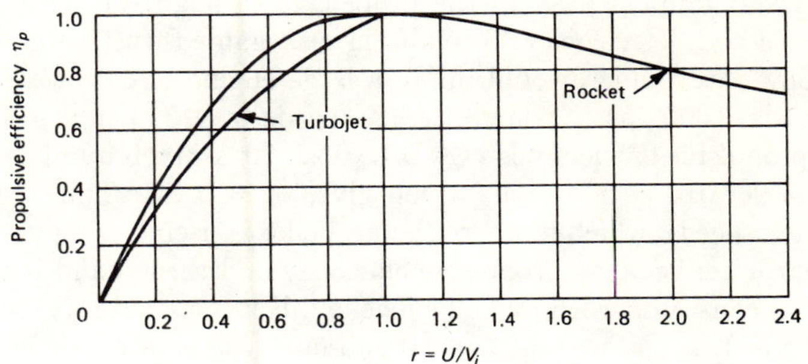

Fig. 2.6 Propulsive efficiency.

2.8 Propulsive Efficiency

The propulsive efficiency η_p requires some analysis. Now an efficiency can be defined as output over input or, alternatively, as output over output plus losses. This latter way provides a clearer understanding in this case, as the loss is simply the kinetic energy corresponding to the absolute velocity of the propellants after discharge from the vehicle, i.e. $V_2^2/2_{g_c}$. This residual kinetic energy is dissipated by viscous effects in the atmosphere. The absolute velocity V_2 is the vector sum of the relative (jet) velocity and the forward speed, or in scalar terms, $V_j - U$. The output propulsive energy E_p is the product of thrust and forward speed, F_nU. For simplicity we shall take the simple expressions for thrust, assuming complete expansion (equivalent to using V_{je}) and constant mass flow for the time being.

For airbreathing engines then,

$$\eta_p = \frac{\dot{m}(V_j - U)U}{\dot{m}(V_j - U)U + \dot{m}(V_j - U)^2/2}$$

$$= \frac{2U}{2U + V_j - U} = \frac{2U}{U + V_j} = \frac{2U/V_j}{1 + U/V_j} = \frac{2r}{1 + r} \tag{2.9}$$

where $r = U/V_j$.

We can also obtain η_p directly from the ratio of useful output over input, the latter being the increase in kinetic energy of the propellant. Thus

$$\eta_p = \frac{\dot{m}(V_j - U)U}{\dot{m}(V_j^2 - U^2)/2} = \frac{2U(V_j - U)}{(V_j - U)(V_j + U)} = \frac{2U}{U + V_j} = \frac{2r}{1 + r} \tag{2.10}$$

Figure 2.6 shows a plot of η_p vs $r = U/V_j$. η_p is obviously zero at standstill and then increases, but at a decreasing rate until it reaches a value of unity at $r = 1$. So it is necessary for the jet velocity to equal the forward speed for 100% propulsive efficiency. However, referring to the expression for thrust, when $V_j = U$ the thrust is zero and hence $\eta_p = 1$ is meaningless. The thrust is greatest at standstill and then decreases linearly with U, so that there has to be a balance between obtaining a useful amount of thrust and using the propulsive energy efficiently. $U/V_j > 1$ has no meaning as the thrust would be "negative."

Turbojets and ramjets pass all the propellant through the power plant and thus must have a high jet velocity V_j to obtain reasonable thrust without requiring a high mass flow rate (which would imply a large engine). A typical value of V_j may be taken as 2,000 fps. At low aircraft speeds, say 300 mph = 440 fps for example, the propulsive efficiency is very low, 0.22. At a Mach number of unity at 35,000 ft, U = 968 fps = 660 mph, which gives $\eta_p = 0.484$. Thus turbojets are inefficient at low speeds, which is where the propeller is useful. A value of $\eta_p \approx 0.5$ is a good compromise between propulsive efficiency and thrust, and it will be seen shortly that it is consistent with overall efficiency. Hence an aircraft cruising speed of $M = 0.85$, which is the present typical subsonic value, is consistent with efficiency. However there is a case for obtaining better propulsion efficiency if possible and this will be investigated later.

Looking at the case of the rocket, the thrust is $\dot{m}V_j/g_c$ and hence

$$\eta_p = \frac{\dot{m}V_jU}{\dot{m}V_jU + \dot{m}(V_j - U)^2/2} = \frac{2V_jU}{V_j^2 + U^2} = \frac{2U/V_j}{1 + U^2/V_j^2} = \frac{2r}{1 + r^2} \qquad (2.11)$$

This is also shown in Fig. 2.6. The efficiency is somewhat higher than that for an airbreathing engine at lower values of U/V_j but also has a value of unity at $r = 1$. In this case however, the thrust does not go to zero, because it is constant at any speed with a value $\dot{m}V_j/g_c$. Hence there is still thrust with forward speed U greater than V_j, so that $U/V_j > 1$. The efficiency declines from unity, as it must do to have a proper meaning, but at a lower rate for $r > 1$ than for $r < 1$. η_p is still about 0.5 for $r \approx 3.7$. Rockets may attain very high speeds and still have a good propulsive efficiency.

2.9 Thermal Efficiency

Considering turbojets and ramjets as engines, their useful output is that at nozzle exit immediately before discharge, that is, thrust + exit loss. The input is the sum of the chemical energy in the fuel available for combustion and its kinetic energy, since it is already traveling at the vehicle speed. The available chemical energy is the heating value of the fuel, ΔH_c, Btu/lb. For simplicity in analysis, the propellant flow rate will be considered as constant through the engine and taken as $\dot{m}_a$, the air flow rate, that is, $\dot{m}_f$ is considered negligible compared with $\dot{m}_a$. Thus

$$\eta_{th} = \frac{\dot{m}_a(V_j - U)U + \frac{1}{2}\dot{m}_a(V_j - U)^2}{\dot{m}_f(g_c\Delta H_c + \frac{1}{2}U^2)}$$

Dividing through by $\dot{m}_a$ and substituting $f = \dot{m}_f/\dot{m}_a$, the fuel-air ratio, then this reduces to

$$\eta_{th} = \frac{V_j^2 - U^2}{f(2g_c\Delta H_c + U^2)} = \frac{1 - U^2/V_j^2}{(fU^2/V_j^2)[(2g_c\Delta H_c/U^2) + 1]}$$

which with $U/V_j = r$ reduces to

$$\eta_{th} = \frac{1 - r^2}{fr^2[(2g_c\Delta H_c/U^2) + 1]} \tag{2.12}$$

Similarly for a rocket, with ΔH_p the heating value of the propellants (fuel + oxidizer),

$$\eta_{th} = \frac{\dot{m}V_jU + \frac{1}{2}\dot{m}(V_j - U)^2}{\dot{m}(g_c\Delta H_p + \frac{1}{2}U^2)}$$

which can be reduced to

$$\eta_{th} = \frac{1 + r^2}{r^2[(2g_c\Delta H_p/U^2) + 1]} \tag{2.13}$$

For a given fuel, fuel-air ratio and forward speed, the thermal efficiency increases continuously as r decreases, which is the opposite of propulsive efficiency. The overall efficiency $\eta_0 = \eta_{th}\eta_p$ must then be examined to see the combined effect (taking the transmission efficiency as unity).

2.10 Overall Efficiency

For a turbojet or ramjet, using Eqs. 2.9 and 2.12,

$$\eta_0 = \eta_p\eta_{th} = \frac{2r}{1 + r} \cdot \frac{1 - r^2}{fr^2[(2g_c\Delta H_c/U^2) + 1]} = \frac{2(1 - r)}{fr[(2g_c\Delta H_c/U^2) + 1]} \tag{2.14}$$

For a given fuel, fuel-air ratio and jet velocity (i.e., a given engine at a particular condition), then η_0 is a function only of forward speed U. For $\Delta H_c = 18{,}000$ Btu/lb (a typical value for a hydrocarbon fuel) and with $U = 700$ mph, $2g_c\Delta H_c/U^2 \gg 1$. Hence resubstituting $r = U/V_j$, and simplifying, we can get

$$\eta_0 \propto UV_j - U^2$$

and

$$\partial\eta_0/\partial U = V_j - 2U$$

Setting this equal to zero,

$$U/V_j \approx \tfrac{1}{2} \tag{2.15}$$

for maximum overall efficiency of a turbojet or ramjet.

With $\Delta H_c = 18{,}000$ Btu/lb, $f = 0.02$, and $V_j = 2000$ fps, η_0 has a maximum at $U = 1000$ fps (≈ 680 mph) and the value is only 0.11 [see Fig. 2.7(a)]. Overall efficiencies of turbojet engines are very low, due in large part to the high temperature at which the propellant gas is discharged to the atmosphere.

For a rocket, using Eqs. 2.11 and 2.13,

$$\eta_0 = \eta_p\eta_{th} = \frac{2r}{1 + r^2} \cdot \frac{1 + r^2}{r^2[(2g_c\Delta H_p/U^2) + 1]} = \frac{2}{r[(2g_c\Delta H_p/U^2) + 1]}$$

$$= \frac{2V_j}{(2g_c\Delta H_p/U) + U} \tag{2.16}$$

For a given propellant and jet velocity,

$$\partial\eta_0/\partial U \propto -2V_j[-(2g_c\Delta H_p/U^2) + 1]$$

Setting this equal to zero, then for maximum efficiency,

$$U = (2g_c\Delta H_p)^{1/2} \tag{2.17}$$

that is, there is a particular optimum flight speed that depends on the propellant. It is useful to plot η_0 vs U for various values of V_j taking a particular value of ΔH_p. With $\Delta H_p = 6130$ Btu/lb of propellants (stoichiometic mixture of H_2 and O_2), $2g_c\Delta H_p = 3.065 \times 10^8$ (ft/sec)^2 and the square root of this is 17,500 fps. Figure 2.7(*b*) shows η_0 vs U for several values of V_j. For $V_j = 17{,}500 = U$, when the propulsion efficiency is unity, the overall efficiency is unity. This means that all the energy contained in the propellants as ΔH_p is comprised in V_j and this would require complete expansion to zero pressure. Actual values of V_j for this propellant would be about 10–12,000 fps and thus η_0 would be much lower. This example demonstrates the two factors entering into overall efficiency, utilization of the whole chemical energy available in the propellant and the effect of the U/V_j value.

As mentioned earlier, values of efficiency such as these are seldom used directly. This is because in propulsion, the mission is controlling. Thus in commercial air transport, the prime factor in development was speed, resulting in the present condition of the maximum subsonic cruising speed of 550–600 mph ($\approx$0.85 Mach number). As it happens, with $U/V_j = 0.5$ then the optimum jet velocity is about 1700–1800 fps, which is lower than that of most simple jet engines although not greatly so. Considerations of engine size and weight are major factors in the overall performance of an aircraft and these tend to overshadow considerations of efficiency in the particular sense of η_p and η_0. However, there is certainly a case for improvement of η_p and this has resulted in the development of the turbofan which we will discuss later.

For rockets, there has been little consideration of efficiency to date, the prime motive being to lift a given load off the surface of the earth and to give the vehicle a particular velocity for its mission, either as an orbiting satellite or to escape to space.

The importance of propulsion efficiency is then not as a quantitative parameter but as a guide to the use of classes of propulsion units for various duties. It serves to delineate types of engine somewhat in the manner of specific speed of turbomachines. Thus with $V_j \approx 2000$ fps, it is obvious that turbojets are not suited for low-speed flight. Development of jet engines for higher combustion temperatures to obtain more thrust for a given size leads to higher jet velocities and these are compatible with higher forward speeds. For rockets, the vehicle speed is continually varying and for long space missions, propellant consumption may be crucial. Hence the question of designing for a varying jet velocity to produce an optimum efficiency may be important in relation to the reliability and complexity of such an engine. This too we will analyze later.

2.11 Specific Power

Engine power as such has no meaning for jet propulsion but we can calculate a "thrust power" or jet power. This is simply net thrust times forward speed, F_nU. Because U is variable, thrust power does not have much meaning per se as shaft power of a propeller engine has meaning. A propeller engine converts shaft power

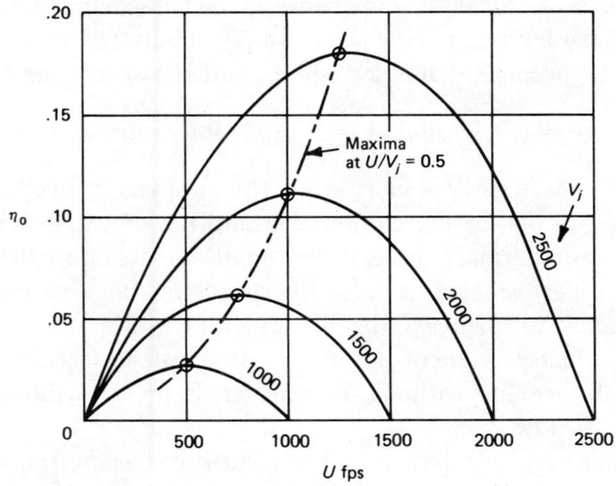

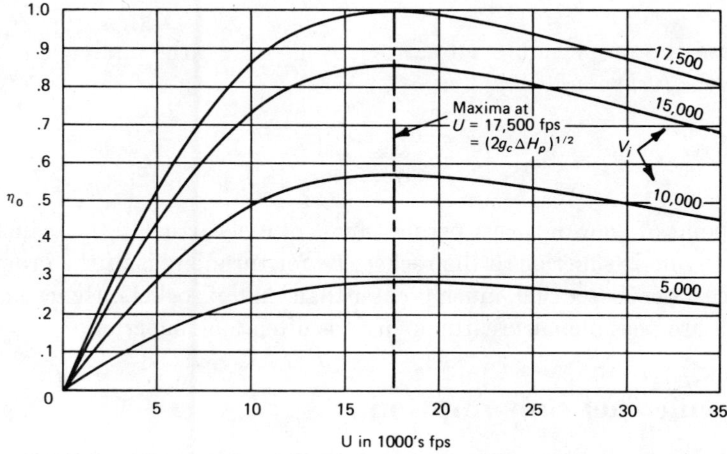

Fig. 2.7 (*a*) Overall efficiency—airbreathing engines.
(*b*) Overall efficiency—rockets.

to thrust via the propeller and so an equivalence can be made. A useful expression is that 1 lb thrust is equivalent to 1 hp at 375 mph ($375 \times 88/60 \times 1/550 \approx 1$).

The enormous power capabilities of jet engines can be demonstrated in this manner. Thus an engine delivering 20,000 lb of thrust at 600 mph has a thrust horsepower of 32,000. Imagine this as a shaft power engine!

2.12 Specific Thrust and Specific Impulse

The thrust per unit flow rate of propellant is a useful criterion of performance, as it represents the "intensity" of the engine design. For airbreathing engines, the propellant is air, so that the *specific thrust* is

$$F_s = [\dot{m}_a(V_{je} - U)/g_c\dot{m}_a] = (V_{je} - U)/g_c \quad \text{lbf/(lbm/sec)} \tag{2.18}$$

i.e., lb thrust per lb/sec of air flow (sometimes called air specific impulse). Turbojets at sea level stationary have a specific thrust of 50–100 lb/(lb/sec).

For rockets, the propellant flow rate is $\dot{m}_p$ and the *specific impulse* is given by

$$I_s = (\dot{m}_p V_{je}/g_c\dot{m}_p) = V_{je}/g_c \quad \text{lbf/(lbm/sec)} \tag{2.19}$$

Specific impulse is mostly a matter of the propellant itself for the normal chemical rocket, as it operates by combustion and nozzle expansion with no compressor or turbine. Furthermore the specific impulse or its equivalent, jet velocity, is one of the major performance criteria for rocket missions as will be developed later. It is customary to "cancel" the lbf and lbm in Eq. 2.19 and quote I_s in seconds. This is perhaps convenient but at the same time unfortunate, as "seconds" does not have much meaning without the context. It must certainly be thought of as force per unit mass flow rate.

Liquid-propellant rockets have typical values of specific impulse of 200–450 sec, the latter for H_2–O_2 under vacuum conditions. Solid propellants now in general use have a maximum value of I_s of about 250 sec, but this is being improved gradually.

Because the propellant is all carried by a rocket, the equivalent of I_s for a turbojet is the *fuel specific impulse*,

$$I_{fs} = \frac{F_n}{\dot{m}_f} = \frac{\dot{m}_a F_s}{\dot{m}_f} = \frac{F_s}{f} \tag{2.20}$$

From the point of view of thrust per unit mass of fuel carried in the vehicle, the airbreathing engine is superior to the rocket. I_{fs} for turbojets is of the order of 2500–5000 lb thrust per lb/sec fuel, much greater than that of rockets. Hence in situations where both are possible, airbreathing engines often look superior.

2.13 Specific fuel consumption

Although thermal efficiency was seen to lack importance as a measure of merit in its own right, its concomitant of fuel consumption is a very commonly used parameter.

For airbreathing engines, it is used as *specific fuel consumption*, SFC, defined as

$$\text{SFC} = \dot{m}_f/F_n \tag{2.21}$$

and given in lb/hr of fuel per lb of thrust. It can be seen to be the inverse of fuel specific impulse I_{fs}. Conversely, the inverse of rocket specific impulse is specific fuel consumption (but quoted as (lb/hr)/lb thrust, not as lb/sec). Design sea-level stationary values of SFC for turbojets are about 0.7–0.8 (lb/hr fuel)/lb thrust.

Range

Although pertaining more particularly to a vehicle rather than to the propulsive means, range is a parameter into which the engine performance enters. For level flight, constant velocity, and steady flow, the thrust of the vehicle equals the drag, i.e., $F = D$. Writing the drag as $L/(L/D)$ where L is the lift and putting the lift equal to the vehicle weight $W = Mg/g_c$, then

$$F = D = \frac{L}{L/D} = \frac{W}{L/D} = \frac{Mg/g_c}{L/D}$$

The propulsive output is FU and this is equal to the product of overall efficiency and energy input, which for an airbreathing engine is the chemical energy, i.e., $\dot{m}_f \Delta H_c$. Hence

$$FU = \eta_0 \dot{m}_f \Delta H_c = MgU/(g_c L/D) \tag{2.22}$$

The fuel is part of the vehicle mass m and $\dot{m}_f = -dm/dt$. The instantaneous mass m is a function of time and distance, i.e., $m = m(s, t)$

$$\therefore dm = (\partial m/\partial s)\,ds + (\partial m/\partial t)\,dt$$

and

$$dm/dt = (\partial m/\partial s)(ds/dt) + (\partial m/\partial t)$$

For uniform steady flow, $ds/dt = U$, $\partial m/\partial t = 0$ and $\partial m/\partial s = dm/ds$ hence

$$dm/dt = U(dm/ds)$$

Substituting in Eq. 2.22,

$$-\eta_0 U \frac{dm}{ds} \Delta H_c = \frac{mgU}{g_c L/D}$$

and

$$ds = -\frac{\eta_0 \Delta H_c L/D}{g/g_c} \frac{dm}{m}$$

For a particular vehicle in steady flight, η_0, ΔH_c and L/D are constant, hence we can integrate and get

$$s_2 - s_1 = \frac{-\eta_0 \Delta H_c L/D}{g/g_c} \ln \frac{m_2}{m_1}$$

and with $m_1 = m_0$ at $s_1 = 0$ the initial condition, and $m_2 = m$ at $s_2 = s$ any subsequent condition, then

$$s = \frac{\eta_0 \Delta H_c L/D}{g/g_c} \ln \frac{m_0}{m} \tag{2.23}$$

This is a form of Breguet's *range formula.* It shows that range is directly dependent on the overall efficiency of the propulsive unit and the aerodynamic characteristics of the vehicle as the lift-drag ratio. The logarithmic form of the mass ratio is an important feature which we will discuss in detail in the rocket analysis.

Problems

2.1 An airfoil of chord c is placed in an incompressible flow and yields a flow pattern with a wake having a velocity deficiency as shown. The plane of the wake where measured is sufficiently far downstream so that the static pressure may be considered as uniform at the ambient value. The velocity distribution can be approximated as follows:

$$\frac{u}{U_\infty} = 1 - 0.01\left(1 + \cos\frac{2\pi y}{b}\right) \quad \text{for} \quad -\frac{b}{2} < y < \frac{b}{2}$$

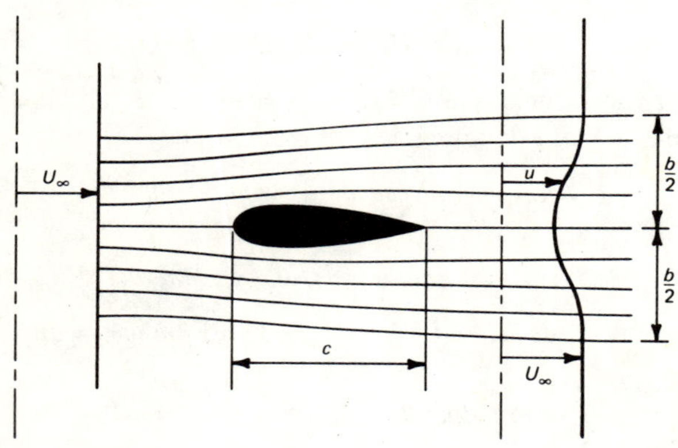

If $b/c = 2.5$, find the drag coefficient C_D per unit span (width) of the airfoil. State carefully the assumptions in your analysis.

2.2 The thrust for a stationary turbojet has been given in Eq. 2.5c as $\dot{m}V_j/g_c$, that is, with no inlet air velocity component. Yet a turbojet on the ground with the engine running has a significant air velocity at the intake and it is sometimes quite a problem to prevent the ingress of foreign matter (including people). Explain this apparent paradox.

2.3 A rocket with a convergent nozzle of 4-in. throat diameter gives a thrust of 3000 lb when tested at ground level. Find the required combustion chamber stagnation pressure, assuming it is greater than 30 psia. Use $k = 1.3$.

2.4 A rocket has a nozzle designed to provide complete expansion of the gases when run at ground level. If operated in space, show that the fractional increase of thrust is $1/kM^2$, where M is the nozzle-exit Mach number. Assume that the mass flow rate and the gas pressure and temperature remain the same.

2.5 It was suggested (G.M. Dusinberre, 1958) that for a choked, convergent nozzle of an airbreathing engine, the propulsive efficiency might better be expressed using the energy input (i.e., as denominator in the efficiency expression) as the sum of the change of kinetic energy of the fluid and the work of a nonexpanding engine working between pressure levels p_j and p_a, that is,

$$\eta_p = \frac{\text{thrust}}{(V_j^2 - U^2)/2g_c + V_j A_j (p_j - p_a)/\dot{m}}$$

Discuss the significance of this definition, compare it with other possibilities and assess its usefulness.

2.6 Show that for a turbojet with a propulsive efficiency of 70%, a flight Mach number of 0.8 in the stratosphere means a specific thrust of 20.6 lb/(lb/sec).

2.7 A turbojet engine has a mass flow rate of 150 lb/sec at ground level and has a convergent nozzle of throat area 3.12 ft², which always operates in a choked condition with a discharge temperature of 1400°F. Find the thrust (a) at the stationary condition at ground level and (b) at $M = 0.85$ at 30,000 ft. Find the propulsive efficiency for (b). Take $k = 1.33$ and assume the mass flow rate varies directly with atmospheric density.

CHAPTER 3

Flow in Diffusers, Combustors, and Nozzles

For airbreathing engines, the problem of inducing the correct quantity of air at a level of velocity and distribution suitable for optimum engine performance can be complex. Apart from the aerothermochemistry of combustion, which will not be taken up here, gas dynamic aspects of combustion as the heating of a gas stream impose some limitations in performance. For all types of propulsion engines using expansion of the working medium for attaining a jet velocity, the nozzle is highly important, as any loss here is directly reflected in the V_j of the thrust expression. These aspects of diffusers, combustors and nozzles, as they are pertinent to propulsion performance, are discussed in this chapter.

3.1 Review of Flow Parameters

A few of the essential relationships for gas flow will be given first, to establish nomenclature and use.

1. Equation of continuity.

$$m = A\rho V \tag{3.1a}$$

or

$$\frac{dA}{A} + \frac{d\rho}{\rho} + \frac{dV}{V} = 0 \tag{3.1b}$$

2. Equation of motion.

One-dimensional steady flow, reversible, the simple Euler equation.

$$(dp/\rho) + (VdV/g_c) = 0 \tag{3.2}$$

3. Steady-flow energy equation.

$$h_1 + V_1^2/2g_c + Q_{12} = h_2 + V_2^2/2g_c + W_s \tag{3.3}$$

where h = enthalpy, Q = heat transferred in and W_s = shaft work out. This excludes gravitational potential energy and electromagnetic energy.

In terms of stagnation values, using subscript zero to indicate a stagnation property,

$$h_{0_1} + Q_{12} = h_{0_2} + W_s$$

For a perfect gas, which is our constant assumption here, $h = c_pT$, where c_p = specific heat at constant pressure, hence,

$$c_{p_1}T_{0_1} + Q_{12} = c_{p_2}T_{0_2} + W_s \tag{3.4}$$

4. Mach number.

$$M = V/a = V/(g_c kRT)^{1/2} \tag{3.5}$$

5. Stagnation properties.

$$T_0 = T + (V^2/2g_c c_p) = T\{1 + [(k-1)/2]M^2\} \tag{3.6a}$$

$$p_0 = p(T_0/T)^{k/k-1} = p\{1 + [(k-1)/2]M^2\}^{k/k-1} \tag{3.6b}$$

It is sometimes useful shorthand to denote the dynamic temperature as θ_v, that is,

$$\theta_v = (V^2/2g_c c_p) = [(k-1)/2]TM^2 \tag{3.7}$$

Another shorthand symbol is used later, substituting $\epsilon = (k-1)/k$ as exponent in the isentropic pressure–temperature relationship, i.e. $T_2/T_1 = (p_2/p_1)^\epsilon$.

6. Choking flow.

Choking flow in a duct occurs when the Mach number reaches unity and this can occur only at the minimum cross section, that is, at the *throat*. Choked states are denoted by a superscript asterisk, e.g., p^*, T^* and A^*. A useful relationship for the mass flow rate at the choked condition is as follows:

$$\dot{m} = A\rho V = A(p/RT)[M(g_c kRT)^{1/2}]$$

Substituting Eqs. 3.6(a) and (b), then

$$\dot{m} = \frac{Ap_0M}{T_0^{1/2}}\left(\frac{g_c k}{R}\right)^{1/2}\left(1 + \frac{k-1}{2}M^2\right)^{-[k+1/2(k-1)]} \tag{3.8a}$$

For $M = 1$ and for given values of k and R,

$$\dot{m} \propto A^* p_0/(T_0)^{1/2} \tag{3.8b}$$

Hence for a given adiabatic flow, $A^* \propto 1/p_0$, i.e., the choking area is inversely proportional to the stagnation pressure, or for a given area, the mass flow rate is proportional to the stagnation pressure.

3.2 Inlet Flow

The engine or device inside the control region (nacelle or cowling) acts as a variable throttling device, varying the amount of air that enters. Flight velocities are invariably much higher than values usable by the engine without great loss and the problem is to achieve *diffusion* of the air, i.e., reduction of velocity and rise of static pressure. The amount of this recovery of pressure is most important, as at high speeds the pressure regain can be large. For example, using Eqs. 3.6a and b at a Mach number of 0.85, a typical cruising speed for subsonic transport aircraft, the temperature ratio T_0/T is seen to be 1.1445 and the corresponding pressure ratio p_0/p is 1.602. That is, if the air were brought completely to rest from $M = 0.85$, it would have a pressure 1.6 times the ambient pressure. This is known as the *ram* effect and the pressure attained by the air after diffusion is the *ram* pressure. At supersonic velocities the ram ratio is very high, p_0/p for $M = 2.7$ being 23.25. Stagnation pressure is defined by an isentropic process and diffusion is never reversible so that the actual pressure attained is lower than the ideal.

Diffusion can be a very inefficient process, because the inevitable boundary layer is acted upon by the increasing pressure in the flow direction and tends to separate with resulting loss of available energy. In a propulsive device, diffusion may occur *externally*, that is before entering the structural inlet, as well as *internally* in the usual manner consequent on change of area. The air entering the physical inlet need not, and indeed seldom does, have the same cross-sectional area out in the free stream as that at the inlet. Thus, as is shown in Fig. 3.1, there is a capture area A_∞ which may be less than or greater than the inlet area A_i. In Fig. 3.1(a), there is external compression of air as $A_\infty < A_i$ and this occurs isentropically, or very nearly so, as there are no solid surfaces to initiate a boundary layer. This would represent the state of affairs at the higher speeds and lower engine mass flow rates. In Fig. 3.1(b), there is external acceleration as $A_\infty > A_i$ and this could occur at maximum engine rpm and very low forward speeds, e.g., at takeoff. The latter situation is most undesirable as the necessary internal diffusion could not be accomplished satisfactorily. External diffusion, though efficient with respect to the engine, can cause external drag by virtue of the induced flow over the outside of the inlet duct, so that some compromise is necessary. However, it is better to have too large an inlet than one too small so that the internal diffusion load is lessened.

3.3 Subsonic, Incompressible Flow

Subsonic diffusion has been made the focus of an intensive study in recent years, before which knowledge of it was sparse and unsystematic. Although much still remains to be understood, some of the main features for incompressible flow are becoming clear and some reliable quantitative information has been made available. Much of the recent work has been done by Kline and his fellow workers at Stanford.[1] They distinguished four regimes of flow: (1) smooth, well-behaved flow; (2) transitory stall, unsteady in space and time; (3) fully developed stall, with the main flow along one wall only; and (4) jet flow, with separation from both walls. Pressure recovery for straight-walled diffusers is not necessarily optimized in region 1, as apparently some stall occurs even in efficient diffusion. The variables of geometry are divergence angle (rate of area change), area ratio (total area change), and shape (rectangular, conical, etc.). Flow variables are entry boundary-layer thickness, Reynolds number, degree of turbulence, entry velocity distribution and, in compressible flow, entry Mach number. Performance may be expressed as loss of stagnation head or pressure, or as pressure recovery defined as a pressure recovery coefficient, that is, $C_{pr} = p_2 - p_1/(\rho V_1^2/2g_c)$. A diffuser may be required for a

[1] For example, S. J. Kline, D. E. Abbott and R. W. Fox, "Optimum Design of Straight-Walled Diffusers," *Trans. ASME Series D, J. Basic Eng.*, 81, 321 (1959); also A. T. McDonald and R. W. Fox, "An Experimental Investigation of Incompressible Flow in Conical Diffusers," *Int. J. Mech. Sci.*, 8, 125 (1966).

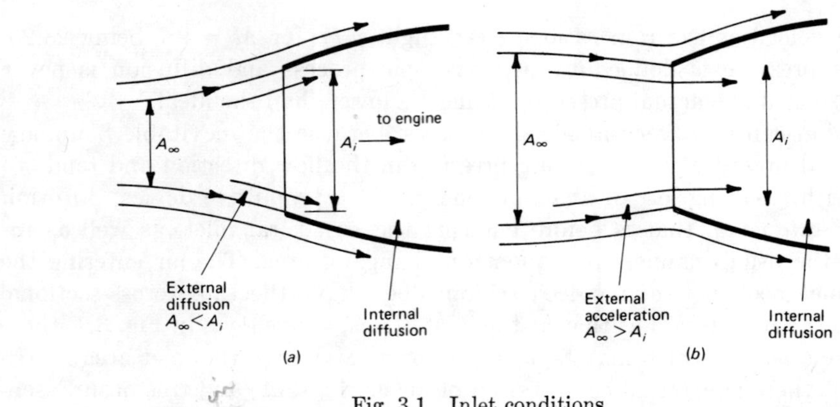

Fig. 3.1 Inlet conditions.

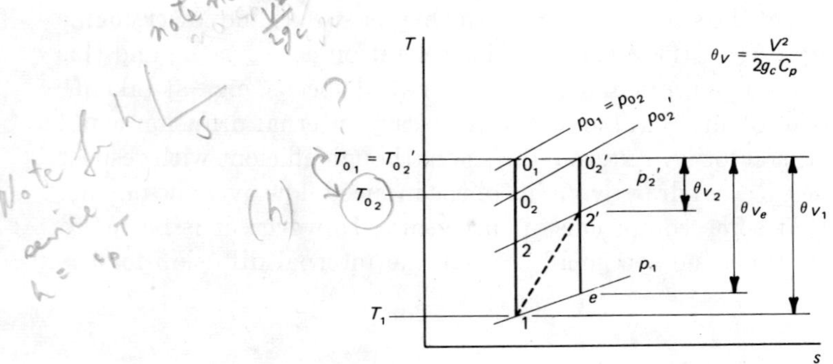

Fig. 3.2 Diffuser performance.

number of design desiderata, namely (1) minimum loss of pressure for a given pressure rise, (2) maximum recovery (a) for a given area ratio, (b) for a given length ratio and (c) for any possible geometry (best optimum solution).

With the above variables and performance requirements, it is not possible to summarize data adequately. It must suffice to say that allowable divergence angles of the diffuser must be kept small, 8–20° included angle dependent on the area ratio and other factors. Very small divergence angles mean long diffusers for a given effect and thus a high wall friction loss. Large angles imply separation and severe loss due to jet flow. Using the incompressible flow criterion of C_{pr} defined previously, it should be possible with an area ratio not greater than 2 to obtain a value of C_{pr} close to 0.9.

3.4 Performance Parameters—Compressible Flow

For compressible flow, the simple pressure recovery coefficient is not satisfactory and we must look further for useful definitions of performance.

Figure 3.2 shows the ram compression process on a T-s diagram. State 1 is that of the ambient atmosphere with stagnation state 01 corresponding to a forward velocity U or Mach number M_1. The air is not completely stagnated and has a discharge velocity V_2, represented on the T-s diagram by θ_{V_2}. The ideal or reversible adiabatic process is represented by 1–2 for the static state and 1–01 for the stagnation state, so that $p_{0_2} = p_{0_1}$. The actual process with loss is 1–2′ or 1–02′ with the same stagnation temperature because the process is adiabatic, but with a lower stagnation pressure due to increase of entropy.

Various parameters are used or have been proposed for expressing the diffuser performance. Some are efficiencies in the sense of energy ratios, some are not.

I. Stagnation pressure ratio parameter.

$$r_0 = (p_{0'_2}/p_{0_2}) = (p_{0'_2}/p_{0_1}) \tag{3.9}$$

This is the ratio of actual to ideal stagnation pressure after diffusion. The ratio is simple to comprehend and simple to use but does not express energy quantities directly.

II. Isentropic or adiabatic efficiency.

This is the ratio of ideal enthalpy change $h_{0_2} - h_1$ to the actual enthalpy change for the same stagnation pressure, $h_{0'_2} - h_1$. Thus

$$\eta_{is} = \frac{h_{0_2} - h_1}{h_{0'_2} - h_1} \approx \frac{T_{0_2} - T_1}{T_{0'_2} - T_1}$$

neglecting the very small differences of specific heat in diffuser processes. Using the expressions

$$T_{0_2}/T_1 = (p_{0'_2}/p_1)^{(k-1)/k} \quad \text{and} \quad T_{0'_2} - T_1 = \theta_{V_1} = [(k-1)/2]T_1M^2,$$

then

$$\eta_{is} = \frac{(p_{0'_2}/p_1)^{(k-1)/k} - 1}{[(k-1)/2]M^2} \tag{3.10}$$

or alternatively,

$$p_{0'_2} = p_1\{1 + \eta_{is}[(k-1)/2]M^2\}^{k/(k-1)} \tag{3.11}$$

where M is the flight Mach number corresponding to U_1. Since

$$\frac{p_{0'_2}}{p_1} = \frac{p_{0'_2}}{p_{0_1}} \cdot \frac{p_{0_1}}{p_1} \quad \text{and} \quad \frac{p_{0_1}}{p_1} = \left(1 + \frac{k-1}{2}M^2\right)^{k/(k-1)}$$

then

$$\frac{p_{0'_2}}{p_{0_1}} = \left(\frac{1 + \eta_{is}[(k-1)/2]M^2}{1 + [(k-1)/2]M^2}\right)^{k/(k-1)} = r_0 \tag{3.12}$$

or

$$\eta_{is} = \frac{\{1 + [(k-1)/2]M^2\}r_0^{(k-1)/k} - 1}{[(k-1)/2]M^2} \tag{3.13}$$

III. Kinetic energy efficiency or energy ratio.

It can be argued that since the purpose of the propulsion engine is to provide a high-velocity jet of gas by expansion, a more suitable criterion is based on the energy available for expansion after diffusion. Thus in Fig. 3.2 this energy is represented by $h_{0'_2} - h_e$, the ideal enthalpy change from the actual diffusion state back to the original pressure. Thus

$$\eta_k = \frac{h_{0'_2} - h_e}{h_{0_1} - h_1} \approx \frac{T_{0'_2} - T_e}{T_{0_1} - T_1} = \frac{T_{0_1} - T_e}{T_{0_1} - T_1}$$

This can be transformed as follows:

$$\frac{T_{0_1} - T_e}{T_{0_1} - T_1} = \frac{T_{0_1}[1 - (T_e/T_{0_1})]}{T_{0_1} - T_1} = \frac{(T_{0_1}/T_1)[1 - (p_1/p_{0'_2})^{(k-1)/k}]}{T_{0_1}/T_1 - 1}$$

and with

$$\frac{p_1}{p_{0'_2}} = \frac{p_1}{p_{0_1}}\frac{p_{0_1}}{p_{0'_2}} \quad \text{and} \quad \frac{T_{0_1}}{T_1} = 1 + \frac{k-1}{2}M^2,$$

then

$$\eta_k = 1 - \frac{2}{(k-1)M^2}\left[\frac{1}{r_0^{(k-1)/k}} - 1\right] \tag{3.14}$$

and

$$p_{0'_2}/p_{0_1} = r_0 = \{1 + (1 - \eta_k)[(k-1)/2]M^2\}^{-k/(k-1)} \tag{3.15}$$

From Fig. 3.2 it is seen that η_k can be expressed directly as $\theta_{V_e}/\theta_{V_1}$, the ratio of kinetic energies of expansion and ideal diffusion.

A comparison of the isentropic and kinetic energy efficiencies η_{is} and η_k in terms of the stagnation pressure ratio r_0 is shown in Fig. 3.3, with Mach number as the variable plotted as abscissa. For $r_0 = 0.9$ and similar values associated with high efficiency as encountered in subsonic flight, there is little difference between η_{is} and η_k. It is seen that there is a good spread of values of η_{is} and η_k over the range of Mach number 0.5–1.0 where high values of r_0 are relevant. For supersonic Mach numbers, where shock losses tend to make lower values of r_0 more general, there is a greater difference between η_{is} and η_k. For hypersonic Mach numbers, say $M > 3$, values of η_k are high, even for very low values of r_0 and hence this parameter by itself does not provide a good spread of values, e.g., at $M = 5$, a change of pressure recovery from 0.2 to 0.5 is reflected by a change of η_k from only about 0.88 to 0.95.

IV. Process efficiency.

Because many favor the η_k type of parameter, the factor of its insensitivity at high Mach numbers has led to the suggestion of a new parameter called the *process efficiency*, K_D. The lowest practical value of η_k may be taken to be that when

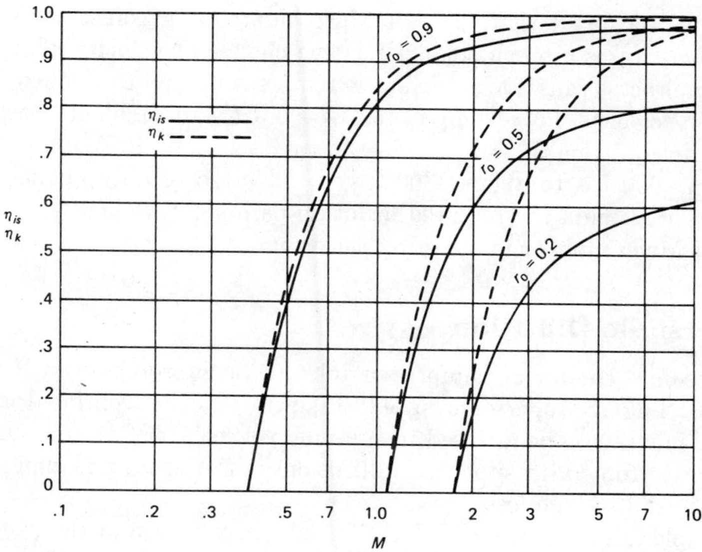

Fig. 3.3 Diffuser efficiency vs Mach number.

there is no rise in static pressure, that is, in Fig. 3.2, state 2′ becomes state *e* and the process occurs at constant pressure. In this case, $V_e^2 = V_2^2$ and $\eta_k = V_2^2/V_1^2$. Now the process efficiency K_D is defined so that it is zero for $\eta_{k\min} = V_2^2/V_1^2$, and equal to unity at $\eta_k = 1$, i.e. expanding the range of useful η_k. This can be given by the expression

$$K_D = \frac{\eta_k - (V_2/V_1)^2}{1 - (V_2/V_1)^2} \tag{3.16}$$

or

$$\eta_k = K_D + (1 - K_D)(V_2/V_1)^2 \tag{3.17}$$

Although this has been established by argument based on convenience, it can also be shown that it has thermodynamic significance of its own. Thus if we form a parameter based on *changes* of kinetic energy rather than on absolute values, we have

$$\frac{V_e^2 - V_2^2}{V_1^2 - V_2^2} = \frac{(V_e^2/V_1^2) - (V_2^2/V_1^2)}{1 - (V_2^2/V_1^2)} = \frac{\eta_k - (V_2/V_1)^2}{1 - (V_2/V_1)^2} = K_D \tag{3.18}$$

Furthermore, K_D was formerly used as an experimental value in Eq. 3.17 for η_k, as it was found to correlate test values in reasonable fashion. Thus its use is satisfactory from a practical and an analytical background, although it has not come into general use. A full discussion of this and other possible diffusion parameters is given in the quoted reference.[2]

[2] E. T. Curran and M. B. Bergsten, "Inlet Efficiency Parameters for Supersonic Combustion Ramjet Engines, Report No. APL TDR 64-61, Research and Development Division, Air Force Systems Command, June 1964.

It should be noted that at high Mach numbers, a considerable change of Mach number causes a much smaller relative change of velocity. This is because the diffusion process causes a rise in static temperature and consequent increase in the acoustic velocity. For example, for $r_0 = 0.5$ at an ambient temperature of 400°R, a diffusion from $M = 5$ to $M = 3$, a reduction of 40%, entails a velocity change from 4900 fps to about 4300 fps, a reduction of only about 12%. Thus one must have in mind the required diffusion parameter, reduction of velocity or reduction of Mach number in assessing performance.

3.5 Supersonic Diffusion

Turbojets require the air at compressor inlet to be subsonic, even if the vehicle itself is traveling at supersonic speed. Likewise the conventional ramjet also requires the inlet air to be diffused to subsonic velocities before combustion. Thus there is a real problem in supersonic diffusion as the usual phenomena of shock diffusion can lead to high loss.

If a simple diffusing duct or "Pitot" intake in the form of the usual subsonic inlet is used, shock waves must occur at some place to effect the transition from supersonic to subsonic flow. The location of the shock depends on the fraction of the design engine mass flow which is required at a given condition. At engine conditions requiring mass flow rate $\dot{m} < \dot{m}_D$, the design flow rate, then the capture area A_∞ is less than the inlet area A_i. Because a supersonic flow cannot "bend," as downstream conditions cannot be signaled upstream with $V > a$, then there must be a transition to subsonic flow via a normal shock wave so that the unrequired air can turn and be spilled over the leading edge outside the engine duct. This is shown in Fig. 3.4(a). When $\dot{m} = \dot{m}_D$, then no spillage is required and a normal shock occurs right at the inlet plane. Oblique shocks occur externally at the leading edge, [Fig. 3.4(b)]. When greater than design flow is required, oblique shocks occur internally at the edge of the inlet, followed by a normal shock at a Mach number lower than that corresponding to M_∞ [Fig. 3.4(c)].

Figure 3.5 shows the extent of the loss due to a normal shock. The highest line in the figure shows the pressure ratio $p_{0_{2'}}/p_1 = p_{0_1}/p_1$ for completely reversible adiabatic diffusion, i.e. no shock and $\eta_{is} = 1$. The next two lines show this pressure

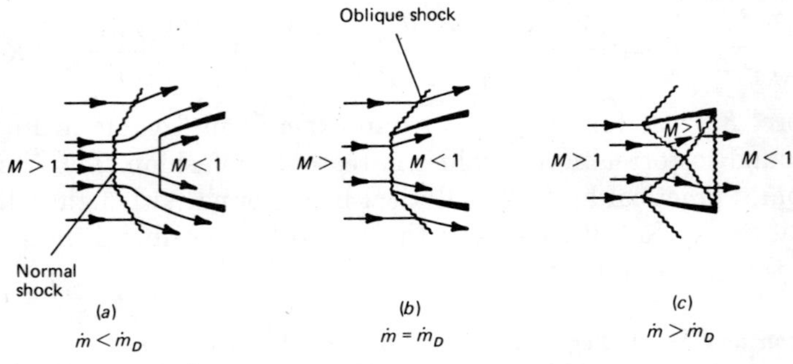

Fig. 3.4 Supersonic diffusion—Pitot intake

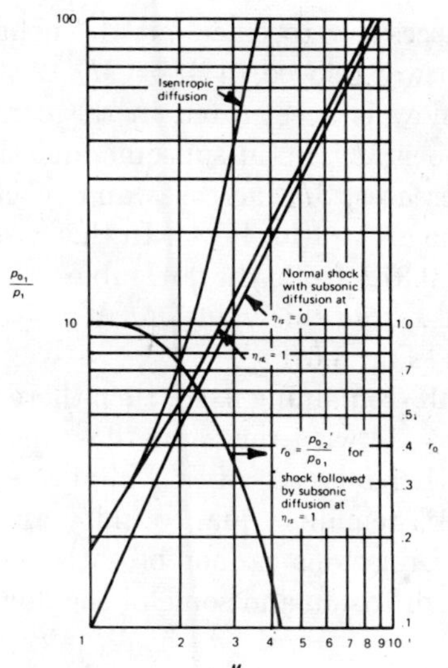

Fig. 3.5 Normal shock loss vs Mach number.

ratio with shock, one for the subsequent subsonic diffusion being reversible ($\eta_{is} = 1$) and one for $\eta_{is} = 0$. The lowest curve represents the stagnation pressure ratio $r_0 = p_{0_{2'}}/p_{0_1}$ for normal shock with subsequent diffusion at $\eta_{is} = 1$. It is seen that at low supersonic velocities $1 < M < 1.5$, shock loss is relatively low and the subsonic diffusion process is very important. As Mach number increases however, r_0 decreases rapidly and there is a very considerable loss due to shock, with the subsonic diffusion not being very important. At $M > 3$, the loss becomes excessive.

If the performance with normal shock is intolerable, then we must look for alternative solutions. An obvious thought is to design the intake for reversible supersonic diffusion, that is, with initially a *decreasing* area leading to a *throat* and then *increasing* area for subsonic diffusion. This is the convergent-divergent shape of a supersonic nozzle but with the flow reversed in direction. Nozzles of this type can have a good performance so possibly diffusers might do so too. There is however a major difficulty in reaching the design state from the starting condition and to understand this, the performance of such a diffuser will be analyzed carefully. It will be carried out as a quantitative example as actual values will aid in comprehension.

Let us suppose that we have designed a convergent-divergent inlet for a Mach number M_D of 2.0. From the relationships for reversible adiabatic flow,[3] the area ratio, i.e. inlet area to throat area A_i/A^*, is 1.6875 or $A^*/A_i = 0.593$. For the design situation, the throat area A_t is the area A^* where the Mach number is unity. Thus the diffuser receives air at $M_\infty = 2$, the capture area A_∞ is equal to A_i, there is diffusion to the throat where $M = 1$, followed by further subsonic diffusion with increasing area to some desired Mach number less than unity. This is shown in Fig. 3.6(a).

[3] For example, J. H. Keenan and J. Kaye, "Gas Tables," Wiley, New York, 1948.

Now, however, it is necessary to trace out the behavior from the stationary state through increasing forward speed to $M = M_D$. At very low subsonic values of M_∞ [Fig. 3.6(*b*)] the flow is accelerated in the convergent part of the inlet but the throat Mach number M_t is still subsonic and the flow decelerates in the divergent part. As M_∞ increases, M_t increases and reaches a value of unity at a value of M_∞ dictated by the area ratio A_i/A_t. In this case with $A_i/A_t = 1.6875 = A_i/A^*$, then $M_i = M_\infty = 0.372$ from the air tables. Thus the throat is *choked*, $M_t = 1$, and cannot increase from this value. At $M_\infty = 0.372$ the capture area A_∞ is exactly A_i the inlet area [Fig. 3.6(*c*)].

As M_∞ increases, with M_t remaining fixed, then there must be external diffusion from A_∞ to A_i, so at $M_\infty = 1$, with reversible flow inside and out, $A_\infty = A_t$, as shown in Fig. 3.6(*d*). Further increase of M_∞ implies supersonic flight, although M_∞ is still less than M_D. M_t remains equal to unity and because $A_i/A_t = A_i/A^*$ is fixed, the isentropic flow relations cannot be satisfied. A_∞ is too small for M_∞ and there must be external diffusion and some of the flow must be spilled. Because

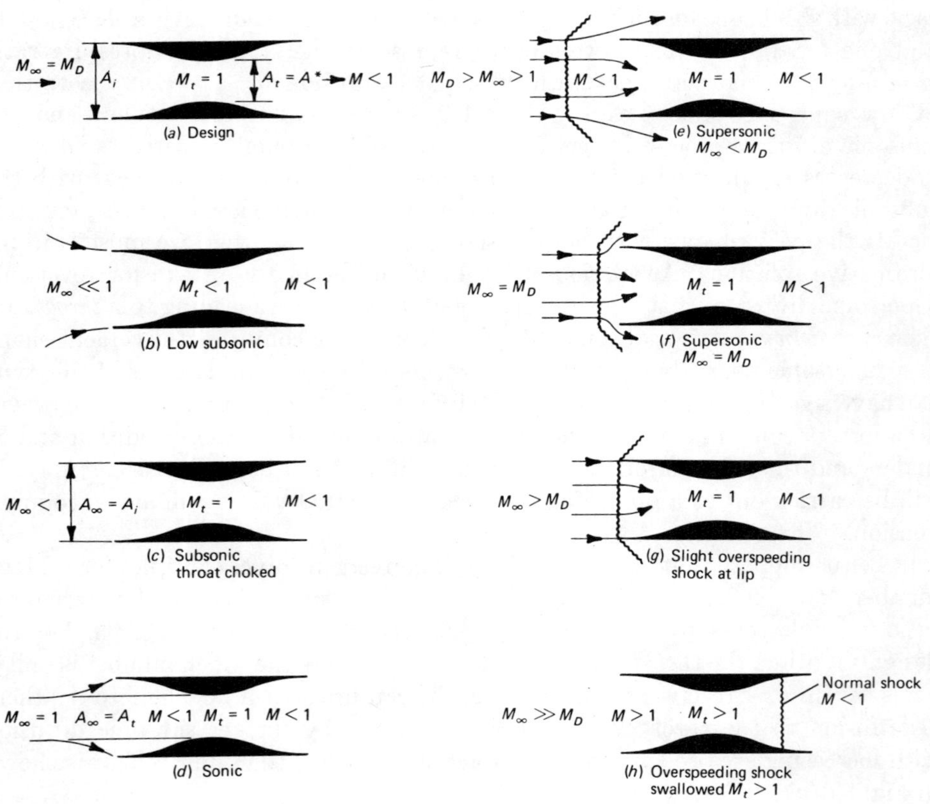

Fig. 3.6 Convergent-divergent diffuser conditions

the flow is supersonic the streamlines cannot bend, and therefore a normal shock occurs ahead of the inlet. The subsonic flow behind the shock can then spill around the inlet, to provide just enough internal flow to choke the throat at the fixed area ratio A_i/A^*. This is shown in Fig. 3.6(*e*).

As M_∞ goes on increasing, the downstream subsonic Mach number behind the shock decreases and less spillage is required, so the shock moves closer to the lip of the inlet. When $M_\infty = M_D = 2.0$, the downstream subsonic Mach number is 0.577 and the corresponding area ratio A_i/A^* for isentropic flow is 1.217, or $A^*/A_i = 0.822$. But the actual area ratio A_i/A_t is 1.6875 or $A_t/A_i = 0.593$, which is lower than required or, expressed otherwise, with a fixed inlet area A_i, the throat area A_t is too small to "swallow" the flow. Thus even at $M_\infty = M_D$, the shock remains in front of the lip and spillage must occur [Fig. 3.6(*f*)]. Another way of looking at it is to note that because of the shock at supersonic M_∞, there is an increase of entropy and loss of stagnation pressure, so that the ideal reversible flow relations cannot be satisfied.

The result is then a normal shock at $M_\infty = M_D$, just as for the simple Pitot intake, and a corresponding loss. Is there any way in which the converging-diverging diffuser could "swallow" the flow and diffuse the supersonic flow without shock? One method might be to make the throat large enough to suit the condition where the shock is right at the lip, whence a slight further acceleration would allow the shock to be swallowed and supersonic flow prevail in the convergent section. This would mean that the area ratio A_i/A_t would have to be 1.217, i.e. an oversize throat for the same inlet area. Then at $M_\infty = M_D = 2.0$, the shock would be swallowed and the flow would decelerate supersonically in the decreasing area. However, the throat is too big for $M_t = 1$ and $A_t/A^* = (A_t/A_i)(A^*/A_i) = 0.822/0.593 = 1.387$ and the corresponding supersonic Mach number for this area ratio is 1.539. Hence the flow is diffused from 2.0 at the inlet to only 1.539 at the throat and then accelerates again in the diverging portion. A normal shock would then occur somewhere in this diverging portion depending on the engine conditions and again there would be a loss, albeit smaller than that for external shock as it could be made to take place at a Mach number less than $M_D = 2.0$.

Another possibility might be to increase the forward speed M_∞ beyond M_D to the point where the downstream Mach number of the bow shock is just that corresponding to $A_i/A^* = 1.6875$, i.e. $M_i = 0.372$ as for low subsonic flight speed. For the given example, this cannot be done because the lowest downstream Mach number even for infinite M_∞ is 0.378. For design Mach numbers only just above unity, overspeeding is a possibility, e.g. for $M_D = 1.3$, $A_i/A^* = 1.0663$ and the corresponding subsonic Mach number is 0.743, which in turn for normal shock requires a supersonic Mach number of about 1.39. So in this case overspeeding to $M_\infty = 1.39$ would allow swallowing of the shock. Obviously overspeeding is an undesirable practice requiring either an oversize engine to attain the necessary speed or some sort of diving maneuver.

Although an oversize throat of fixed area is undesirable, a variable-area nozzle would be satisfactory, as the area ratio could be made large enough to swallow the shock in the starting operation, and then contracted so that the shock in the divergent part would move closer to the throat and decrease in value and, at exactly the design condition, move into the throat at $M = 1$ and disappear. This type of variable-area inlet is difficult mechanically but can be provided by axial movement of a central plug in nonparallel walls.

Another possible solution is to use a perforated inlet,[4] shown diagrammatically in Fig. 3.7. As a fixed diffuser will not swallow the flow because of the reduction of stagnation pressure due to shock, then if the area cannot be changed, the reduction of required mass flow through the throat is another possibility. Thus in the accelerating stage, the perforations in the convergent part of the inlet spill some of the flow (*a*), and with the discontinuity of mass flow, the bow shock can

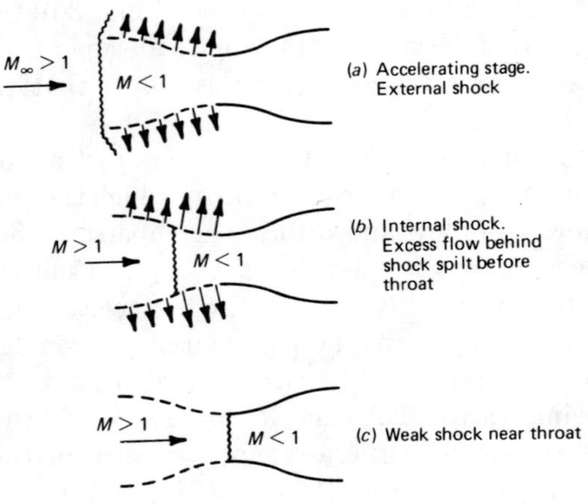

Fig. 3.7 Perforated diffuser.

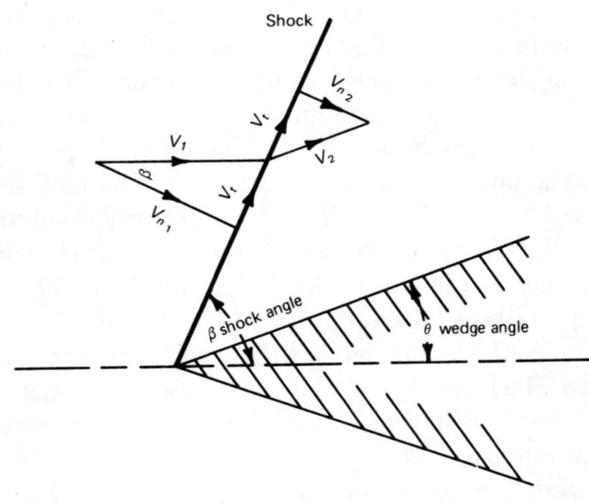

Fig. 3.8 Oblique shock.

[4] J. C. Evvard, and J. W. Blakey, "The Use of Perforated Inlets for Efficient Supersonic Diffusion," NACA TN 3767, 1956.

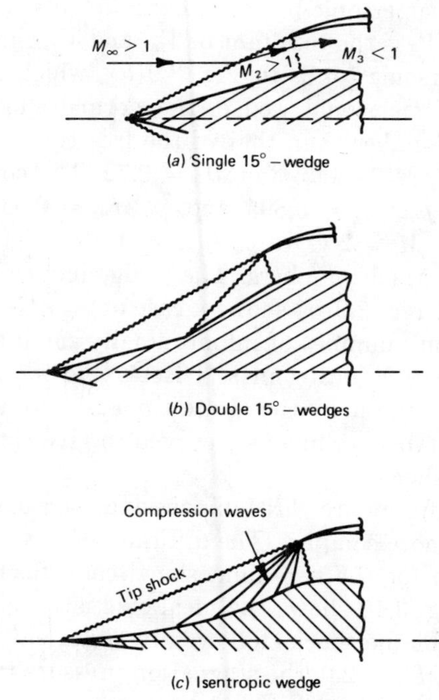

Fig. 3.9 Oblique shock diffusers.

be made to move into the inlet (*b*). The static pressure behind the shock is much higher than in front and bleeding off the downstream air can cause the shock to move toward the throat, becoming weaker and weaker (*c*), and then when it passes to the throat it disappears, when supersonic flow is established upstream and the bleed ports can be closed.

None of these supersonic diffusers with normal shock is particularly satisfactory from the point of view of either performance or complexity. A solution for many circumstances lies in the use of *oblique* shock (Fig. 3.8). If the absolute velocity V is resolved into the components, V_t and V_n, parallel to and normal to the shock, respectively, only the normal component is subject to shock. Quantitatively, the oblique shock behaves according to the normal shock of component M_n and thus for a given freestream Mach number M_∞, the loss of stagnation pressure is less than that for a normal shock at M_∞ to a degree varying with the shock angle. For example at $M_\infty = 2.0$ with the shock angle $\beta = 45°$, then $V_{n1} = V_1 \sin \beta$ and $M_{n1} = M_1 \sin \beta = [(2)^{1/2}/2]M_1 = 1.414$. For a Mach number of 2.0, the ratio of stagnation pressures after and before shock is 0.721. For $M = 1.414$, it is 0.955, so that a considerable loss is avoided by making the shock oblique.

Oblique shocks are caused by some change of conditions, such as surface direction or pressure distribution, which the supersonic stream cannot propagate upstream. For diffusers, an oblique shock is caused by positioning a wedge (in the two-dimensional case) or a spike (in the three-dimensional case), axially ahead of the inlet duct as shown in Fig. 3.9(*a*). The flow following this shock may still be, and nearly always is, supersonic but at a lower Mach number. This is because although V_{n2} is subsonic, V_2 is the resultant of V_{n2} and V_t, the two components resulting usually in a supersonic total. In Fig. 3.9(*a*), which shows one change of direction due to a two-dimensional wedge, the normal shock at the duct inlet occurs at a lower Mach number than that of the freestream. Thus for $M_\infty = 2.0$ and a wedge angle of 15°, $M_2 = 1.45$ and $M_3 = 0.72$. The corresponding losses of stagnation pressure are $p_{0_2}/p_{0_\infty} = 0.895$ and $p_{0_3}/p_{0_2} = 0.945$, giving $p_{0_3}/p_{0_\infty} = 0.845$. A normal shock at $M = 2$ gives $p_{0_3}/p_{0_\infty} = 0.721$. The reduced loss is even more apparent at higher Mach numbers. The wedge-induced oblique shock need not be limited to one and Fig. 3.9(*b*) shows a case of two 15° angles of turn.

In the limit, an infinite number of infinitesimally small angles could be used, giving a reversible process, resulting in an "isentropic wedge" external diffuser [Fig. 3.9(*c*)]. In practice, the necessary finite tip angle produces a shock but the rear part of the wedge can be continuously curved to give continuous compression waves rather than finite shocks.

More often, such diffusers are three-dimensional and axisymmetric, in which case the flow problem is more complex. The diffusion however is more efficient and less loss is obtained than for the two-dimensional case. Such diffusers are called after Oswatitsch[5] and Fig. 3.10 shows the ideal stagnation pressure ratio p_{0_n}/p_{0_∞} for one, two, and three oblique shocks followed by a normal shock. It is seen that for one oblique shock at $M_\infty = 2.0$, the stagnation pressure ratio is just about 0.9, compared with 0.845 for the two-dimensional case quoted above. Such efficiencies are not necessarily obtained under actual conditions because of boundary layer effects. Because of the large static pressure change across a shock, conditions are conducive to boundary layer separation.

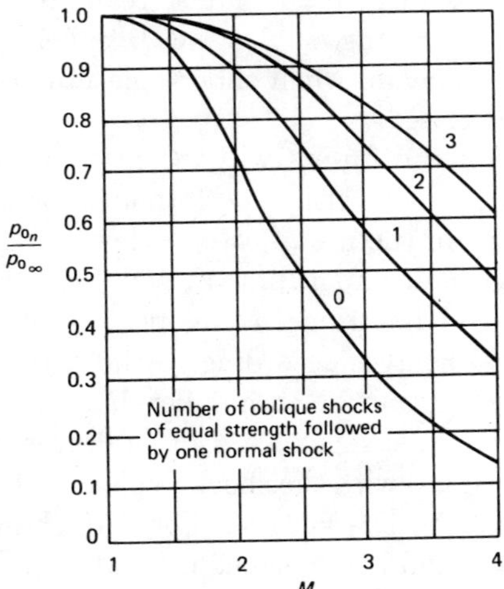

Fig. 3.10 Stagnation pressure ratio for n oblique shocks of equal strength followed by a normal shock (after Oswatitsch).

[5] K. Oswatitsch, "Pressure Recovery for Missiles with Reaction Propulsion at High Supersonic Speeds," NACA TM 1140, 1947 (translation).

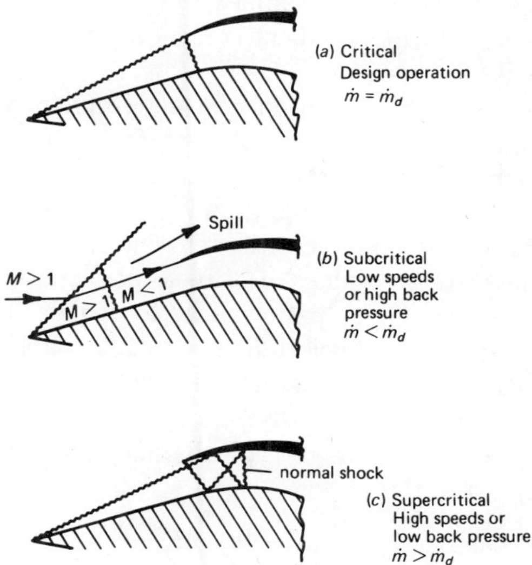

Fig. 3. 11 Modes of operation of spike diffusers.

As for the wedge, the axisymmetric spike can be continuously curved to give ideally an isentropic diffusion by infinitesimal compression waves rather than by shock. In practice however it has limitations, as when in yawed flow, shock may occur and it may be preferable to have the shocks definitely located by finite changes of direction.

These external-compression diffusers can operate exactly in the manner indicated only at the actual design point and when this is not the case, the shocks are of different strengths, of different angles, and sometimes at different locations. Figure 3.11(a) shows a simple one-step diffuser operating at the design point, with the oblique shock at the cowl lip and the normal shock emanating from this point. This is known as the *critical* mode of operation. When less flow is required or at lower forward speeds, the shock angle β (Fig. 3.8) increases, the oblique shock does not intercept the lip, and the normal shock occurs externally. This is called *subcritical* flow [Fig. 3.11(b)]. At higher speeds and flow rates, the oblique shock may strike inside the duct, with the normal shock well downstream. This is *supercritical* flow [Fig. 3.11(c)] where the criterion is that the normal shock be inside the duct, even if the oblique shock is not impinging internally.

In subcritical operation, flow instability may develop resulting in diffuser "buzz." This is defined as unstable, subcritical operation associated with fluctuating internal pressures and a shock pattern oscillating about the diffuser entrance. The fluctuation of the shock has an upstream limit when it detaches from the nose or spike as a bow shock and air is expelled from the inlet. Then the shock reattaches and moves downstream to a limiting position and air is again swallowed. This cycle is unstable and repeats, giving intermittent flow to the engine, which may cause compressor stall or the combustion to become highly inefficient, even to the point of extinction. Buzz is a complex phenomenon, not easy to explain absolutely. Somewhat empirical methods of control are possible, with theory as a guide but not in a detailed manner.

It will be realized that only some of the major phenomena of supersonic diffusion have been touched upon here. Diffusers for transport aircraft become complicated, with variable ramps, cross-sectional areas and spill ports, and automatic control is required for a range of engine and ambient conditions. Considerable detail and a full bibliography are given in the reference.[6]

With respect to quantitative values of diffuser performance, a standard used for evaluating and comparing engine performance in vehicles for specific missions is that of the Aircraft Industries Association. This A.I.A. standard, whose use may be thought of as comparable to that of the standard atmosphere in providing a common base for evaluation, is obviously subject to considerable variations in practice but, nevertheless, is relevant to typical situations. It is given in terms of the stagnation pressure ratio, thus,

$$r_0 = p_{0'2}/p_{01} = 1 - 0.1(M_\infty - 1)^{1.5} \tag{3.19}$$

This is shown plotted in Fig. 3.12, together with the corresponding values of isentropic efficiency η_{is} calculated from Eq. 3.12 and with the normal shock loss for comparison.

Equation 3.19 gives zero pressure recovery for $M \approx 5.6$ and hence is pessimistic at the higher Mach numbers. A military standard for $1 < M < 5$ is given by

$$r_0 = 1 - 0.075(M - 1)^{1.35} \tag{3.20}$$

and for $M > 5$,

$$r_0 = 800/(M^4 + 935) \tag{3.21}$$

This is also given in Fig. 3.12.

3.6 Combustion

The combustion of fuel with oxidant is a very complex problem, combining chemistry, fluid mechanics and thermodynamics, sometimes called aerothermochemistry. Here we shall only touch upon the major limiting features of combustion in relation to propulsion performance and not on the principles and methods of combustion phenomena in general.

The purpose of combustion is the release of chemical energy to the working substance to obtain an increase of internal energy as temperature. Pressure is then the agent by which this thermal energy is converted to directed bulk kinetic energy for effective propulsion. In airbreathing engines, the fuel is a very small and often negligible part of the propulsive fluid but in rockets it is the order of half the total. Hence in the former, the fuel can be chosen wholly for its combustion, storage, handling and cost qualities, while for the latter there can be very important pro-

[6] "Handbook of Supersonic Aerodynamics," Sec. 17, "Ducts, Nozzles and Diffusers," NaVWeps Report 1488 (Vol. 6), Supt. of Documents, U.S. Government Printing Office, Washington, D.C., Jan. 1964.

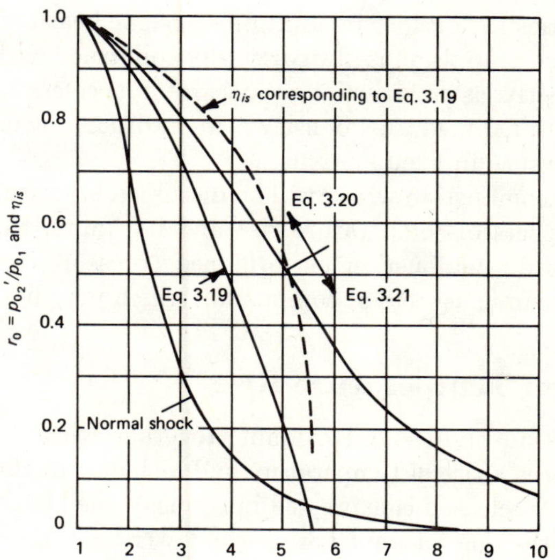

Fig. 3.12 Nominal standard performance of supersonic diffusers.

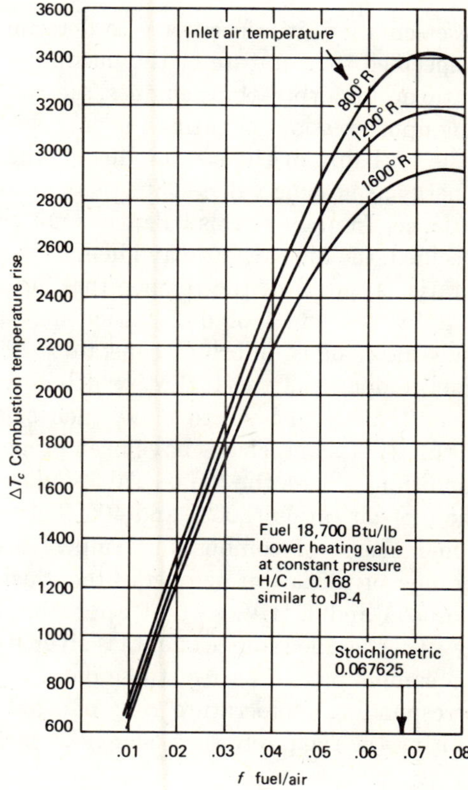

Fig. 3.13 Combustion temperature rise vs fuel-air ratio (data from Huntley).

pulsion characteristics to consider in addition—specific heat and molecular weight, for example. Because propulsion is always vitally concerned with mass of the whole system, either directly as an inertial property for acceleration or indirectly as weight in a gravitational field, fuel density is an additional consideration normally not of prime importance in fixed systems.

The aspects of combustion which will be discussed here are concerned with the thermodynamic aspects of temperature rise and its limits, the general effects of ambient conditions on combustion performance (pressure, temperature and velocity), and gas dynamic aspects of combustion which may limit propulsive effect.

3.7 Combustion Temperature Rise

The chemical reaction of fuel and oxidant, together with the presence of inert components, produces a rise of temperature and a change in chemical composition. The chemical energy released is expressed in terms of the heating value, which has several forms, and the one relevant here is the lower heating value L.H.V. at constant pressure. The use of the lower value implies that the final temperature at discharge from the propulsive device is above the condensation temperature of the water vapor in the products of combustion if the mixture originally contained hydrogen and oxygen.

For performance calculation, it is necessary to determine the amount of fuel required to obtain a specified temperature or the maximum possible temperature, i.e. the fuel-oxidant ratio or, for airbreathing engines, the fuel–air ratio. In principle, this is found by setting up expressions equating the enthalpy of entering reactants, say air and fuel, to the enthalpy of the exit products. The entering enthalpy includes the chemical energy as heating value. Thus there must be an energy balance and a mass-species balance. In practice this balance is not easy to make because if the temperature is specified, the fuel–air ratio and hence the products are unknown, while if the fuel–air ratio is specified, the temperature is unknown. The problem is further complicated by the effect of dissociation at high temperatures and, hence, requires consideration of equilibrium constants. Each calculation has to specify a particular fuel composition and L.H.V., which may vary for hydrocarbon fuels. Since it would be tedious to have to make such calculations ab initio for each and every condition, tables and charts have been prepared for "typical" fuels used in "typical" conditions. Thus the "Gas Tables" (see p. 49) give enthalpy values for two specific fuel–air ratios (200% and 400% of the theoretical air) and elsewhere one may find charts of combustion temperature rise plotted against fuel–air ratio for a range of initial air temperatures. Such charts presuppose a particular fuel composition and L.H.V. so that approximate corrections have to be made for differences. One detailed source of data is given by a NACA publication[7] and Fig. 3.13 shows a sample plot of part of these data. Up to a fuel–air ratio of about 0.04, with corresponding temperature rises of 2200–2400°F, the lines are slightly curved, exhibiting the relatively slow change of specific heats and compo-

[7] S. C. Huntley, "Ideal Temperature Rise due to Constant-pressure Combustion of a JP-4 Fuel," NACA RM E55G27a, 1955.

sition. At high temperatures however, the rate of increase of temperature rise with f decreases rapidly, a maximum is reached, and then the slope becomes negative. This is due to dissociation, with the energy addition that goes to vibration of the molecules being sufficient to split them into their component atoms, with decrease and eventual cessation of increase of molecular translational energy known to us as temperature. Thus there is a *maximum flame temperature* for a given fuel and this occurs not at the stoichiometric fuel–air ratio but at greater values (rich mixtures), as excess fuel tends to suppress dissociation. Figure 3.13, although quite inadequate for numerical work, shows the order of f required for a given temperature. For example, a turbojet of 13/1 pressure ratio with a compressor efficiency of 0.85 with an inlet temperature of 60°F has a combustion inlet temperature of about 1200°R. For a turbine inlet (= combustion discharge) temperature of 2000°F, a combustion temperature rise of 1260°F, the fuel–air ratio is 0.02. This is only about 0.3 of stoichiometric so that a considerable quantity of excess air is required.

3.8 Combustion Performance

The combustion reaction process results from the collision of molecules of fuel and oxidant, with the collision requiring an energy level above a certain *activation energy E* so that the molecules are broken down into simpler elements, with subsequent collisions leading to the final products of combustion. The process is one of reaction kinetics and even the simplest of combustion reactions involving only two chemical elements can require several intermediate steps. However, the general chemical kinetic approach is useful in obtaining an overall view of the effect of controlling parameters.

In a gas reaction involving collision of two molecules, a bimolecular reaction, the rate of reaction can be expressed from the kinetic theory as

$$\dot{r} \propto C_0 C_f \rho^2 \sigma^2 T^{1/2} M^{-3/2} e^{-E/RT} \tag{3.22}$$

where

$\dot{r}$ = mass rate of reaction per unit volume of conversion of one component
C_0 = fractional concentration of oxygen
C_f = fractional concentration of fuel
ρ = mixture density
σ = measure of molecular diameter
T = temperature of mixture
M = mean molecular weight
E = energy of activation
R = gas constant

The first six terms on the right-hand side represent the number of collisions occurring and the last term, $\exp(-E/RT)$, is the fraction of collisions with sufficient energy to result in reaction. This last term is very important as it is exponential and is seen to depend on a constant for a given mixture, the activation energy, and the temperature of the mixture. A small change in temperature can make a large change in the value of the term.

Now considering the terms individually, the concentrations C_0 and C_f are a function of temperature, because as the reaction proceeds the concentrations are reduced as the temperature increases. Density ρ is proportional to pressure and inversely to temperature, while σ and M are fixed for a given mixture. Thus Eq. 3.22 can be written as

$$\dot{r} = Ap^n f(T) \tag{3.23}$$

where exponent n generalizes the expression for other than bimolecular reactions. The value of n is composed of a weighted mean of some sort made up of the order of the various intermediate reactions contributing to the final state, e.g. one, two or three. The experimentally adduced value for stoichiometric reactions is about 1.8 and this shows that combustion may be difficult at high altitudes where the absolute pressure is very low, even after ram and engine compression. The effect of temperature is felt mostly from the exponential expression and it is found that combustion proceeds very slowly at mixture strengths, and hence flame temperatures, not far removed from the stoichiometric and is impossible at some reduced value. Thus for the example of combustion temperature rise given in the preceding section, reaction at a uniform temperature of 2000°F proceeds at such a slow rate that it is virtually nonexistent as combustion in the manner required for engines. Thus the combustor has to be designed so that the reaction itself occurs close to the stoichiometric by mixing only a portion of the air with the fuel, adding the remainder as diluent after combustion of the fuel is complete to reduce the mixture temperature to the required value for the turbine inlet.

A property of a combustible mixture at a given state is the flame speed or rate of propagation of the flame. As a property, it is valid only for a quiescent mixture or in laminar flow, because turbulence (small-scale) or eddying introduce independent variables. Flame speed has a very low value, even in what are considered to be fast-reacting mixtures such as hydrogen-air, being of the order of a few feet per second. This gives notice that some special arrangement for combustion is necessary, as throughflow velocities in any propulsion engine are higher by one or more orders of magnitude. The practical solution is to introduce the fuel into regions

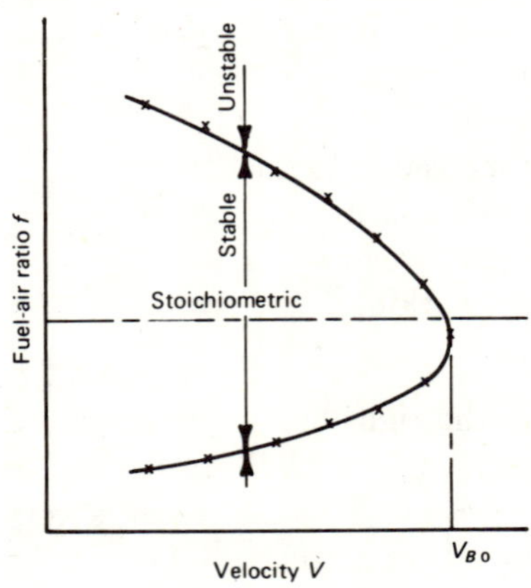

Fig. 3.14 Stability diagram.

where the air velocity is very low and turbulence is high. For example, a bluff body in the flow acting as a baffle creates a wake and a region of reversed flow due to separation. Here the flame can stabilize to provide a combustion region and to feed hot products to preheat the incoming mixture. This stabilizing zone is a key feature of combustors and the characteristics of the combustion can be shown on a stability diagram such as Fig. 3.14. For a quiescent mixture, the fuel–air ratios between which the mixture will react are known as the limits of inflammability and are a function of temperature and pressure only. For a flowing mixture, the velocity becomes the important parameter. The flame is possible, i.e. stable, only between certain rich and lean mixture strengths at a particular value of mixture velocity (or air velocity if the fuel is injected into the air without previous mixing). At fuel–air ratios outside of these, the flame blows off the baffle and is extinguished. The higher the velocity, the narrower is the range of stable flame with respect to mixture strength and at some value of velocity, the peak blow-out velocity V_{Bo} is reached.

A very simple model yields useful information on combustion behavior and design. It is hypothesized that at the peak blow-out velocity point, the rate of flow of fluid into the reaction zone associated with the baffle just exceeds the reaction rate possible and the flame is extinguished. This mass rate of flow $\dot{m}$ is taken to be proportional to the peak blow-out flow velocity V_{Bo}, the area of the reaction zone d^2 where d is a characteristic dimension of the baffle, and the density of the mixture, which for a given inlet temperature is proportional to the pressure p. Hence $\dot{m} \propto V_{Bo}d^2p$. The reaction rate is the rate per unit volume $\dot{r}$ times the reaction zone volume, taken as proportional to d^3. Hence for the blow-out condition,

$$V_{Bo}d^2p \approx \dot{r}d^3$$

and

$$V_{Bo}p \approx \dot{r}d$$

From the reaction rate expression (Eq. 3.23), for a given mixture and temperature level, $\dot{r} \propto p^n$, thus

$$V_{Bo}/p^{n-1}d = \text{constant} \tag{3.24}$$

This simple analysis resulting in Eq. 3.24 has proven very reasonable in correlating test data from a number of experiments. The value of the exponent of the pressure is found to be a little less than unity, agreeing quite well with the value of $n = 1.8$ mentioned previously as representative of the overall order of a hydrocarbon reaction. Thus it would appear as though stability in the sense of ability to sustain combustion is approximately proportional to the product pd, implying poor performance at high altitude and for small-scale combustors. For operation at very low pressures, then a few large combustors are better than a number of smaller ones.

From the previous summary, we may see that (1) the combustion zone should operate around the stoichiometric mixture strength regardless of the overall mixture strength, (2) a region of low velocity or reversed flow must be provided within the overall throughflow region, (3) low pressure and low initial temperature have a deleterious affect on combustion, and (4) the scale of the combustor can be important.

The size of combustors can be approximated by values of allowable combustion intensity or heat release rate. Because of the effect of pressure noted previously, the parameter is given as Btu/(hr) (ft^3) (atm) and values of the order of 10×10^6 are in use for turbojets. Sometimes, following the result of Eq. 3.23, the pressure dependence is given as $(atm)^{1.8}$. Although volumetric heat release rate is an important parameter, the area of the combustion flow path is also useful and values of up to about 10×10^6 Btu/(hr) $(ft)^2$(atm) are again typical. Lower values of these combustion intensities are very desirable as they are easier to realize and result in less arduous conditions for the combustor to withstand.

Because combustion intensities have gradually been increased as engines have developed, it is of interest to know if there is some maximum value possible. Experiments in a "perfectly stirred reactor",[8] that is, in an experimental spherical combustor incorporating elaborate arrangements to provide the most intense mixing, yielded a value of about 300×10^6 Btu/(hr) (ft^3) (atm), some 30 times the present practical limit. This represents a chemical limit, that is, a completely homogenous chemical reaction with the restraints of physical mixing removed. There would appear to be room for improvement then, but it must be borne in mind that intense mixing requires "high" turbulence and the corresponding pressure loss cannot be tolerated.

3.9 Choking Effect of Heat Addition

Addition of heat to a gas flowing in a constant area duct causes changes in pressure, temperature and velocity, regardless of whether this energy addition is caused by heat transfer through the walls or by direct combustion of fuel. To investigate this effect, the three conservation equations are invoked for the particular conditions:

(1) Continuity

$$\dot{m} = A\rho V$$

and for constant area

$$\dot{m}/A = \rho V = G = \text{constant} \tag{3.25}$$

(2) Momentum (assuming frictionless flow)

$$pA + (\dot{m}V/g_c) = \text{constant}$$

$$p + \frac{(\dot{m}/A)V}{g_c} = \text{constant}$$

[8] J. P. Longwell and M. A. Weiss, "Heat Release Rates in Hydrocarbon Combustion," Joint Conference on Combustion, I. Mech. E. and ASME, 1955.

and

$$p + (\rho V^2/g_c) = \text{constant} \tag{3.26}$$

(3) Energy

$$c_pT_1 + (V_1^2/2g_c) + Q = c_pT_2 + (V_2^2/2g_c)$$

or

$$T_{0_1} + (Q/c_p) = T_{0_2} \tag{3.27}$$

These equations can be combined to show changes of properties as heat is transferred. Thus for a given initial state, a subsequent state can be calculated and the properties plotted in a number of ways. One of the most significant is to introduce entropy, as entropy is a function of any two other properties for a simple, compressible substance, and then to plot temperature T against entropy s. For a given value of the specific mass flow rate G, a curve results as shown in Fig. 3.15. This is known as a *Rayleigh* line and the process of heat addition to a fluid flowing in a frictionless, constant-area duct is known as a Rayleigh process. For each value of s up to a certain limiting value there are two values of T. Analysis shows that the upper value of T corresponds to a lower value of velocity, and the lower one to a higher value, and correspondingly, subsonic and supersonic values of Mach number. State y would appear to be a limiting state and this can be investigated as follows. The assumption of frictionless flow allows us to use the Euler equation

$$(dp/\rho) + (VdV/g_c) = 0$$

and introducing Eq. 3.25, we can get

$$dp/d\rho = G^2/g_c\rho^2 = V^2/g_c \tag{3.28}$$

Thus along a Rayleigh line of constant G, $V = (g_c dp/d\rho)^{1/2}$ anywhere. At point y, $ds = 0$, and instantaneously the process is adiabatic. Under these conditions of isentropic flow, since there is internal reversibility (no friction),

$$V = (g_c dp/d\rho)_{\text{isen}} = a,$$

the acoustic velocity. Hence at y, $M = 1$.

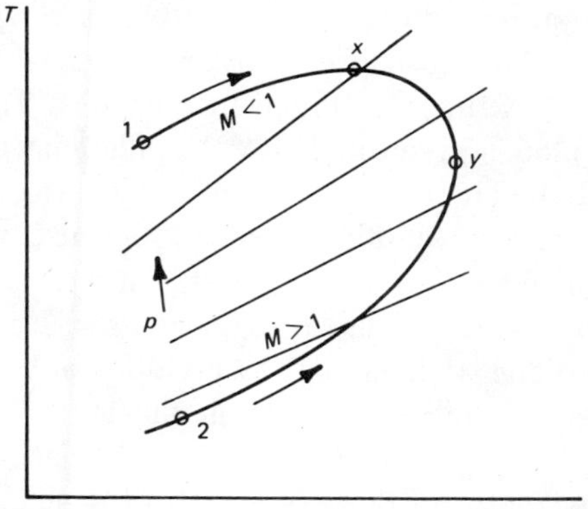

Fig. 3.15 Rayleigh line.

From the second law of thermodynamics, heat addition can cause only an increase of entropy and thus changes of state by heating can proceed only from left to right on the curve. So, starting at point 1, adding heat causes an increase in velocity and Mach number from a subsonic value until point y is reached, where $M = 1$. Further addition of heat at this point under the same conditions is impossible because the entropy cannot decrease by this means. Similarly, starting from point 2, addition of heat causes a reduction of an originally supersonic Mach number until $M = 1$ is reached. What happens if further heat is added at point y, something which we know is physically possible? The only thing that can happen is for the mass flow rate to change, i.e. to move to another G-line which will allow an increase of entropy by addition of heat at a lower specific mass flow rate.

Thus there is a possible *choking* effect with heat addition at constant area, that is, at a given initial Mach number, there is a limiting value of T_0^*/T_{0_i} at which the Mach number becomes unity. This means that at a given inlet temperature, the energy input as heat $c_p(T_0^* - T_{0_i})$ is limited and this is a gas dynamic effect, not a physicochemical one. To obtain high energy inputs then, the initial Mach number must be kept low, i.e. there must be adequate diffusion if the approach Mach number is high, as in the case of ramjets for example.

Using the conservation equations underlying the process, together with the definition of Mach number and the equation of state for a perfect gas, ratios of upstream and downstream properties can be formulated in terms of their respective Mach numbers. For example, from Eq. 3.26,

$$p_1 + (\rho_1 V_1^2/g_c) = p_2 + (\rho_2 V_2^2/g_c)$$

With $\rho = p/RT$ and $V^2 = M^2(g_c kRT)$, then this becomes

$$p_1/p_2 = (1 + kM_2^2)/(1 + kM_1^2)$$

Using the downstream state 2 as a reference state (*) where $M_2 = 1$, then with the upstream state as p and T at any Mach number M,

$$p/p^* = (1 + k)/(1 + kM^2) \tag{3.29}$$

This expression can be tabulated once and for all for a range of values of M, with a fixed value of k. A similar procedure relating the variables can be followed for T/T^*, T_0/T_0^* and p_0/p_0^*. Values are given in the "Gas Tables" previously mentioned and a skeleton plot is given in Fig. 3.16. The curve for T_0/T_0^* shows the choking effect at Mach 1. The curve for p/p^* shows a continuous decrease with increasing Mach number as we might expect, because this implies increasing velocity and hence momentum, which requires a force that can come only from pressure. Note however that as discussed earlier and seen in Fig. 3.15, $M = 1$ can be approached by increasing M from subsonic values but must be approached by decreasing M in the supersonic region. Thus the static pressure p *decreases* in subsonic flow and *increases* in supersonic flow with heating. On the other hand, the stagnation pressure p_0 always decreases with heating. The loss is not great in subsonic flow, where p_0/p_0^* has a maximum value of about 1.27, but is large at high supersonic numbers, e.g., $p_0/p_0^* = 3.42$ for $M = 3$.

The curve for T/T^* exhibits a maximum for M a little less than unity and this is clearly seen in the Rayleigh process (Fig. 3.15). Heating increases T from low subsonic values but eventually a state is reached in which the density is decreasing so rapidly that the velocity increases sufficiently fast so as to cause the static temperature to fall. Using Eq. 3.28,

$$V^2/g_c = dp/d\rho$$

and dividing through by kRT gives

$$V^2/g_ckRT = M^2 = (kRT)^{-1}(dp/d\rho) = (kp/\rho)^{-1}(dp/d\rho) \tag{3.30}$$

From the equation of state,

$$dp/p = (d\rho/\rho) + (dT/T)$$

and at state x, $dT = 0$, hence $dp/p = d\rho/\rho$ and Eq. 3.30 becomes

$$M^2 = k^{-1} \tag{3.31}$$

Thus at $M = (1/k)^{1/2}$ the gas has its maximum temperature in a heating process at constant area. (But note that stagnation temperature, which represents heat input, continually increases.)

We see from this analysis that choking is possible with heat addition, that initial Mach number should be low for maximum heat addition, and that a penalty is paid in decrease of stagnation pressure. These effects can be mitigated by increasing the area in the flow direction but this cannot readily be generalized and often it is not possible or desirable to increase area in an actual design for reasons of overall size of the engine.

3.10 Nozzles

The function of the nozzle is to expand the hot gases down to ambient pressure, transforming thermal energy to directed kinetic energy to produce thrust. The assumptions made for initial analysis are that the flow is adiabatic and reversible (isentropic) with varying area. Using the basic relations for continuity, momentum, and energy for these conditions shows that a convergent-divergent form is needed for the maximum velocity if the available pressure ratio exceeds a certain value and that in the supersonic flow regime, isentropic flow occurs only for a particular receiver (back) pressure for a nozzle of given area ratio with given entry conditions.[9] The behavior is recapitulated by Fig. 3.17, which plots the variation of pressure along the axis of a nozzle, which is equivalent to plotting it against area or area ratio.

When the back pressure is slightly lower than the upstream stagnation pressure p_0, the flow is subsonic throughout, accelerating to the throat where the velocity is a maximum and decelerating thereafter so that the nozzle exit pressure p_E just

[9] For a review of basic nozzle flow, see for example A. H. Shapiro, "The Dynamics & Thermodynamics of Compressible Fluid Flow," Vol. I, Ronald Press, New York, 1953 or D. G. Shepherd, "Elements of Fluid Mechanics," Harcourt, Brace & World, New York, 1965.

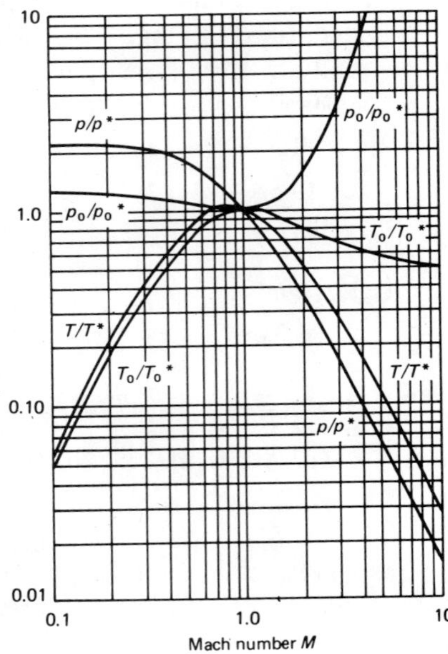

Fig. 3.16 Variation of properties with Mach number—Rayleigh process.

matches the back pressure p_B. In this regime, shown as (*a*) in Fig. 3.17, the mass flow varies with the pressure ratio p_B/p_0. As pressure p_B is lowered, the velocity increases everywhere. At a certain value of p_B/p_0, the Mach number reaches unity at the throat but is still subsonic thereafter, Fig. 3.17(*b*). The value of p_B/p_0 is a function of the *area ratio* A_E/A_t (A/A^* for isentropic flow). The mass flow is then the *choking* mass flow and remains the same for all lower values of p_B. As the throat is now choked, further reductions of back pressure p_B cannot yield an isentropic solution with $p_E = p_B$. Flow continues, of course, and in the diverging part of the nozzle a normal shock occurs, at a strength and location such that the succeeding subsonic flow is diffused to the exact pressure $p_E = p_B$. Three such possible states are shown as *c* in Fig. 3.17.

With further reduction of p_B, the normal shock moves downstream until it occurs right at the nozzle exit plane, with the pressure at the downstream side of the shock just equal to p_B. This is shown as state *d*. Following this, lowering of p_B produces oblique shocks at the nozzle exit, as shown at *e*. The pattern of reflected shocks, expansion waves, and compression waves required to adjust the pressure is shown in Fig. 3.18(*a*). This situation is known as *overexpansion*.

When the correct back pressure to satisfy continuity and the isentropic expansion relations is reached, as at *f* in Fig. 3.17, then the flow is supersonic throughout and expansion occurs continuously to $p_B = p_E$ at the exit. This is called *complete expansion*. Further lowering of back pressure requires a further expansion of the gas from p_E and this occurs through a fan of Prandtl-Meyer type expansion waves, *g* of the figure. Reflection of expansion waves from the jet boundary outside the

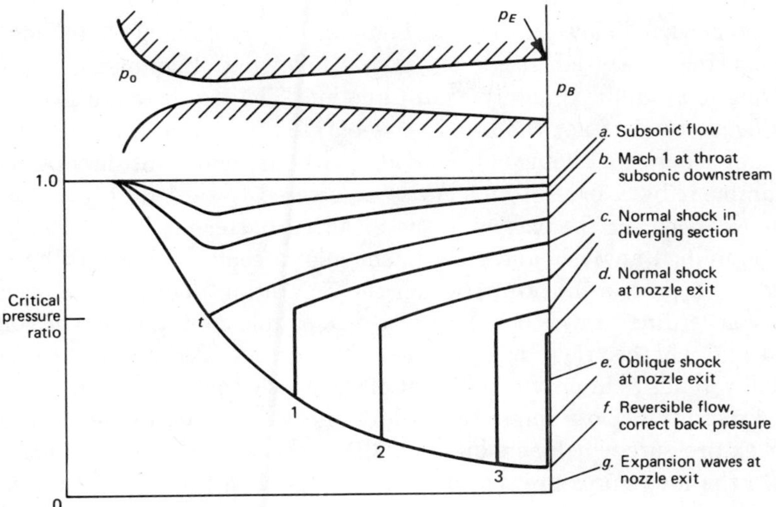

Fig. 3.17 Convergent-divergent nozzle performance.

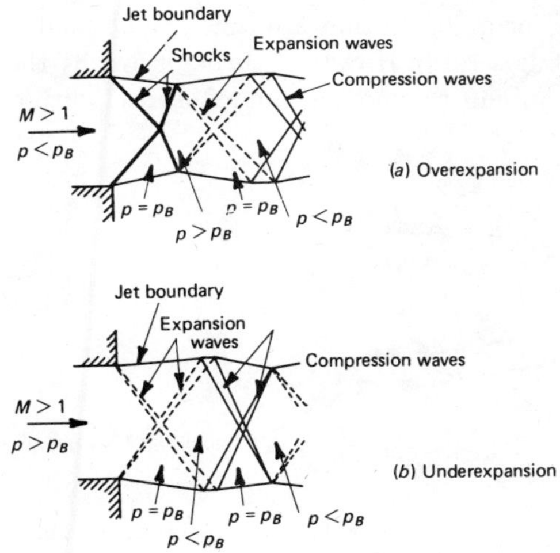

Fig. 3.18 Discharge flow patterns, over- and underexpansion.

nozzle requires compression waves, as shown in Fig. 3.18(*b*). This state of affairs is known as *underexpansion* and is the common case for rockets at high altitude and out in space.

The picture presented above is for idealized conditions. The external flow patterns shown in Fig. 3.18, while essentially seen as the "shock diamonds" characteristic of nozzle flow, are modified considerably by viscous effects, both at the jet boundary and internally in the jet stream. In particular, the normal shock at exit, of Fig. 3.17(*d*), does not occur as shown. At this value of back pressure and for values somewhat below it, the wall boundary layer influences the flow to the extent of causing separation inside the nozzle following oblique shocks such that the flow follows an approximately axial direction. The flow pattern is one of "jet flow" analogous to (but not the same as) separated flow in a diffuser (Fig. 3.19). The region of separation, being subsonic and without significant directed velocity, is at the ambient back pressure p_B. As p_B is lowered toward f, at some value the oblique shocks occur at the exit as in the idealized pattern of Fig. 3.19. A combination of simplified analysis and experiment allows some estimate to be made of the location and pressure ratios at which the separation phenomenon occurs, but values vary according to whether the nozzle is a simple conical shape or contoured (see later) and whether there is an external flow or not. Perhaps surprisingly, the separated flow tends to improve the thrust of the nozzle compared with the idealized pattern, as there is less loss across the oblique shocks than across the normal shock and the back pressure which acts on the walls in the separated zone is higher than that due to the lower pressure which would occur when the flow is overexpanded in this region, i.e. expressing the thrust as $\dot{m}V_E/g_c + (p_E - p_B)A_E$, if p_E is less than p_B, the pressure term becomes negative. Some of this gain is lost from the viscous effects of separation, but nevertheless this is one case in fluid flow for which separation produces a beneficial effect.

A conical type of nozzle is the simplest to construct and the half-angle of divergence may be 20° or more, so that a nozzle can be made much shorter than a diffuser for equivalent area ratio. However, thrust is lost by the nonparallel streamlines at discharge, because of the momentum associated with the radial component

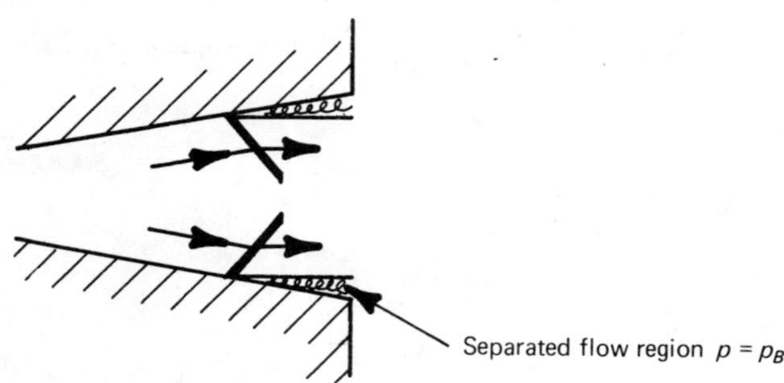

Fig. 3.19 Separated flow, supersonic nozzle.

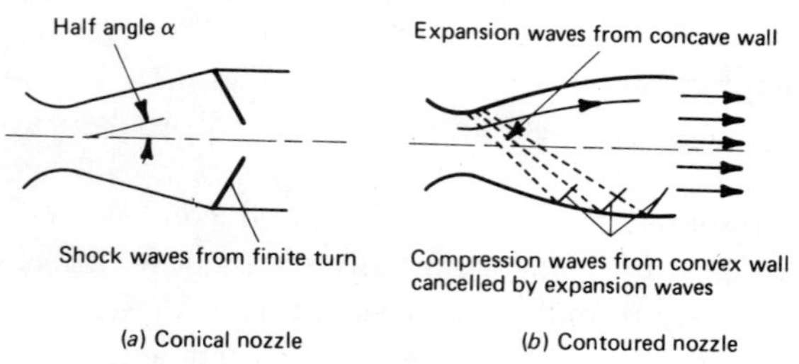

Fig. 3.20 Comparison of conical and contoured nozzles.

of velocity. It can be shown that the loss can be given by a coefficient λ modifying the one-dimensional thrust, i.e., $F_{act} = \lambda F_{calc}$, with

$$\lambda = (1 + \cos \alpha)/2 \tag{3.32}$$

where α is the half-angle of the conical nozzle. The loss is small but significant, e.g. $\lambda \approx 0.97$ for $\alpha = 20°$, and therefore there is a case for trying to make the exit flow parallel to the axis. If this were done by simply adding a parallel portion as shown in Fig. 3.20, then the finite turn angle would cause a shock emanating from the corner and hence much larger losses. If, however, the nozzle is shaped or *contoured* similar to Fig. 3.20(*b*), then apart from friction, a loss-free nozzle with parallel, axial discharge can be made. Full understanding and design requires a knowledge of the method of characteristics in gas dynamics, but the basic concept can be understood as follows. It is the convex portion of the nozzle at the downstream end which causes compression waves that can lead to shock, the finite turn of Fig. 3.20(*a*) being the ultimate development of this. However, the concave portion of the nozzle immediately following the throat produces expansion waves that are isentropic and hence without loss. If then the geometry is made such that the expansion waves from the concave part reach the walls in the convex region, their lowered pressure neutralizes the increased pressure of the compression waves, which hence are not formed. The expansion waves turn the flow continuously and reversibly until at discharge the streamlines are parallel.

Contoured nozzles have the disadvantage of being expensive to manufacture as the shape has to be exact and, moreover, the off-design performance has to be considered. To produce the axial discharge, the nozzle length is considerably greater than that of the corresponding conical nozzle and the frictional loss due to this extra length may nullify the gain due to parallel flow. Hence, even a contoured nozzle may be cut off or truncated leaving a small nonparallelism.

The length of a nozzle, conical or contoured, is a very important design item, because although a large area ratio is desirable so as to reduce the discharge pressure p_E to make it as close as possible to the back pressure p_B, the resulting length may increase the frictional loss appreciably at the very high velocities involved and, perhaps more importantly, increases the weight and overall diameter. Since rockets usually operate in a near-vacuum, a compromise in area ratio always has to be made. Combining the continuity equation with the relationships for reversible adiabatic flow of a perfect gas can result in expressions relating pressure ratio p/p_0, discharge Mach number M and A/A^*, where the last term is the ratio of discharge area to throat area. Thus

$$\frac{p}{p_0} = \left[1 + \left(\frac{k-1}{2}\right) M^2\right]^{-k/(k-1)} \tag{3.6b}$$

$$\frac{A}{A^*} = \frac{1}{M}\left[\frac{2 + (k-1)M^2}{k+1}\right]^{(k+1)/2(k-1)} \tag{3.33}$$

$$= \frac{[(k-1)/2]^{1/2}[2/(k+1)]^{(k+1)/2(k-1)}}{(p/p_0)^{1/k}[1 - (p/p_0)^{(k-1)/k}]^{1/2}} \tag{3.34}$$

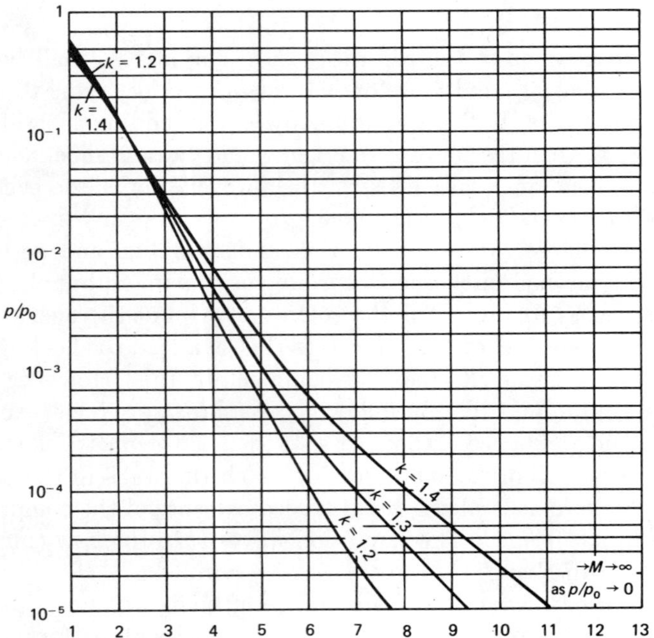

Fig. 3.21 Pressure ratio vs Mach number for isentropic expansion.

p/p_0 vs M is shown in Fig. 3.21 and it will be seen that there has to be a very low back pressure for hypersonic values of M. For example, for a rocket combustion chamber (stagnation) pressure of 1000 psia and p/p_0 of 10^{-5}, which corresponds to an altitude of about 100,000 ft, M is about 7.6 (for $k = 1.2$). Because the nozzle discharge temperature tends to zero as the pressure ratio becomes very small, the Mach number tends to a value of infinity. Figure 3.22 shows p/p_0 vs A/A^* and this is an important figure because a choice of area ratio then determines the expansion ratio and hence the thrust. For the same example as before, a stagnation pressure of 1000 psia and p/p_0 of 10^{-5} and hence altitude of 100,000 ft, the area A/A^* is about 4000, which for a circular nozzle means a diameter ratio of just over 63. Thus the virtual impossibility of attaining complete expansion at high altitudes is apparent.

Perhaps a more useful measure of thrust capability is given by exhaust velocity V_E, rather than exhaust Mach number M. A useful dimensionless parameter for discharge velocity is V_E/a_0, where a_0 is the acoustic velocity corresponding to the stagnation temperature T_0. Putting $M^2 = V_E^2/a_E^2 = V_E^2/g_c kRT_E$ and $a_E^2/a_0^2 = T_E/T_0 = (p_E/p_0)^{(k-1)/k}$ in Eq. 3.6b and rearranging,*

* This relationship can also be obtained from the simple Euler equation $dp/\rho + VdV/g_c = 0$ and the isentropic relationship p/ρ^k = constant, integrating between limits of $p = p_E$ at $V = V_E$ and $p = p_0$ at $V = 0$.

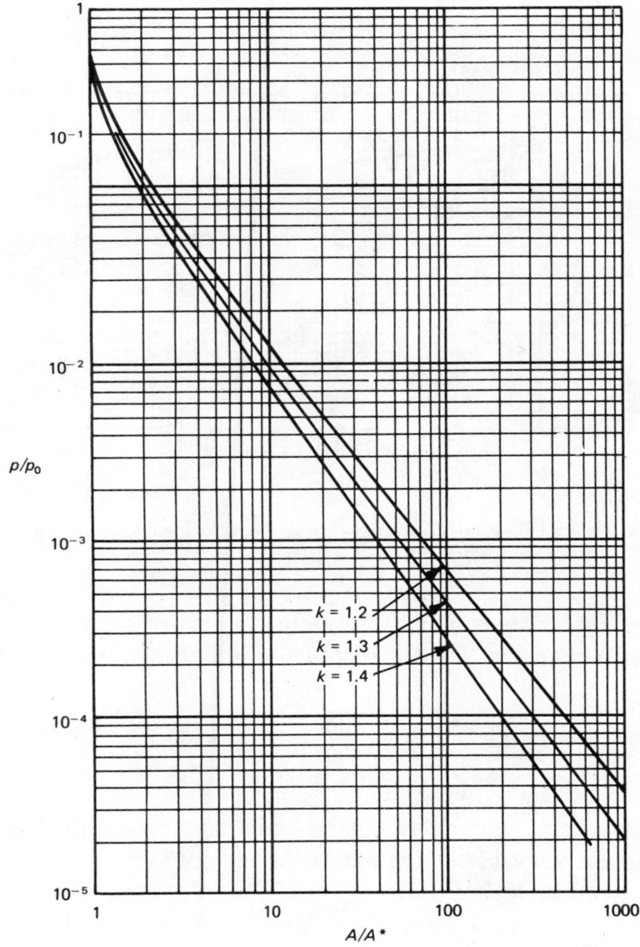

Fig. 3.22 Pressure ratio vs area ratio for isentropic expansion.

$$\frac{V_E}{a_0} = \left\{\frac{2}{k-1}\left[1 - \left(\frac{p_E}{p_0}\right)^{(k-1)/k}\right]\right\}^{1/2} \tag{3.35}$$

This is shown in Fig. 3.23 and for the quoted example V_E/a_0 is about 2.92. Hence with T_0 of 6000°R and $a_0 \approx 4000$ fps, the exhaust velocity is about 11,700 fps or $I_s = 362$. Limiting values of V_E/a_0 are reached when $p/p_0 \rightarrow 0$ and $V_E = 2g_c c_p T_0)^{1/2}$.

Because of the large area ratio required for the pressure ratios involved, complete expansion for rockets is seldom possible and hence the thrust must include the pressure term $A_j(p_j - p_\infty)$. The momentum term is $\dot{m}V_j/g_c$, where V_j is the nozzle discharge velocity V_E corresponding to the area ratio and hence pressure ratio. There is then some loss of thrust, as the pressure term does not give the same value as for complete expansion and it is necessary to know to what this may amount. As an indication, we can calculate the thrust if no divergence beyond the

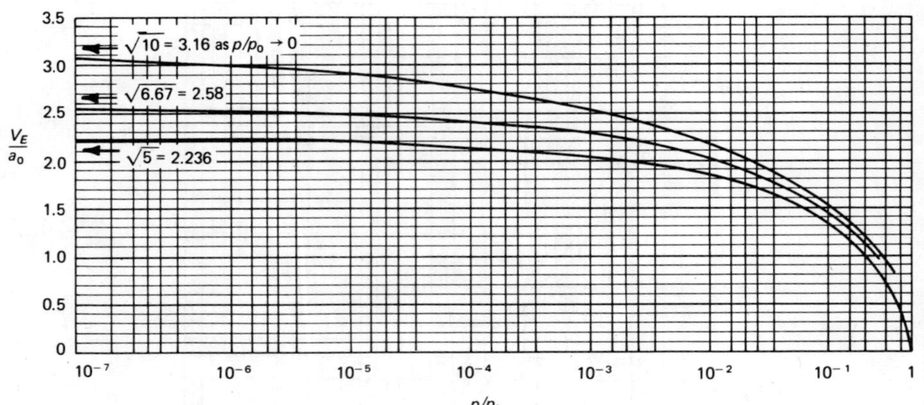

Fig. 3.23 Dimensionless jet velocity vs pressure ratio for isentropic expansion

throat is used, i.e. for a convergent nozzle only, and express it as a fraction of the thrust for complete expansion. This development is left as a problem for this chapter and the result is

$$\frac{F_c}{F_e} = \frac{1 + k^{-1}(1 - \{(k+1)/[2 + (k-1)M^2]\}^{k/(k-1)})}{M\{(k+1)/[2 + (k-1)M^2]\}^{1/2}} \tag{3.36}$$

where F_c = thrust for a convergent nozzle only and F_e = thrust for complete expansion to exhaust velocity equivalent to Mach number M. This is shown in Fig. 3.24. Up to about $M = 2$, the reduction of thrust is comparatively small, but the rate then increases up to $M \approx 5$ when the rate decreases and the value of the factor flattens out. As $M \to \infty$, Eq. 3.36 approaches a limiting value of $(k^2 - 1)^{1/2}/k$, which has values of 0.7, 0.64 and 0.55 for k = 1.4, 1.3 and 1.2, respectively.

The expression (3.36) is valid for rockets at all speeds and for airbreathing engines in the static case, and the ratio F_c/F_E is greater for any forward speed of the latter. For many turbojets, in which the pressure ratios are such that the fully expanded Mach number is not greatly in excess of unity, then the loss due to a convergent nozzle only is small and the resulting simplicity and lowered weight is a considerable advantage. For rockets, with their much greater pressure ratio, some degree of divergence is necessary and therefore the values of Fig. 3.24 are pessimistic, as they are based on a convergent nozzle only. Thus it is seen that although there is a loss of thrust for incomplete expansion, it is not as serious as might have been expected.

For the underexpanded nozzle, the jet boundary spreads out at nozzle discharge, as shown to a small degree in Fig. 3.18(b). This is a consequence of the Prandtl-Meyer expansion of a supersonic flow and the "pluming" of the jet increases as the pressure ratio p_B/p_E decreases.

It can be shown that the Prandtl-Meyer angle, ν, which is the angle through which an initially sonic stream is turned to reach a Mach number M, is given by

$$\nu = \left(\frac{k+1}{k-1}\right)^{1/2} \tan^{-1}\left[\frac{(k-1)(M^2-1)}{k+1}\right]^{1/2} - \tan^{-1}(M^2-1)^{1/2} \qquad (3.37)$$

The maximum deflection then occurs when the expansion occurs into a vacuum and $M \to \infty$. Thus

$$\nu_{\substack{\max \\ M\to\infty}} = \left(\frac{k+1}{k-1}\right)^{1/2} \frac{\pi}{2} - \frac{\pi}{2} = 130.5° \qquad \text{for} \quad k = 1.4$$

Thus theoretically the flow pattern appears as in Fig. 3.25(a), but in practice this will not occur because of viscous effects and the fact that as $M \to \infty$, the temperature goes toward zero absolute. Hence the gas is no longer ideal and its properties cannot be considered those of a continuum. However, considerable

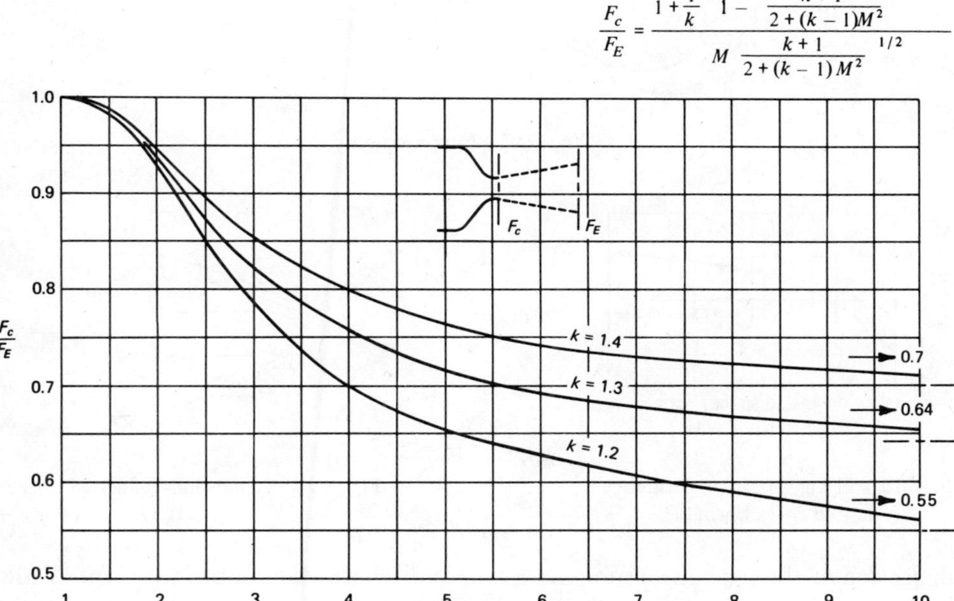

Fig. 3.24 Ratio of thrust for nozzle flow, convergent and fully expanded.

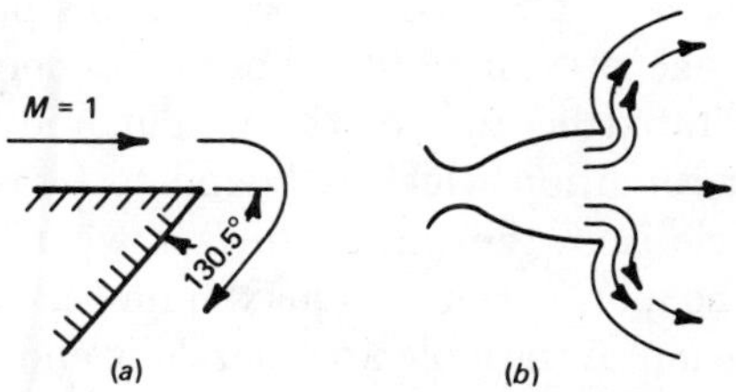

Fig. 3.25 Pradtl-Meyer expansion flow round corners.

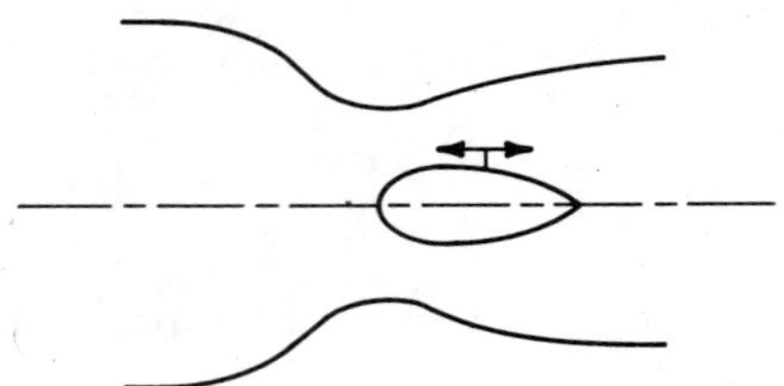

Fig. 3.26 Variable-area nozzle.

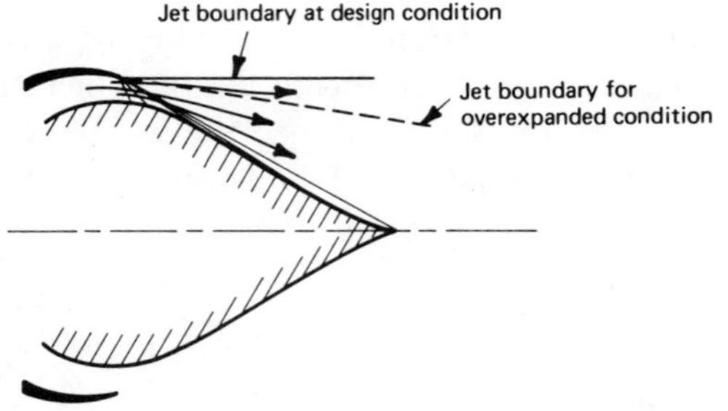

Fig. 3.27 Plug nozzle.

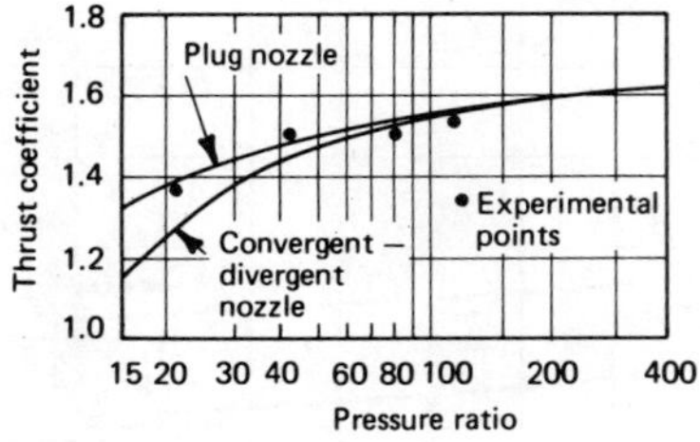

Fig. 3.28 Plug nozzle performance (adapted from Berman).

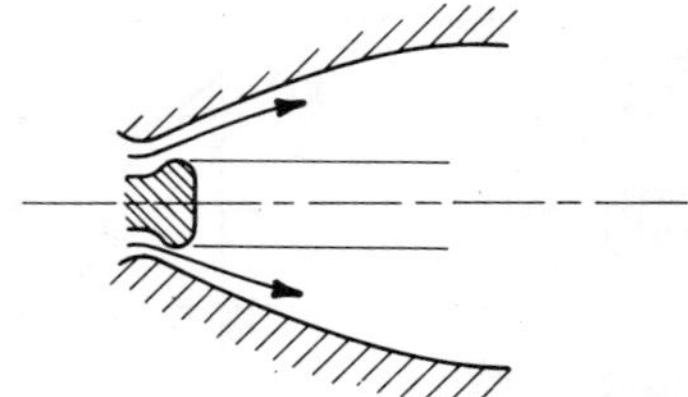

Fig. 3.29 Expansion-deflection nozzle.

deflection of the jet flow occurs and a nozzle discharge flow takes the form of Fig. 3.25(*b*). This flow pattern causes some difficult heat transfer problems at the discharge, particularly when a cluster of rockets is used, as then the jets interfere mutually.

It has been seen that the area ratio of a nozzle can be correct for fully expanded reversible flow only for one pressure ratio. Loss in underexpansion is relatively small and has to be tolerated because of the weight and size limitation. In the case of overexpansion, a condition which is bound to be encountered during the passage of a rocket vehicle from the earth's surface where the back pressure is a maximum, the loss may be more serious, as maximum thrust is usually required in the ascent phase of a mission. A variable area nozzle would alleviate this situation, analogous to the starting operation of a convergent-divergent diffuser, and one method of doing this is shown in Fig. 3.26. The central plug can be moved axially to provide variable throat areas and hence area ratios. For turbojets with afterburning, the range of required area is large and a variable area nozzle is almost mandatory. A more simple form can be used that dispenses with the central plug but has an outer contour that has two positions, one for ordinary operation and one for afterburning operation. For the short time involved and with the temperatures involved ($\approx 5000°F$), a variable central plug for rockets is difficult and costly.

However, there is a form of fixed plug nozzle that has considerable possibilities. The central plug is retained but the surrounding outer cone is removed from the throat onwards, as shown in Fig. 3.27. In supersonic flow, there is a free-jet boundary at the outside as in an underexpanded nozzle, and thus the flow can adjust itself to varying back pressure without a solid boundary imposing a continuity requirement leading to shock recompression. The inner cone provides the necessary surface to take the thrust. At the design point, the flow at the boundary would be parallel to the axis, while at conditions lower than this the outer boundary would be inclined downwards with some loss of thrust due to nonaxial flow. At back pressures below that for the design area ratio, i.e. underexpansion, the plug nozzle has just about the same efficiency as the conventional convergent-divergent form, but in the overexpanded condition, its performance is much better (Fig. 3.28).[10] (The thrust coefficient C_F is a measure of performance discussed in Chapter 7.) Such nozzles have been used but there are difficulties in geometries and in cooling the central plug at gas temperatures of the order of 5000°F.

Plug nozzles of this form tend to be long in comparison with conical nozzles if designed for isentropic flow to give uniform axial flow with no overexpansion at the tip. They can be shortened by cutting off the tip with small effect on performance and 40% shortening has been attained without significant loss. This represents a considerable saving in weight and reduces the cooling problem.

The idea of having a free-jet boundary to accommodate varying pressure ratios is used again in the *expansion-deflection* nozzle (Fig. 3.29), which is really the opposite of the plug nozzle. The gas is contained by the outer cone but the inner flow can assume its own boundary according to the flow conditions.

Another class of nozzles is called the ejector type. Apart from any thrust augmentation, which is difficult to achieve without excessive length and weight, the main idea of the ejector nozzle is to induce an external flow of air which controls the primary flow through the central nozzle. Thus one use is to promote the flow separation that occurs in overexpansion, a condition which yields better performance than nonseparated flow due to the wall pressure distribution as mentioned earlier. Figure 3.30 shows diagrammatically a blow-in-door ejector nozzle.[11] In the static condition, the doors are open, the small amount of air injected keeps the main flow separated, and the nozzle acts essentially as a convergent nozzle. At subsonic-transonic speeds, a large amount of air is induced with the minimum

[10] Adapted from K. Berman, "The Plug Nozzle: A New Approach to Engine Design," *Astronautics*, 5, 4 (April 1960).

[11] D. Nisdal and J. J. Horgan, "Thrust Nozzles for Supersonic Transport Aircraft," ASME Paper No. 63-AHGT-73, 1963.

possible change of velocity and this keeps an outer shroud of air around the main jet, which thus suffers little loss as it is protected by an aerodynamic boundary even though the flow is nominally overexpanded. At supersonic speed at the design condition, the doors are closed and a rear flap opened to give the correct area ratio and the nozzle acts as a properly designed convergent-divergent nozzle.

A complicating factor in nozzle design is that the static behavior and the flight behavior can be different. This is because external flow over the nozzle can produce expansion or shock patterns that change the local pressure conditions at the nozzle discharge. This seems to be particularly true for the otherwise satisfactory external-diffusion plug nozzle, in which the outer flow is controlled by the ambient pressure. On the other hand, the blow-in-door ejector type behaves well under static or flight conditions. Thus it is important that a nozzle be tested under the proper simulated flight conditions.

3.11 Effect of Real Gas Flow

Up to this point, the flow has been considered mostly to be that of a perfect gas, usually with a constant specific heat. At the extremes of temperature and pressure which may be encountered in aerospace operations, this is no longer true and for exact calculations the properties of real gases must be introduced. Furthermore, exhaust products may include liquid or solid particles, sometimes adventitiously as carbon particles and residual ash, and sometimes deliberately as additions of powdered metals to increase the heating value of a fuel. Here we shall simply discuss the major effects qualitatively, as exact quantitative analyses are complex and often uncertain.

One of the most important manifestations is that of high temperature and low pressure resulting in dissociation of the gas molecules into atoms and radicals. This has already been mentioned in the discussion of combustion and it also has

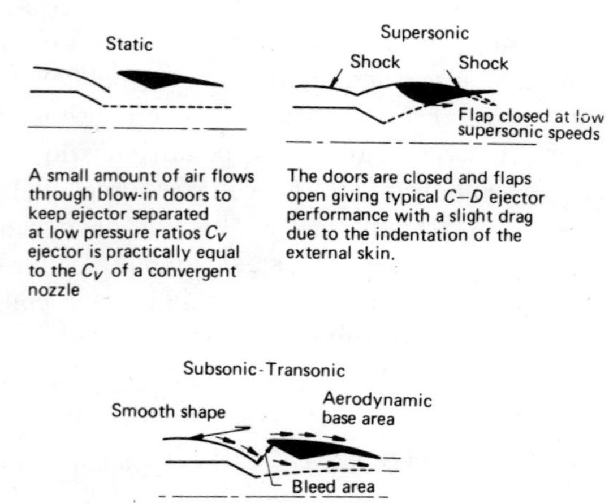

Fig. 3.30 Ejector-type nozzle (from Nisdal & Horgan).

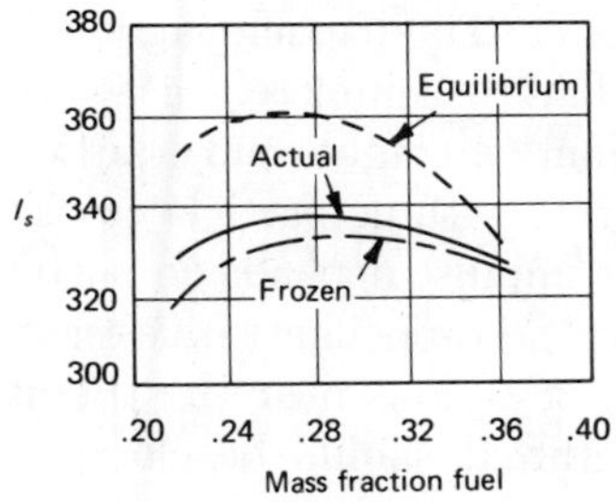

Fig. 3.31 Effect of dissociation on nozzle flow.

an important effect in nozzle flow. As the gases expand, the temperature and the pressure both decrease, with the former having the most effect on dissociation, or rather in this case reassociation, of the atoms and radicals to molecules. The energy release of reassociation increases the temperature of the gas and hence its kinetic energy. However, the recombination process takes a finite time and the residence time in the nozzle of a given particle is of the same order, so that it is a function of the particular species present, the temperature and pressure levels, and the length of nozzle which decides the actual energy transfer process. Thus calculations have to be for specific cases for a precise answer. However, one can estimate the range of possible performance by assuming on the one hand that the reactions occur instantaneously, i.e. the flow is in chemical and thermodynamic *equilibrium** at any moment, and on the other hand that no chemical reaction occurs at all during passage through the nozzle, i.e. the flow is *frozen* at the initial state. Actual flow is somewhere between these two states and Fig. 3.31 shows a typical result, representative of the process in a JP4-liquid oxygen rocket. Two points are apparent: first, that performance as represented by specific impulse can ideally be up to some 10% better for equilibrium flow than for frozen flow and, second, that there can be a particular value of fuel-oxidizer ratio which yields maximum performance and this value is a function of the degree of equilibrium attained. In this case, the experimental performance is closer to the frozen flow line but other reactants, notably H_2–O_2, more nearly approach equilibrium flow, the criterion being relaxation time of the species. Prediction of performance in nozzle flow is very important and digital-computer methods are available for the very complex reactions of many propellant combinations.

3.12 Particle Flow (Condensed Flow)

Small solid and liquid particles in nozzle flow cause loss of thrust dependent on the amount present and the size of the particles. This loss is due to the lower velocity of the particle relative to the gas (particle lag), to the thermal (internal) energy stored in the particles, and to the drag between solid and gas giving a loss of stagnation pressure similar to wall friction. Many assumptions have to be made but these appear to be valid for the size of particle normally encountered, which is a few microns. Drag is assumed to occur according to Stokes Law and likewise

* This state of affairs in nozzle flow is sometimes called *shifting equilibrium*.

the corresponding heat transfer. The particle lag can be important, with the heat transfer factor less so. Similar to completely gaseous flow, it is possible to distinguish extremes of operation, an equilibrium state with particle acceleration and heat transfer and a frozen state with no particle acceleration and no heat transfer. The overall effect on specific impulse depends on particle size, with greater loss for larger sizes. The loss with typical propellants investigated is of the order of 5–10%. The particle lag effect is most serious near the throat, where gas acceleration is greatest, but although the throat should be elongated and made more gradual than usual, any improvement effected is only modest.

Problems

3.1 Calculate the velocity ratio V_2/V_1 for the diffusion of air from $M_1 = 5$ to $M_2 = 3.85$ with $T_1 = 400°R$.

3.2 For an initial Mach number of 5.3, find the final Mach number for the final velocity to be one-half its initial value.

3.3 For a diffusion process of air initially at $T_1 = 400°R$ from $M_1 = 5$ to $M_2 = 3$ resulting in a stagnation pressure ratio r_0 of 0.6, find values of the isentropic efficiency, the kinetic energy efficiency, and the process efficiency.

3.4 For a flight Mach number of 2.23 at an altitude of 40,000 ft where the atmospheric pressure is 2.73 psia, find the final stagnation pressure and Mach number for a diffuser system that has a kinetic energy efficiency of 0.935 and yields a final velocity of 800 fps.

3.5 Show that for a diffusion process with stagnation pressure ratio r_0, the ratio of isentropic efficiency to kinetic energy efficiency is given by

$$\eta_{is}/\eta_k = r_0^{(k-1)/k}$$

3.6 A diffuser has an outlet/inlet area ratio of 1.67 and model tests show that its efficiency is such that $r_0 = 0.92$. If the engine intake (at the diffuser outlet) admits air at $M = 0.33$, find (a) the diffuser inlet Mach number, and (b) the maximum possible engine intake Mach number.

3.7 A combustor has inlet air stagnation conditions of $T_0 = 600°R$ and $p_0 = 50$ psia at a Mach number of 0.4. The combustor has a constant area of 0.1 ft^2. Fuel is burned and the process is assumed to be equivalent to heat addition with a constant specific heat of 0.24 Btu/lb °R and $k = 1.4$.
   (a) Find the necessary rate of heat addition in Btu/sec to cause choking.
   (b) If the inlet stagnation temperature and pressure are kept constant and the heat added is 1400 Btu/sec, find the resulting inlet Mach number and mass flow rate.

3.8 A constant-area combustor has an inlet Mach number of 0.25 and a stagnation temperature of 720°R. The exit stagnation temperature is 2575°R and the ratio of exit to inlet stagnation pressure is 0.71. Find the loss of stagnation pressure due to friction and turbulence in the combustor as a fraction of inlet stagnation pressure.

3.9 A convergent nozzle operates continuously with a stagnation pressure of 25 psia and a stagnation temperature of 2000°F. Assume $k = 1.31$.
(a) What is the exit temperature at ground level?
(b) At what altitude does the nozzle choke?
(c) What is the nozzle discharge temperature and pressure at 100,000 ft?
(d) What is the specific gross thrust (i.e., nozzle only) at 110,000 ft?

3.10 A convergent-divergent nozzle has an area ratio (outlet/throat) of 2 and an inlet stagnation pressure of 10 psia. What is the location of the normal shock (as an area ratio) and the discharge Mach number if the ambient back pressure is 6.23 psia.

3.11 Show that for a stationary convergent nozzle the ratio of thrust F_c to the thrust F_e for a nozzle giving complete expansion to Mach number M_E is

$$\frac{F_c}{F_e} = \frac{1 + k^{-1}(1 - \{(k+1)/[2 + (k-1)M_E{}^2]\}^{k/(k-1)})}{M_E\{(k+1)/[2 + (k-1)M_E{}^2]\}^{1/2}}$$

and that as $M_E \to \infty$, $F_c/F_E \to (k^2 - 1)^{1/2}/k$.

CHAPTER 4

Airbreathing Engines

4.1 The Basic Cycle

As stated previously, we shall be dealing only with the constant-pressure, steady-flow type of airbreathing engine, the normal turbojet, turboshaft, and ramjet engines. For these the basic cycle is that of Joule or Brayton, consisting of reversible adiabatic compression, heat addition at constant pressure, reversible adiabatic expansion, and heat rejection at constant pressure, as shown in Fig. 4.1. Figure 4.1(a) shows the cycle on a p–v diagram and the relatively large area representing the compression work should be noted. This underlines the necessity for high compressor efficiency in distinction to the Rankine cycle where for a steam turbine the pump work is very small, as the working substance is in the liquid phase. Figure 4.1(b) shows the Brayton cycle on a T–s diagram.

The actual engine, apart from losses, only simulates the cycle because heat addition is by internal combustion of fuel and the products are discharged to atmosphere, with fresh air being taken into the compressor to complete the "cycle." The individual processes are not reversible, with inefficiencies of compression and expansion giving increases of entropy and the combustion process requiring a drop of pressure for mixing as well as having losses due to incomplete chemical reaction. The exhaust process entails only a small loss of pressure in the usual case. These are shown in Fig. 4.2. Superscript primes are used where necessary to indicate actual states in contrast to ideal states.

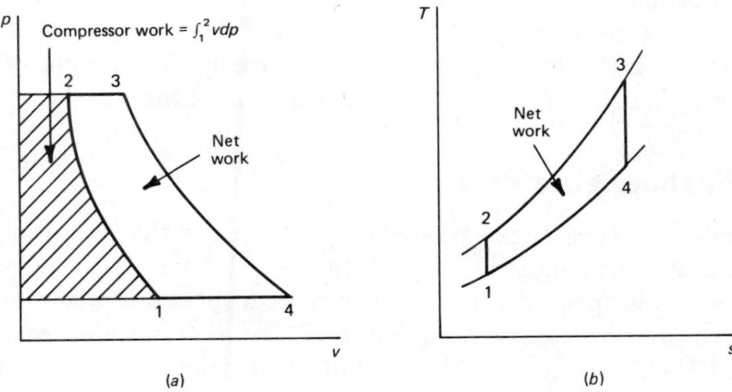

Fig. 4.1 Joule or Brayton cycle: (a) p-v diagram; (b) T-s diagram.

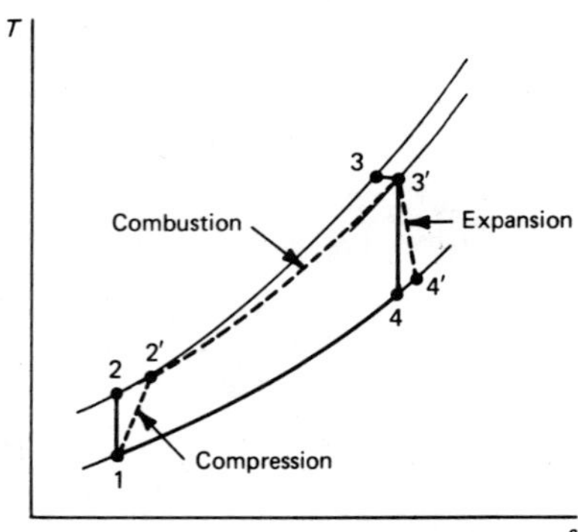

Fig. 4.2 Actual engine processes for Brayton cycle.

The turboengines, as their name implies, have a turbine in the expansion process. For the turbojet, this turbine is required to supply only just the power needed for compression. As the expansion process has an initial temperature much higher than that for compression, the expansion ratio necessary is less than the compression ratio and hence further expansion to the initial pressure is possible and this represents the useful output. For the turboshaft engine, this additional output is obtained through another turbine which drives the propeller, fan, rotor or other propulsion device, while for the simple turbojet, the expansion occurs through a plain nozzle to obtain thrust directly. The ramjet engine has no compressor or turbine, as the compression is obtained by the "ram" effect of forward speed, that is, almost stagnating air taken into the inlet duct at the vehicle speed U. The ramjet is obviously a much simpler engine but it is not operable until some finite speed sufficient for reasonable compression is obtained. The fact that the combustion gases do not have to pass through a turbine allows higher temperatures to be used in greatly simplified fashion.

The turboengines all have in common a diffuser (ram) process, a compressor process, combustion, and a turbine process which supplies the compressor work. Hence we will analyze these common processes and then discuss the utilization of the remaining expansion for the different types of engine. The ramjet will be taken up separately, although its processes are similar to some of those in the turboengines.

4.2 The Turboengine Processes

Figure 4.3 shows the turboengine processes. (Note that in this figure and following T–s diagrams, actual temperatures are distinguished from ideal temperatures by means of superscript "prime.") The air is initially at pressure p_1 and temperature T_1 and reaches pressure p_{0_1} and temperature T_{0_1} by ram diffusion consequent on forward speed U or Mach number M. The diffusion is not reversible and while the stagnation temperature is the same for both real and ideal flow, the actual pressure denoted here by p_{0_1} is lower than the ideal pressure in the ratio r_0 as discussed in

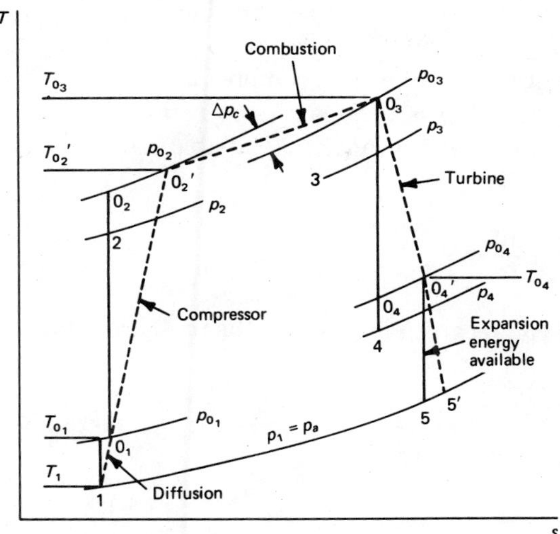

Fig. 4.3 Turboengine processes.

Chapter 3. The compressor process is from p_{0_1} to p_{0_2}, with the discharge temperature being $T_{0'_2}$ consequent on increase of entropy. The combustion process raises the temperature to T_{0_3}, with a drop of pressure to P_{0_3}. The gases then expand through the turbine to some pressure p_{0_4} and temperature $T_{0'_4}$ to supply the compressor work. The energy available for useful output is that for expansion from p_{0_4} and $T_{0'_4}$ to the ambient pressure $p_a = p_1$.

Figure 4.4 shows the compression process in greater detail. From the steady-flow energy equation for an adiabatic compression,

$$h_1 + (V_1^2/2g_c) + W_c = h_2 + (V_2^2/2g_c)$$

and for a perfect gas $h = c_pT$ and with $h_0 = h + V^2/2g_c$ then

$$W_c = h_{0_2} - h_{0_1} = c_{p_2}T_{0_2} - c_{p_1}T_{0_1}$$

Now because of the great advantage for analysis, it is convenient to regard the specific heat during compression as constant, hence where c_{p_a} represents an average value for air,

$$W_c = c_{p_a}(T_{0_2} - T_{0_1}) \tag{4.1}$$

In fact c_{p_a} does not vary a great deal in the range of compression except for the highest pressure ratios. The error is just over 1% at 10/1 and is not allowable for precise work, such as specifications, guarantees and the like, when tables of air properties with variable specific heat must be used. However, by treating air as a perfect gas with constant specific heat, many analytical expressions are readily obtained and the practice is common.

Equation 4.1 shows that it is the *stagnation* temperature which is significant and as velocities can be high, it is important to use stagnation properties. The dynamic temperature or temperature equivalent of velocity is $V^2/2g_c c_{pa}$, which for air at atmospheric temperature can be expressed as $(V/110)^2$ with V in fps. Thus a velocity of 550 fps is equivalent to a temperature of 25°F and such a value is not uncommon at entry or discharge from a compressor.

Compressor efficiency η_c is defined as the ratio of the work for a reversible adiabatic compression to that for the actual adiabatic process for the same pressure ratio. Thus from Fig. 4.4,

$$\eta_c = \frac{W_c}{W_{c'}} = \frac{h_{0_2}}{h_{0'_2}} \approx \frac{c_{pa}(T_{0_2} - T_{0_1})}{c_{pa}(T_{0'_2} - T_{0_1})} = \frac{T_{0_2} - T_{0_1}}{T_{0'_2} - T_{0_1}} \tag{4.2}$$

The numerator can be given in terms of the pressure ratio, since the process is a reversible adiabatic, i.e.,

$$T_{0_2} - T_{0_1} = T_{0_1}[(T_{0_2}/T_{0_1}) - 1] = T_{0_1}[(p_{0_2}/p_{0_1})^{(k-1)/k} - 1]$$

Thus

$$\eta_c = \frac{T_{0_1}[(p_{0_2}/p_{0_1})^{(k-1)/k} - 1]}{T_{0'_2} - T_{0_1}} \tag{4.3}$$

Values of η_c for turbojet compressors range from about 0.80 to 0.87 dependent on pressure ratio, with η_c decreasing with increase of pressure. It is sometimes convenient to use a constant value of the polytropic or infinitesimal stage efficiency η_s, which allows for the inherent reduction of adiabatic efficiency with pressure ratio due to the preheat effect instead of a varying value of η_c. It can be shown[1] that

$$(\eta_c)^{-1}[(p_{0_2}/p_{0_1})^{(k-1)/k} - 1] = (p_{0_2}/p_{0_1})^{(k-1)/k\eta_s} - 1 \tag{4.4}$$

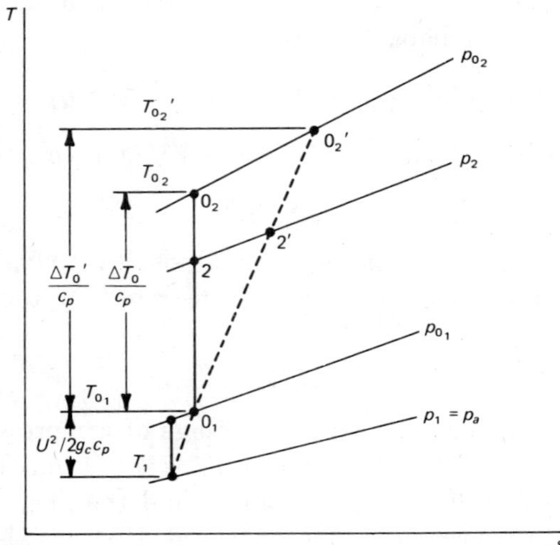

Fig. 4.4 Compression processes.

[1] See, for example, D. G. Shepherd, "Principles of Turbomachinery," Macmillan, New York, 1956.

and hence, writing $p_{0_2}/p_{0_1} = p_{r_c}$ and $(k - 1)/k = \epsilon$,

$$W_{c'} = [(c_{p_a} T_{0_1})/\eta_c](p_{r_c}{}^{\epsilon} - 1) = c_{p_a} T_{0_1}(p_{r_c}{}^{\epsilon/\eta_s} - 1) \tag{4.5}$$

The combustion process from state 02′ to 03 (Fig. 4.3) is subject to a loss of stagnation pressure Δp_c required for mixing and combustion turbulence and this can be 4–5% of p_{0_2}. The value of Δp_c depends on the volume allowed for combustion and hence the throughflow velocity, and on the severity of the conditions for which combustion is required, such as minimum pressure and inlet temperature or range of fuel-air ratio. In addition to the pressure loss in combustion, there may be a loss due to unburned fuel, represented as a combustion efficiency η_{comb} defined as the ratio of the fuel theoretically required for the specified enthalpy rise to the actual fuel required for the same enthalpy rise. Normally η_{comb} is very close to unity, with 98–99% being used as a conservative figure but under certain conditions, notably those near the stability limit discussed in Chapter 3, it may deteriorate considerably.

The turbine expansion takes place from p_{0_3} to p_{0_4}, providing work $W_{t'}$ just equal to the compressor work $W_{c'}$. Although the air mass flow through the turbine is augmented by the fuel flow, it is customary to take it as simply the air flow, i.e., equal to the compressor flow, because invariably some air is bled from the compressor for cooling. Turbine efficiency is defined as the ratio of actual work to the work for the reversible adiabatic process between the same pressures. Thus from Fig. 4.3 and using an average value of specific heat for the gases as c_{p_g},

$$\eta_t = \frac{h_{0_3} - h_{0'_4}}{h_{0_3} - h_{0_4}} \approx \frac{c_{p_g}(T_{0_3} - T_{0'_4})}{c_{p_g}(T_{0_3} - T_{0_4})} = \frac{T_{0_3} - T_{0'_4}}{T_{0_3} - T_{0_4}}$$

$$= \frac{T_{0_3} - T_{0'_4}}{T_{0_3}[1 - (T_{0_4}/T_{0_3})]} = \frac{T_{0_3} - T_{0'_4}}{T_{0_3}[1 - (p_{0_4}/p_{0_3})^{(k-1)/k}]} \tag{4.6a}$$

Denoting p_{0_3}/p_{0_4} as p_{r_t} and $(k - 1)/k$ as ϵ, then Eq. 4.6 may be written as

$$\eta_t = \frac{T_{0_3} - T_{0'_4}}{T_{0_3}[(p_{r_t}{}^{\epsilon} - 1)/p_{r_t}{}^{\epsilon}]} \tag{4.6b}$$

Values of η_t are about 87–90% for turbojets. Expressions in terms of polytropic efficiency corresponding to those for compression are

$$W_t{}' = c_{p_g} T_{0_3} \eta_t \left[\frac{p_{r_t}{}^{\epsilon} - 1}{p_{r_t}{}^{\epsilon}}\right] = c_{p_g} T_{0_3} \left[\frac{p_{r_t}{}^{\eta_s \epsilon} - 1}{p_{r_t}{}^{\eta_s \epsilon}}\right] \tag{4.7}$$

It is obvious that the maximum value of energy available for useful output requires that p_{0_4} should be as high as possible, and from the work expressions we can obtain this. Equating $W_{c'}$ and $W_{t'}$, we have

$$c_{p_a}(T_{0'_2} - T_{0_1}) = c_{p_g}(T_{0_3} - T_{0'_4})$$

and using Eqs. 4.5 and 4.7,

$$c_{p_a}(T_{0_1}/\eta_c)[(p_{0_2}/p_{0_1})^{\epsilon_a} - 1] = c_{p_g} \eta_t T_{0_3}[1 - (p_{0_4}/p_{0_3})^{\epsilon_g}]$$

from which

$$p_{0_4} = p_{0_3}\left\{1 - \frac{c_{p_a}}{c_{p_g}}\frac{T_{0_1}}{T_{0_3}}\frac{1}{\eta_c\eta_t}\left[\left(\frac{p_{0_2}}{p_{0_1}}\right)^{\epsilon_a} - 1\right]\right\}^{k_g/(k_g-1)} \tag{4.8}$$

From this expression we can deduce the following:

1. p_{0_3} should be as high as possible for a given p_{0_2}, so that Δp_c should be as small as possible.

2. c_{p_a}/c_{p_g} should be as small as possible, although actually there is nothing effective that can be done about it. It is controlled by the fluid properties of temperature and composition, and other criteria for the choice of these override any consideration of specific heat values. The effect however is not small, c_{p_g} being some 15–20% higher than c_{p_a}.

3. T_{0_1}/T_{0_3} should be low and expressed in this form as a ratio it shows clearly that the effect of low ambient temperature is as great as high combustion temperature. This is an important feature of turboengine performance, and the atmospheric temperature can make a great difference to takeoff performance, for example. For a turbojet of pressure ratio 15/1 with a combustion temperature of 1700°F and using representative efficiencies, the thrust at takeoff for inlet air temperatures of 0°F and 90°F compared to the thrust at 60°F is about +25% and −10%, respectively. Operation at high ambient temperatures can, therefore, be quite a handicap and sometimes it may be necessary to limit takeoff weight on hot days.

4. The component efficiencies η_c and η_t must be high and a percentage variation of each is as important as the same percentage variation of combustion temperature. Thus in the effect on p_{0_4}, 1% of compressor or turbine efficiency has an effect approximately equivalent to 25°F of combustion temperature.

5. The effect of pressure ratio cannot be determined by inspection, as from the expression, the ratio p_{0_2}/p_{0_1} should be low but $p_{0_3} \approx p_{0_2}$ should be high. However we can get a general answer from a T–s diagram.

Figure 4.5 shows three ideal Brayton cycles, A of very low, B of intermediate, and C of very high pressure ratio, all having the same range of temperature, T_1 to T_3. It is apparent that the extremes of pressure ratio both give a very small work output and that there is some intermediate pressure ratio that gives maximum

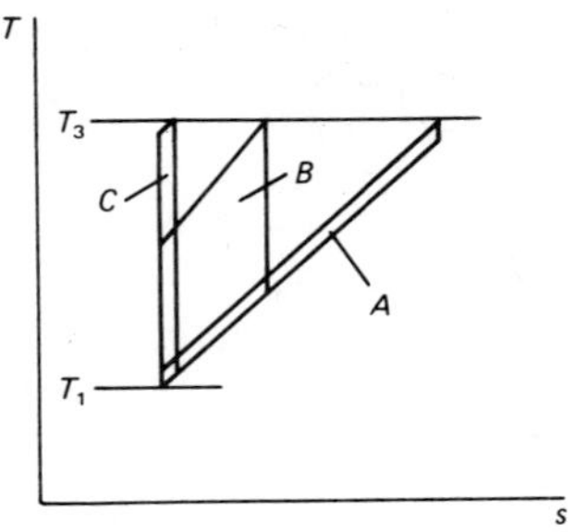

Fig. 4.5 Effect of pressure ratio on the ideal Brayton cycle.

output for a given temperature ratio. This important conclusion is modified, but not changed in kind, by the effect of losses and inefficiencies. Thus, each cycle has an optimum value of pressure ratio for maximum output. It is easy to see from this diagram that an increase of temperature ratio at the same pressure ratio increases the work output or energy available.

Figure 4.5 can also be used to study the problem of efficiency or its inverse, specific fuel consumption. For the ideal cycle, at the low pressure ratios the heat added is at a relatively low average temperature and the second law of thermodynamics tells us that this represents serious irreversibility and hence poor efficiency. As the pressure ratio increases, the temperatures of heat addition and rejection increase and decrease, respectively, until at the highest pressure ratios, the cycle approaches a Carnot cycle with the highest possible efficiency. Thus we conclude that for the ideal cycle, the efficiency increases with pressure ratio. However, for the real cycle this is not so. With increasing pressure ratio, the effect of compressor and turbine efficiencies is such that the actual net work becomes very small, approaching zero at a pressure ratio much less than that of the ideal cycle, while the necessary heat addition remains more nearly the same. The result is that again there is a maximum value of efficiency at some pressure ratio and this is always at a higher pressure ratio than that for maximum work. Thus there has to be a compromise in the cycle parameters between maximum work and minimum SFC. Some typical numerical results for turbojets will be given shortly.

4.3 Cycle Performance

It is useful to generalize the cycle performances of turboengines of various kinds by calculating the energy available over a range of cycle parameters and then showing engine performances of various kinds with the energy available for particular conditions. The range of variables is very large—such as pressure ratio, combustion temperature, component efficiencies and losses, forward speed, and altitude—with each case requiring varying specific heat and adiabatic index. In order to generalize the performance, an analysis is made here with pressure ratio, combustion temperature, forward speed, and altitude as the major variables, with efficiencies and losses being taken at fixed representative values and using fixed mean values of specific heat. Thus some imprecision is inevitable but of a relatively small order compared to the effect of the main variables. The results are correct relatively but inexact absolutely.

The expression for energy available, E_{av}, can be obtained in a manner somewhat similar to that used for the qualitative analysis for p_{0_4} above. This is carried out in Appendix 4.1 with the result that

$$E_{av} = c_{p_n} T_{0_1}[(T_{0_3}/T_{0_1}) - (c_{p_a}/c_{p_g})(p_{r_c}{}^{\epsilon_a/\eta_s} - 1)] \times \left\{1 - \left[\frac{p_1/p_{0_1}}{K_{comb} p_{r_c}[1 - (c_{p_a}/c_{p_g})(T_{0_1}/T_{0_3})(p_{r_c}{}^{\epsilon_a/\eta_s} - 1)/\eta_t]^{1/\epsilon_g}}\right]^{\epsilon_n}\right\} \quad (4.9)$$

with the state symbols as given in Fig. 4.3.

The fixed values to be assigned in this relationship are c_{p_a}, c_{p_g}, ϵ_a, ϵ_g, ϵ_n, K_{comb}, η_s and η_t. The major parameters are p_1/p_{0_1}, T_{0_1}, T_{0_3} and p_{r_c}. The former group are assigned the following values:

$$c_{p_a} = 0.24 \text{ Btu/lb °R}, \quad k_a = 1.4$$
$$c_{p_g} = 0.287 \text{ Btu/lb °R}, \quad k_g = 1.314$$
$$c_{p_n} = 0.275 \text{ Btu/lb °R}, \quad k_n = 1.333$$
$$\epsilon_a = (k_a - 1)/k_a = 1/3.5 = 0.286$$
$$\epsilon_g = (k_g - 1)/k_g = 1/4.184 = 0.239$$
$$\epsilon_n = (k_n - 1)/k_n = 1/4 = 0.25$$
$$K_{\text{comb}} = 1 - \Delta p_c/p_{0_2} = 1 - .04 = 0.96$$
$$\eta_s = 0.88 \text{ (polytropic)}$$
$$\eta_t = 0.90 \text{ (isentropic)}$$

The parameters are taken as follows:

$$p_1/p_{0_1} = (1/r_0)[1 + (k-1)M^2/2]^{-k/(k-1)} \tag{4.10}$$

where

$$\begin{aligned} M &= 0 \text{ at ground level} \\ &= 0.6 \text{ at 15,000 ft}, T_1 = 465.2°\text{R} \\ &= 0.9 \text{ at 35,000 ft}, T_1 = 394.1°\text{R} \\ &= 2.2 \text{ at 35,000 ft}, T_1 = 394.1°\text{R} \end{aligned}$$

and

$$\begin{aligned} r_0 &= 0.98 \text{ at } M = 0.6 \text{ (stagnation pressure ratio, Eq. 3.8)} \\ &= 0.98 \text{ at } M = 0.9 \\ &= 0.87 \text{ at } M = 2.2 \\ T_{0_1} &= T_1(1 + (k-1)M^2/2) \\ T_{0_3} &= 1600°\text{F}, 1800°\text{F}, 2000°\text{F}, 2200°\text{F}, 2400°\text{F} \\ p_{r_c} &= 2\text{–}28 \end{aligned}$$

The choice of Mach numbers is representative, (a) the static (and approximate takeoff) condition for obvious reasons, (b) a medium speed–medium altitude condition, (c) a high subsonic speed at the present-day nominal cruising altitude, and finally (d) a low supersonic speed at the same altitude. Values of energy available E_{av} from Eq. 4.9 for this range of major parameters and for the fixed values of properties and efficiencies are shown in Figs. 4.6 through 4.9.

The figures show the anticipated maximum of useful output at a particular pressure ratio, with this optimum p_{r_c} increasing with increasing combustion temperature T_{0_3}. E_{av} increases rapidly with p_{r_c} to the maximum and then decreases quite slowly except for the $M = 2.2$ case. Here the maximum occurs at a very low compressor pressure ratio, because the ram pressure due to diffusion has a high value ($p_{0_1}/p_1 \approx 9.3$ at $M = 2.2$ for $r_0 = 0.87$). E_{av} decreases more rapidly with increasing p_{r_c} in this case. The energy available increases for a given p_{r_c} and T_{0_3} in the four cases from $M = 0$ to $M = 2.2$. The increase is due to two factors: first the reduced ambient temperature T_1 with altitude, as this implies less compressor work for a given pressure ratio, and second, the increased forward speed that yields a higher pressure p_{0_1} at compressor inlet, this compression not requiring cycle work input.

We can now examine how this energy available E_{av} is utilized in three ways now in common usage, namely the simple turbojet, its development in the turbofan, and in the turboprop. Figure 4.10 shows these possibilities for E_{av}. The upper part

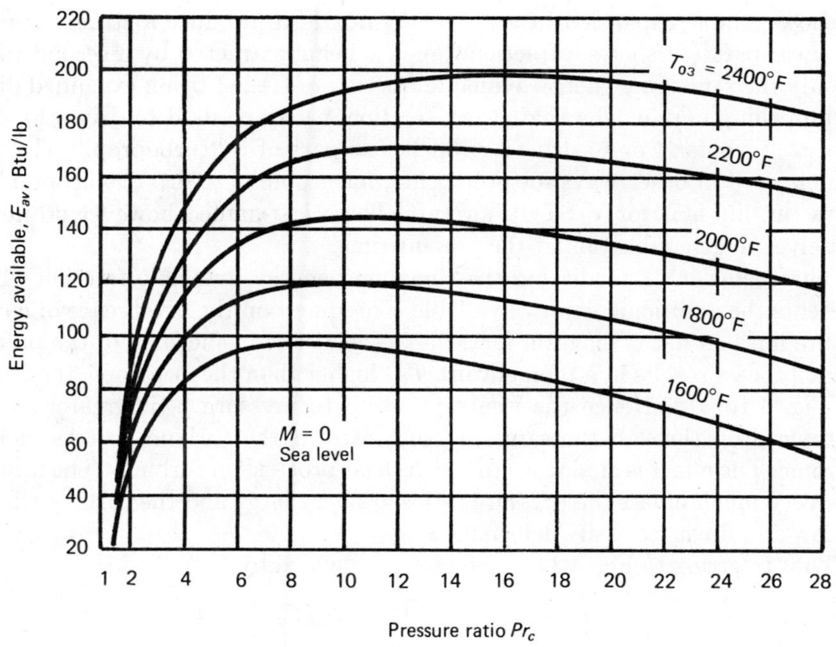

Fig. 4.6 Energy available, $M = 0$, sea level.

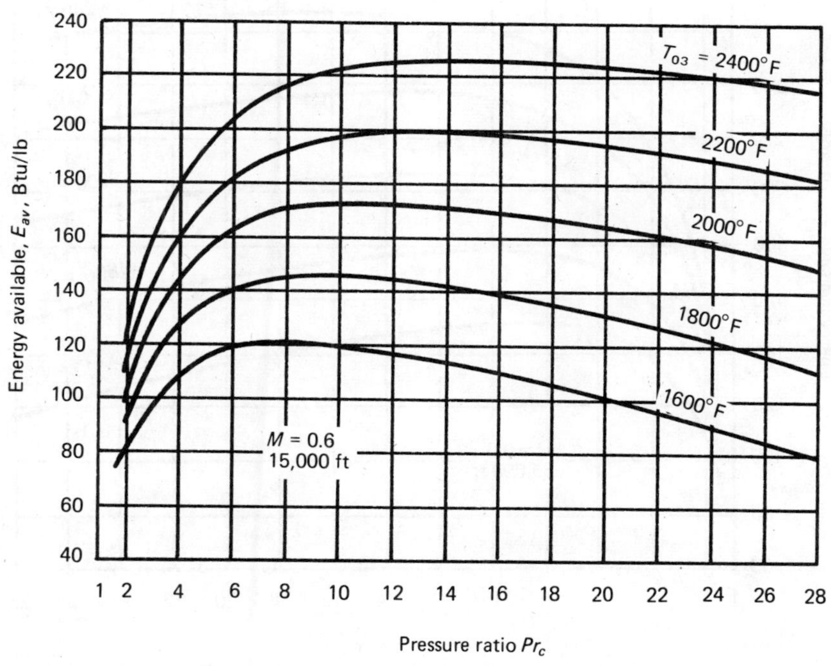

Fig. 4.7 Energy available, $M = 0.6$, 15,000 ft.

(*a*) shows simple expansion in a propelling nozzle to produce a straight turbojet. The lower part (*b*) shows a fraction λ of E_{av} being extracted by a second turbine, and only the remaining energy available fraction $\sigma(1 - \lambda)$ being expanded directly in a propelling nozzle. The shaft work fraction λ may be used to drive the ducted fan of a "turbofan" or to drive a propeller as part of a "turboprop." The turbo-engine is used in other ways for propulsion, for example, to drive helicopter rotors and to run lift fans for V/STOL aircraft. These last-named however do not lend themselves to generalization at the present time.

The coefficient σ modifying the remaining energy available fraction $(1 - \lambda)$ represents the additional energy available consequent on the divergence of constant pressure lines on a T–s diagram for a perfect gas. The inefficiency of the preceding turbine process results in a temperature $T_{04'}$ higher than the isentropic temperature T_{04} [Fig. 4.10(*b*)]. Hence the isentropic drop to pressure p_1 is greater than the isentropic drop through the same pressure ratio if the turbine process had been isentropic. The effect is analogous to the "reheat process" in turbines. The difference is not very much unless the pressure ratios are very large and the turbine efficiency very low and it can be easily determined.

Thus referring to Fig. 4.11, the factor σ is defined by

$$\sigma = \frac{T_d - T_e}{T_b - T_c} = \frac{T_d[1 - (T_e/T_d)]}{T_b[1 - (T_c/T_b)]} = \frac{T_d}{T_b}$$

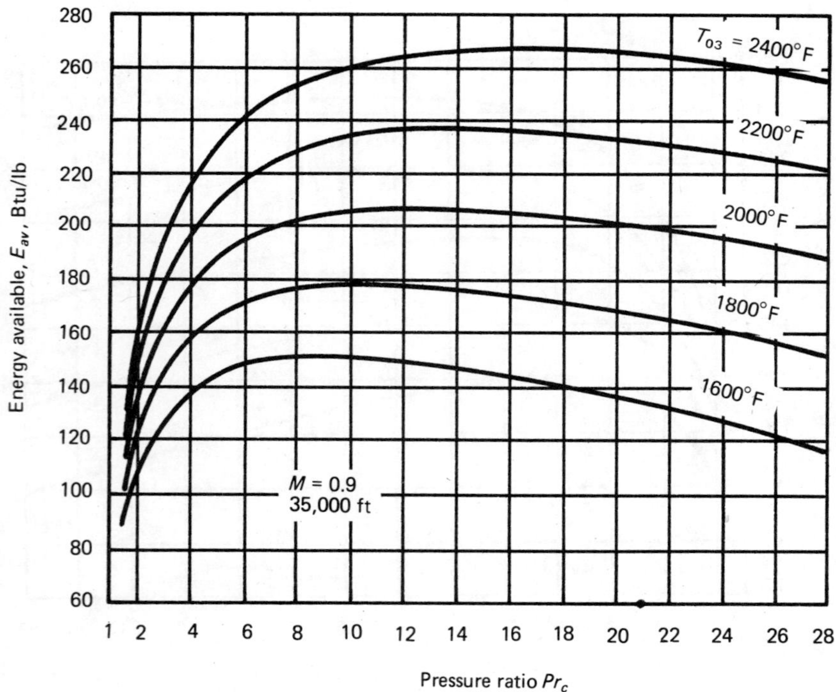

Fig. 4.8 Energy available, $M = 0.9$, 35,000 ft.

since $T_e/T_d = T_c/T_b$ because these are the isentropic temperature ratios for the same pressure ratio p_b/p_c. Now

$$T_a - T_d = \eta_t(T_a - T_b)$$

where η_t is the isentropic efficiency. Hence

$$T_d = T_a - \eta_t(T_a - T_b) = T_a(1 - \eta_t) + \eta_t T_b$$

and

$$T_b = \frac{T_a}{(p_a/p_b)^\epsilon}$$

$$\therefore T_d = T_a(1 - \eta_t) + \eta_t T_a/(p_a/p_b)^\epsilon$$

and so

$$\sigma = \frac{T_d}{T_b} = \frac{T_a(1 - \eta_t) + \eta_t T_a/(p_a/p_b)^\epsilon}{T_a/(p_a/p_b)^\epsilon} = \eta_t + (1 - \eta_t)(p_a/p_b)^\epsilon \quad (4.11)$$

For example, for $\eta_t = 0.87$ and $p_a/p_b = 2$ with $k = 1.33$, then $\sigma = 1.024$. For $\eta_t = 0.7$ and $p_a/p_b = 10$ with $k = 1.33$, then $\sigma = 1.233$. Most values of σ are only a few percent over unity and as its use occurs mainly as the square root, its effect is not great and for initial design values may be considered as unity. (In subsequent analyses, σ has been included for reference, but for numerical work has been taken at a value of unity).

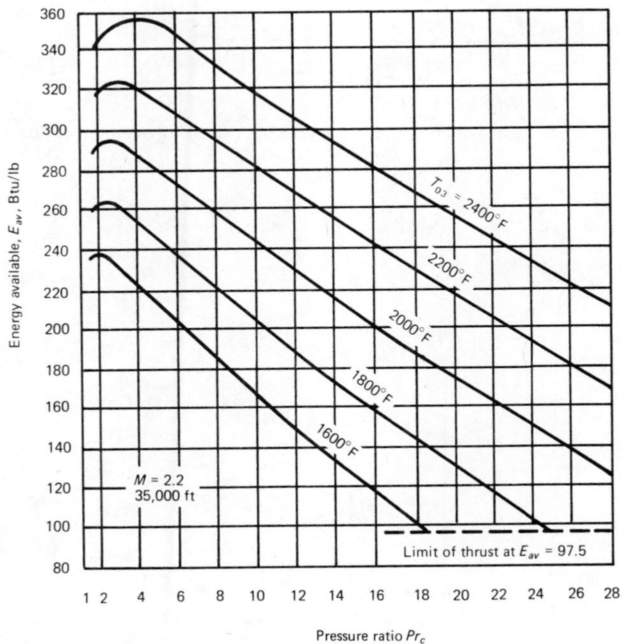

Fig. 4.9 Energy available, $M = 2.2$, 35,000 ft.

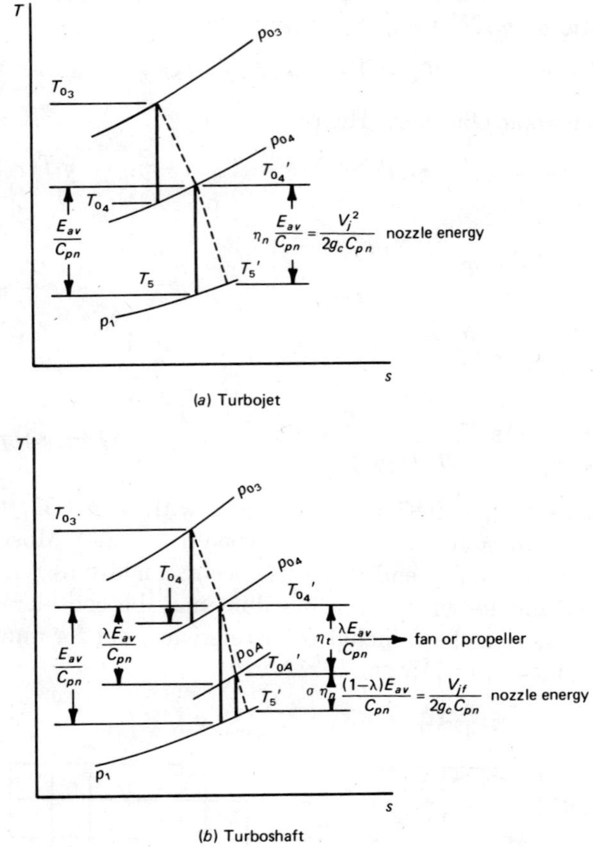

Fig. 4.10 Utilization of E_{av} in turbojet and turboshaft engines.

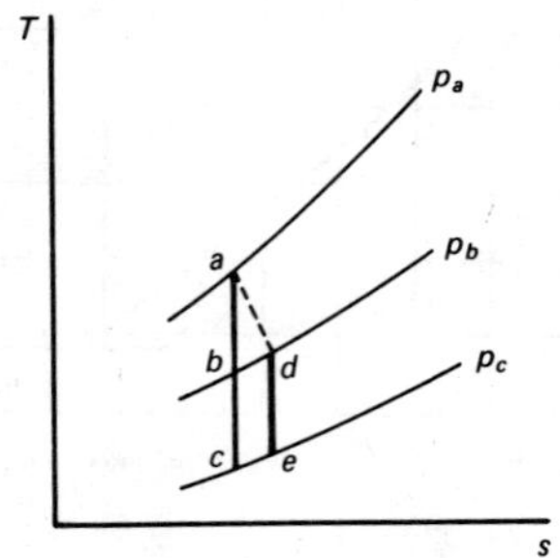

Fig. 4.11 T-s diagram for analysis of correction factor σ due to reheat effect in expansion.

4.4 The Turbojet Engine

For the turbojet, the whole of the energy after the compressor-turbine process is available for thrust in the propelling nozzle. The energy converted to kinetic energy $V_j^2/2g_c$ at the nozzle exit is $\eta_n E_{av}$, where η_n is the isentropic expansion efficiency in the nozzle. This efficiency includes not only the loss due to fluid friction, shock, and so forth, but also covers the case of loss due to underexpansion. That is, η_n includes the effect of a finite nozzle area ratio which, in its extreme form using only a convergent nozzle, is given by the ratio F_c/F_e of Eq. 3.36. Hence V_j, standing for main jet velocity, is in fact an equivalent jet velocity V_{je} of Eq. 2.8, including the incomplete expansion loss and the fluid dynamic loss. In this way, the thrust can be written simply as $\dot{m}(V_j - U)/g_c$. Thus, for specific thrust, lbf/(lbm/sec),

$$F_n = (V_j - U)/g_c = (2\eta_n E_{av}/g_c)^{1/2} - (U/g_c) \qquad (4.12)$$

From the plots of E_{av} at various conditions, the thrust of the simple turbojet can easily be obtained. The value of η_n varies with pressure ratio but assuming that at the design point there is no overexpansion effect and only small underexpansion, a value of 0.94 has been used for demonstration here. Since η_n is defined in terms of actual to ideal kinetic energy, $\eta_n = C_v^2$, where C_v is the velocity coefficient or ratio of actual to ideal velocity given by the nozzle. For $\eta_n = 0.94$, then $C_v \approx 0.97$, which is a reasonable typical value to use.

The specific fuel consumption (SFC) can be calculated using values of fuel-air ratio for a required combustion temperature rise as discussed in Sec. 3.7. Values used here are those of Huntley[2] for JP–4 fuel, a typical turbojet liquid hydrocarbon fuel. The SFC can be obtained thus:

$$\text{SFC} = \frac{\text{lb/hr fuel}}{\text{lb thrust}} = \frac{\text{lb/sec air}}{\text{lb thrust}} \times \frac{\text{lb/sec fuel}}{\text{lb/sec air}} \times \frac{\text{sec}}{\text{hr}} = \frac{3600f}{F_n} \qquad (4.13)$$

Figures 4.12 through 4.15 show values of F_n and SFC against pressure ratio for the four altitude-speed conditions. The shape of the thrust curves follows those of energy available, but is rather flatter because the output is dependent on the first power of velocity rather than the square. Thrust increases with combustion temperature and decreases with forward speed. The latter effect occurs in spite of the decreased ambient temperature, which accompanies the increase of altitude with increase of speed in these figures, and is due to the increasing effect of the inlet drag or momentum, that is the subtractive U/g_c term. At the highest Mach number shown of 2.2, the thrust declines rather rapidly to very small values at low combustion temperatures and pressure ratios greater than the optimum. At high forward speeds, then, the margin of energy available is small. At high altitudes, the aircraft drag is low because of the low ambient density and hence only a small thrust is necessary.

The curves of SFC display two noticeable features. One is that with the exception of those for $M = 2.2$, values of SFC decrease uniformly with increasing

[2] S. C. Huntley, "Ideal Temperature Rise Due to Constant-Pressure Combustion of a JP-4 Fuel," NACA RM E55G27a, 1955.

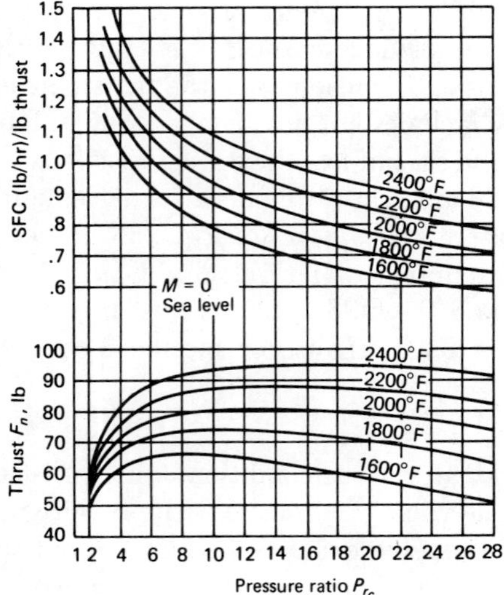

Fig. 4.12 Thrust and SFC for turbojet, $M = 0$, sea level.

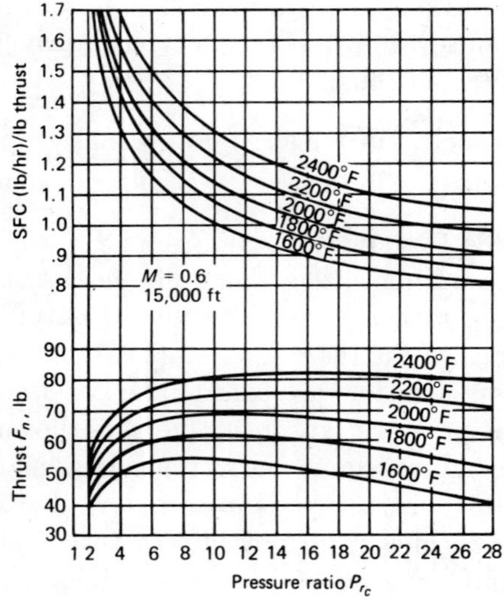

Fig. 4.13 Thrust and SFC for turbojet, $M = 0.6$, 15,000 ft.

pressure ratio. The SFC is the inverse of efficiency, which it was seen had an optimum pressure ratio and thus SFC has a minimum at some value of p_{rc} but this optimum value is greater than 28/1 in the $M = 0$, 0.6, and 0.9 plots. For $M = 2.2$, the minimum is seen clearly at the lower temperatures. Minimum SFC always occurs at a higher pressure ratio than for maximum thrust as discussed in Sec. 4.2.

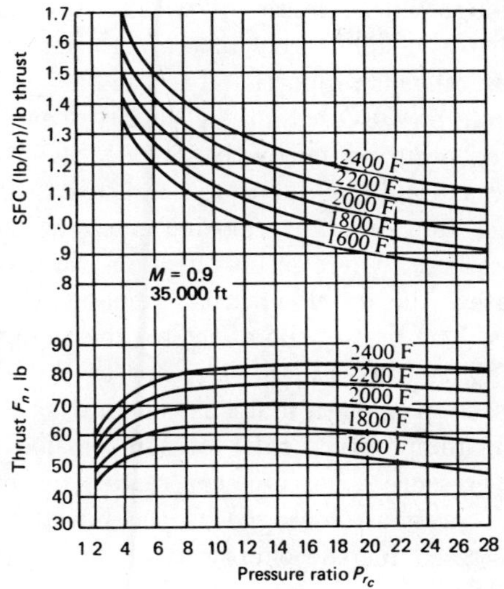

Fig. 4.14 Thrust and SFC for turbojet, $M = 0.9$, 35,000 ft.

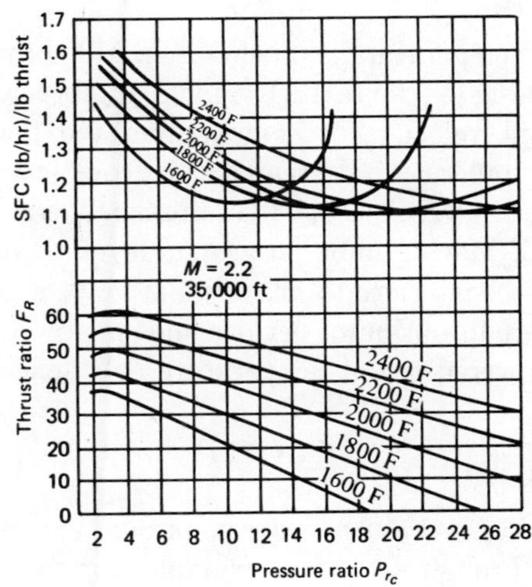

Fig. 4.15 Thrust and SFC for turbojet, $M = 2.2$, 35,000 ft.

The second feature of note is that SFC *increases* with increase of combustion temperature, which is opposite to the usual thermodynamic dictum that efficiency increases with maximum temperature. This may be explained as follows. Increase of fuel input increases *energy* output, but in this case the useful output is not energy but *thrust*. Energy input increases in an amount proportional to increase of V^2, but output increases only proportionately to V and thus the ratio of input to output or SFC increases with fuel input, i.e., increase of combustion temperature. Another way of looking at it is that the stagnation temperature does not change in simple nozzle flow. Thus although thrust increases with combustion temperature, this increase is proportional to the square root of temperature whereas the increase of energy loss to the surroundings is proportional to the first power of temperature.

The effect of altitude on SFC cannot be clearly seen from these figures, as changes of forward speed occur concurrently. It does have two effects, however, one due to pressure decrease and one due to temperature decrease. The compressor of a turboengine behaves as a constant velocity machine and thus mass flow decreases directly with pressure. There is no effect on the cycle, however, so that although thrust decreases, the specific thrust remains the same. Temperature decrease has two effects. The first is the effect on the required compressor work, which is reduced at a given rpm. Energy transfer to the air as enthalpy rise is proportional to compressor speed and, if this is constant as the initial air temperature decreases, then the temperature ratio of compression increases and, correspondingly, so does the pressure ratio. This can readily be seen on a T–s diagram, as the lines of constant pressure converge as the initial temperature decreases (opposite effect as illustrated for the energy available analysis of the preceding section). The increase of pressure is available for expansion work and has been obtained at no cost relative to the process at a higher initial temperature and thus output is increased. Slightly more fuel is required as the compressor delivery temperature is reduced but this is not as great as the increase in thrust and hence SFC improves.

The other effect of temperature is on the internal behavior of the compressor. The latter is usually designed for a high Mach number relative to the blades, i.e., a high relative fluid velocity for compactness and low weight. As altitude increases and ambient temperature decreases, the acoustic velocity decreases (Eq. 3.5) and thus for constant air velocity, Mach number increases. If the compressor is designed to the limit of Mach number at ground level, then such an increase may cause the performance to deterioriate or, alternatively, a safety margin at high altitude may mean less-than-optimum performance at lower altitude. On balance, however, the effect of lowered initial temperature is beneficial.

4.5 The Turbofan Engine

From the relationship $E_{av} = V_j^2/2g_c$, we can obtain values of jet velocity very readily from Figs. 4.7 through 4.10. For example, for $M = 0$ at sea level, with $p_{r_c} = 15$ and $T_{0_3} = 1800°F$, $E_{av} \simeq 115$ Btu/lb yields a jet velocity V_j of about 2320 fps. At takeoff speeds the propulsive efficiency $\eta_p = 2U/(U + V_j)$ is then very low. At $M = 0.6$ and 15,000 ft for the same engine condition, $E_{av} \simeq 142$ Btu/lb and $V_j \simeq 2660$ fps, whence $\eta_p = 1268/3294 \approx 0.39$. At $M = 0.9$ and 35,000 ft, $\eta_p \simeq 0.46$ and at $M = 2.2$, $\eta_p \simeq 0.86$. Only at supersonic speeds, therefore, does the propulsive efficiency reach what might be called a satisfactory value.

This was recognized very early in the development of jet engines and led to the idea of the *ducted fan* or *augmentor*, now called the *turbofan*. The concept is that of using the energy available to provide a greater mass flow at a lower velocity and hence to improve η_p. Instead of using a large fan with very few blades, as in a propeller, energy is used to drive a multibladed fan immediately surrounding the jet engine itself. Thus there are two jet streams, one of hot gas from the turbojet proper and one of cool air from the fan. These can be mixed to give a single stream, which is open to thermodynamic and fluid dynamic loss, although it may be done in order to provide a single exhaust flow which is much more convenient if thrust reversal is required for aircraft braking, and if silencing is necessary.

Even during the earliest development of jet engines during World War II, experimental augmented turbojets were built and run. However, there was a period of about 15 years while jets were establishing themselves as the prime power unit in the air, before the ducted fan became operational on transport aircraft. Now the turbofan appears standard for all large engines for subsonic aircraft. Diagrammatically the turbofan is shown in Fig. 4.16(*a*) and (*b*), the former showing the fan at the back operated directly as an integral unit with the power-extraction turbine, the latter showing the fan at the front to form an integral first stage (or two) with the main jet compressor, and driven via an internal shaft from the extraction turbine. The front fan arrangement is mostly favored now, although there have been rear-fan engines. Thermodynamically there is no difference, the choice being one of mechanical design and manufacture, both as part of the power unit and as part of the aircraft installation. The aft-fan presents a rather complex turbine blade-fan problem, both in stress analysis and in manufacture, but leaves the main turbojet undisturbed until after the compressor turbine. The front fan requires two concentric shafts.

There are three additional parameters introduced by the turbofan although only two are independent. These are (1) the *bypass ratio B* or ratio of augmenter fan mass flow to main jet mass flow, (2) the fan pressure ratio p_{rf} and (3) the energy extraction fraction λ. Because the specific fan work is fixed by the fan pressure ratio, then the total work is proportional to the product of this and the fan

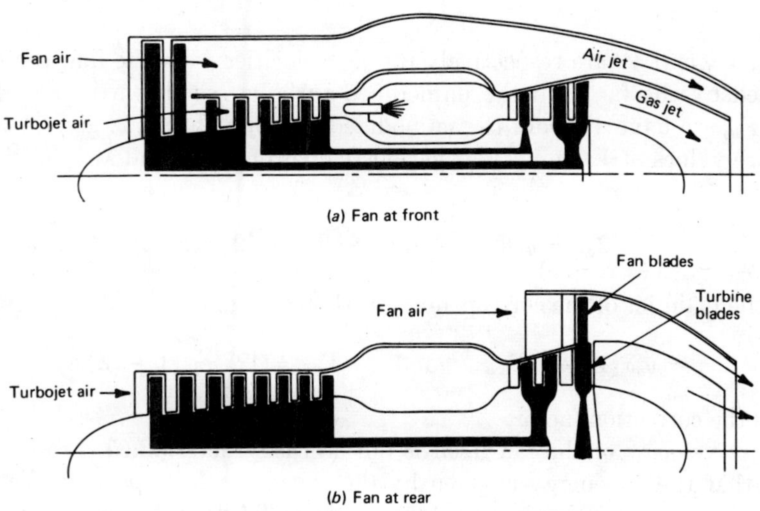

Fig. 4.16 Turbofan arrangements: (*a*) fan at front; (*b*) fan at rear.

air mass flow and, thus, the energy extraction is fixed. We shall wish to investigate the conditions for optimum performance and hence analyze the effect of B, p_{rf} and λ. Additional component efficiencies are those of the energy extraction turbine, which is taken as η_t as for the compressor turbine, and the fan itself. The latter is denoted by η_c and used as a single constant value as pressure ratios are usually low, varying between 1.5 and 2.5. The expansion efficiency is taken as η_n, the same as for the turbojet. Ram compression is also taken as the same process as for the main turbojet, the parameter r_0 yielding the same stagnation pressure and temperature.

Figure 4.17 shows the fan process on a T–s diagram. First of all, energy is fed to the fan as the fraction λ of energy available E_{av}, modified by the extraction turbine efficiency η_t. As specific energy, Btu/lb of *main-jet* air, this quantity must be divided by the by-pass ratio B. Hence, as temperature, the fan energy is $\eta_t \lambda E_{av}/c_{p_a}B$. This energy is given to air at the ram condition p_{0_1} and T_{0_1} and raises its pressure to p_{0_f} at fan discharge. This pressure p_{0_f} is given by the fan energy as modified by the fan efficiency η_c, that is, the isentropic fan temperature rise ΔT_{0_f} is $\eta_c \eta_t \lambda E_{av}/c_{p_a}B$ and then

$$p_{0_f}/p_{0_1} = [(T_{0_1} + \Delta T_{0_f})/T_{0_1}]^{k/(k-1)} \tag{4.14}$$

The propulsion energy available is then that from expansion at p_{0_f} to the ambient pressure p_1, modified by the nozzle efficiency η_n.

The simplest way to envisage turbofan performance is to obtain values as a ratio to that of the unmodified turbojet for the same cycle conditions and based on unit flow through the main jet. Thus we define a *thrust ratio* F_R such that

$$F_R = \frac{\text{turbofan thrust}}{\text{turbojet thrust}} = \frac{\dot{m}_{\text{fan}}(V_f - U) + \dot{m}_{\text{jet}}(V_{jf} - U)}{\dot{m}_{\text{jet}}(V_j - U)}$$

$$= \frac{B(V_f - U) + (V_{jf} - U)}{V_j - U} \tag{4.15}$$

where V_f, V_{jf} and V_j are respectively the jet velocities from the fan, the main jet as modified by the fan, and the unmodified turbojet, as shown in Fig. 4.18. To evaluate F_R, we have V_j from the unmodified turbojet thrust as given previously and require values of V_f and V_{jf} at various values of B, p_{rf} and λ.

For V_j,

$$V_j^2/2g_c = \eta_n E_{av} \qquad \text{and} \qquad V_j = (2g_c \eta_n E_{av})^{1/2} \tag{4.16}$$

For the main jet of the corresponding turbofan engine,

$$V_{jf}^2/2g_c = \eta_n \sigma (1 - \lambda) E_{av} \qquad \text{and} \qquad V_{jf} = [2g_c \sigma \eta_n (1 - \lambda) E_{av}]^{1/2} \tag{4.17}$$

where σ is the correction factor.

The jet velocity of the fan itself is not so easily determined. From Fig. 4.17 it is seen that the jet energy is given by the expansion from p_{0_f} and $T_{0_f'}$ to p_1 at nozzle efficiency η_n, resulting in a final temperature T_x'. It is shown in Appendix 4.2 of this chapter that this energy is given exactly by

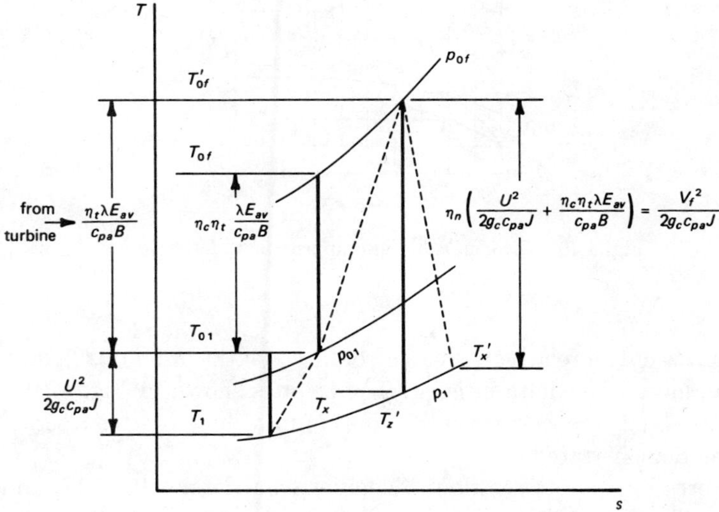

Fig. 4.17 Turbofan process.

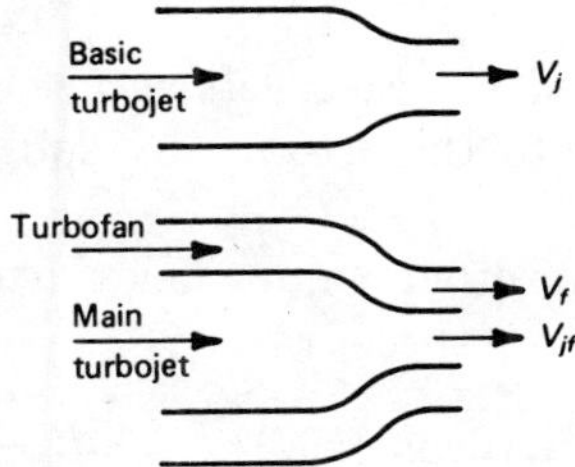

Fig. 4.18 Definition of jet velocities.

$$\frac{V_f^2}{2g_c} = \eta_n c_{pa}\left[T_{0_1} + \frac{\eta_t\lambda E_{\text{av}}}{c_{pa}B}\right]\left[1 - \frac{1}{r_0^{\epsilon_a}[1 + (k-1)M^2/2][1 + (\eta_c\eta_t\lambda E_{\text{av}}/c_{pa}BT_{0_1})]}\right] \tag{4.18}$$

but this expression is very cumbersome, particularly as a derivative is required in the subsequent development. A very close approximation is obtained by taking the energy as the sum of the terms given by the isentropic expansion from p_{0_f} to p_{0_1}, i.e. $\eta_c\eta_t\lambda E_{\text{av}}/B$, plus the kinetic energy corresponding to forward speed U, that is $U^2/2g_c$, the total being subject to the nozzle efficiency η_n. Hence

$$V_f^2/2g_c \approx \eta_n[(\eta_c\eta_t\lambda E_{\text{av}}/B) + (U^2/2g_c)] \tag{4.19}$$

Study of these values on Fig. 4.17 will show that the first term is an underestimate because of the divergence of the constant pressure lines with increased entropy, and that the second term overestimates because it neglects the diffuser efficiency.

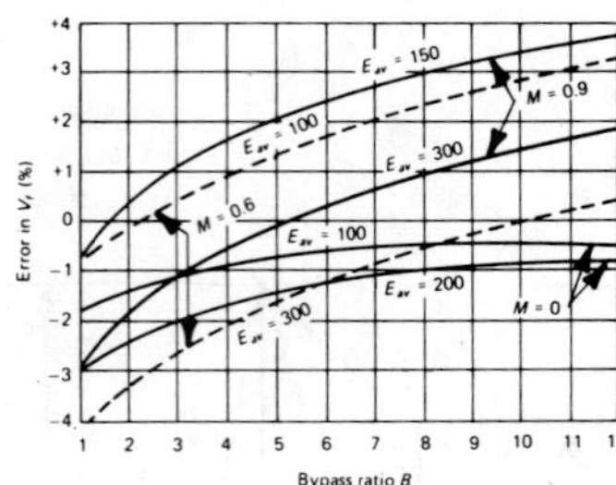

Fig. 4.19 Error in exact and approximate fan velocity.

The net result is only a relatively small error in V_f, of the order of 2% for the bypass ratios of the most immediate interest. The error is shown in Fig. 4.19, which has been calculated for the corresponding optimum values of the fraction λ, these optima being demonstrated shortly.

The expressions for the various velocities from Eqs. 4.16, 4.17, and 4.19 can be substituted in Eq. 4.15, yielding

$$F_R = \frac{B[(2g_c\eta_n\eta_c\eta_t\lambda E_{av}/B) + \eta_n U^2]^{1/2} - BU + [2g_c\eta_n(1-\lambda)E_{av}]^{1/2} - U}{[2g_c\eta_n E_{av}]^{1/2} - U} \tag{4.20}$$

Here as elsewhere, flight speed has been left as velocity U rather than as Mach number, but it can be put into these terms by substituting $U^2/2g_c = kRTM^2/2$ if desired.

Taking λ as the variable, other values being fixed, differentiation yields an optimum value of λ for maximum thrust ratio as

$$\lambda_{opt} = \frac{\eta_c\eta_t B - (\sigma U^2 B/2g_c\eta_c\eta_t E_{av})}{\sigma + \eta_c\eta_t B} \approx \frac{\eta_c\eta_t B - (U^2 B/2g_c\eta_c\eta_t E_{av})}{1 + \eta_c\eta_t B} \tag{4.21}$$

if σ is taken as unity as before.

Values of λ_{opt} plotted against B are shown in Fig. 4.20 for all flight cases, showing (1) the rapid initial rise in λ_{opt} with increase of bypass ratio and subsequent flattening off, and (2) the decreased values of λ_{opt} with increasing Mach number for a given bypass ratio. The latter is the expected result due to increase of propulsion efficiency with increasing forward speed and hence reduced need of turbofan effect for maximum thrust.

For the stationary case, Eq. 4.21 reduces to

$$\lambda_{opt}(\text{stationary}) = \eta_c\eta_t B/(1 + \eta_c\eta_t B) \tag{4.22}$$

showing a very rapid increase with B, independent of E_{av}, tending towards a value of unity ($\lambda_{opt} \approx 0.9$ for $B \approx 10$). Numerical values of λ are not needed directly in the subsequent development as the analytical expression for optimum value is inserted into the thrust expression, but are given here for information.

The optimum value of energy fraction from Eq. 4.21 may then be substituted into Eq. 4.20 for the thrust ratio, yielding

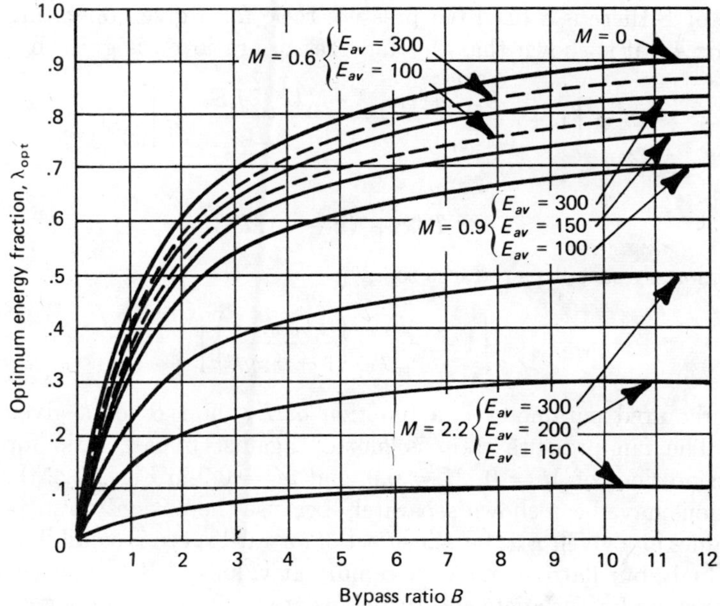

Fig. 4.20 Optimum energy fraction.

$$F_R = (2g_c\eta_n E_{av})^{1/2}\left[\left\{\frac{(\eta_c\eta_t B)^2 - (U^2B^2/2g_cE_{av})}{1+\eta_c\eta_t B} + \frac{U^2B^2}{2g_cE_{av}}\right\}^{1/2} + \left\{\frac{1+(U^2B/2g_c\eta_c\eta_t E_{av})}{1+\eta_c\eta_t B}\right\}^{1/2}\right] - U(1+B) \Big/ (2g_c\eta_n E_{av})^{1/2} \quad (4.23)$$

This relationship, for given component efficiencies and operating condition, is a function of energy available E_{av} and bypass ratio B. Thus a single plot suffices for all values of pressure ratio and combustion temperature, the significant parameter being E_{av}. Figure 4.21 shows F_R vs B for several values of E_{av} for the three operating conditions $M = 0$, $M = 0.6$, and $M = 0.9$. The operating condition of $M = 2.2$ does not appear because the thrust ratio F_R with an unheated turbofan is less than unity except for very high values of E_{av} at low bypass ratios. The losses in the turbofan system outweigh any increase of momentum at this high speed. For the stationary case, $U = 0$, Eq. 4.23 reduces to

$$F_{R|U=0} = (1 + \eta_c\eta_t B)^{1/2} \quad (4.24)$$

a result independent of energy available. Figure 4.21 shows the high values of static thrust ratio at low bypass ratios, demonstrating the usefulness of the turbofan engine for takeoff, which is one of its main advantages.

For a given amount of shaft energy extracted for the turbofan, the product of bypass ratio and specific fan work is fixed. Thus for λ_{opt} at a given E_{av}, then for each value of B there is a fixed fan pressure ratio for a given operating condition. In Appendix 4.3 it is shown that this fan pressure ratio p_{rf} is given by

$$p_{rf} = \left[1 + \frac{E_{av}}{c_{pa}T_{0_1}}\left\{\frac{(\eta_c\eta_t)^2 - (U^2/2g_cE_{av})}{1 + \eta_c\eta_t B}\right\}\right]^{k_a/(k_a-1)} \tag{4.25}$$

where

$$T_{0_1} = T_1[1 + (k - 1)M^2/2]$$

For the stationary case, this reduces to

$$p_{rf}\big|_{U=0} = \left[1 + \frac{E_{av}}{c_{pa}T_{0_1}}\left\{\frac{(\eta_c\eta_t)}{1 + \eta_c\eta_t B}\right\}\right]^{k_a/(k_a-1)} \tag{4.26}$$

Again the required parameter is a function of E_{av} and B for a given operating condition. The fan pressure ratio is plotted against bypass ratio for the three operating conditions of $M = 0$, $M = 0.6$, and $M = 0.9$ in Figs. 4.22 through 4.24. The conditions have been shown separately because there is only a small difference between values at a given E_{av}, i.e. the effect of forward speed is small. For low values of B p_{rf} is high, but flattens out very rapidly at values of B greater than about 6.

These results for turbojets and turbofans are useful in giving a general picture of performance over a wide range of pressure ratio, combustion temperature, and bypass ratio. Although they have general validity, the assumptions made and values of efficiency assumed obviously imply lack of precision, particularly in marginal areas where the balance of useful output depends critically on exact values.

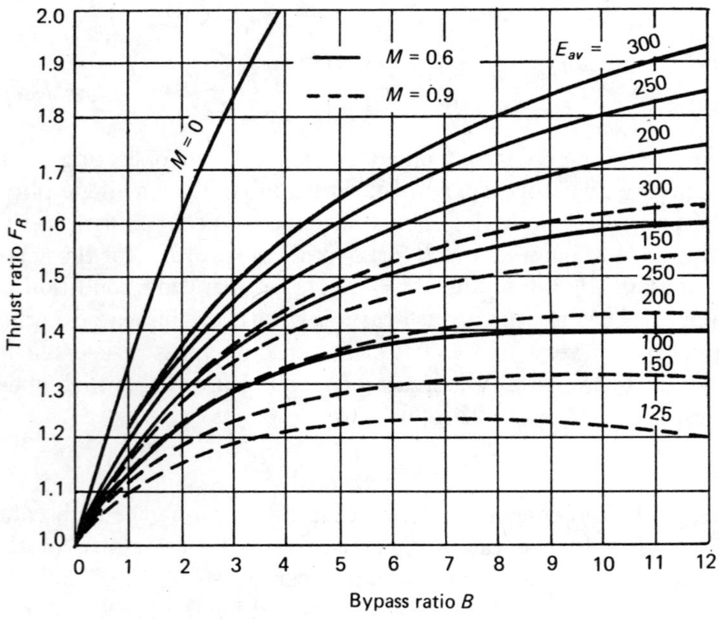

Fig. 4.21 Thrust ratio for turbofan engine.

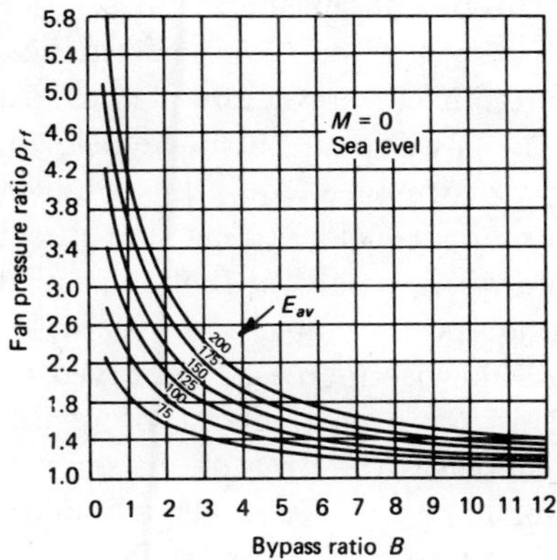

Fig. 4.22 Fan pressure ratio, $M = 0$, sea level.

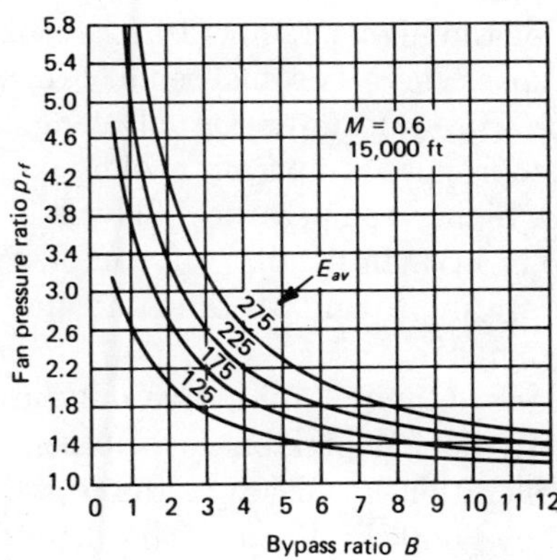

Fig. 4.23 Fan pressure ratio, $M = 0.6$, 15,000 ft.

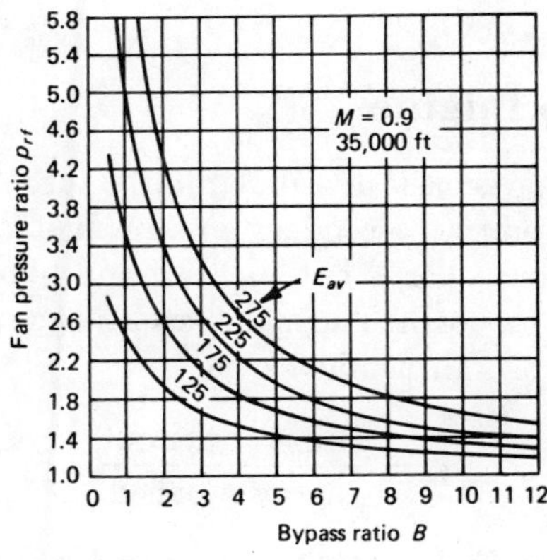

Fig. 4.24 Fan pressure ratio, $M = 0.9$, 35,000 ft.

We may conclude that for subsonic flight, high values of pressure ratio, temperature, and bypass ratio are necessary for the best performance, but must also keep in mind that compromise is often necessary in any particular instance. For propulsion, the overall installation of the power plant in the vehicle and the specific mission are inseparable for optimum performance, i.e., the complete *system* is paramount. Weight and drag are two major considerations that affect the choice or design of power plant. Weight includes the weight of fuel or propellant necessary as well as fixed engine weight. Although performance improves with increase of bypass ratio, the engine weight and size likewise increase. Improved SFC and increased engine weight must be balanced off for a particular type of duty. The increase of size, particularly in diameter, increases the drag of the aircraft, but the degree of this depends on the location, such as "pod" or fuselage mounting, or the aircraft speed. Quite recently, bypass ratios of about unity were favored but development appears now to be for much higher values, of the order of 5, for example.

Combustion temperature is one of the most important parameters and this is steadily increasing as (a) material properties are improved and (b) blade-cooling methods are developed. For the higher temperatures, cooling is almost mandatory and is used in many engines. There is some penalty in power loss to supply the cooling fluid (often compressor air) and in actual blade performance itself, which may have to be slightly compromised to permit cooling passages to be incorporated. On balance, however, the higher temperature allowable more than compensates and the main limitation is in manufacture of complex blade forms. However, techniques and materials are improving all the time and temperatures of 2500°F and more will become commonplace.

Component efficiencies are likely to improve only incrementally, as they are already high. The emphasis on compressor development is to improve range of stability (i.e., a greater toleration of airflow without stalling or choking) and to decrease size for a given flow, both in diameter and length. The former may stem from lower hub ratio (i.d./o.d.) and higher fluid and blade velocities, and the latter from increased work per stage, i.e. fewer required stages. Although progress is slow, supersonic compressors are in development and may be possible for some purposes if range and off-design performance can be improved.

4.6 The Turboprop Engine

While the turboprop engine is basically similar to the turbofan in extracting exhaust energy via a turbine to produce a secondary air stream of lower velocity and greater mass flow, the "fan" or propeller is not enclosed and bypass ratio is not a design parameter as used for the turbofan. The main performance parameter for a propeller is the overall efficiency η_0 as a thrust device, i.e.,

$$\eta_0 = \frac{\text{thrust power out}}{\text{energy supplied}}$$

The thrust power out is the product of propeller thrust F_{prop} and the flight speed U. The energy supplied is the actual extracted energy $\eta_t \lambda E_{\text{av}}$. Both quantities are related to unit mass flow rate of the turbojet. Thus:

$$\eta_0 = (F_{\text{prop}} U/\eta_t \lambda E_{\text{av}})$$

The propeller thrust is then given by

$$F_{\text{prop}} = (\eta_0 \eta_t \lambda E_{\text{av}})/U \tag{4.27}$$

and the residual turbojet thrust F_{jf} as before is $(V_{jf} - U)/g_c$ or

$$F_{jf} = \{[2g_c \eta_n (1 - \lambda) E_{\text{av}}]^{1/2}/g_c\} - (U/g_c) \tag{4.28}$$

The total thrust is then

$$F_{\text{total}} = F_{\text{prop}} + F_{jf} = \frac{\eta_0 \eta_t \lambda E_{\text{av}}}{U} + \frac{[2g_c \eta_n (1 - \lambda) E_{\text{av}}]^{1/2}}{g_c} - \frac{U}{g_c} \tag{4.29}$$

For given efficiencies, energy available, and flight speed, differentiation with respect to λ yields

$$\lambda_{\text{opt}} = 1 - [\eta_n/(\eta_0 \eta_t)^2](U^2/2g_c E_{\text{av}}) \tag{4.30}$$

The thrust ratio F_R may be formed as before,

$$F_R = \frac{F_{\text{prop}} + F_{jf}}{F_{jm}} = \frac{(\eta_0 \eta_t \lambda E_{\text{av}}/U) + \{[2g_c \eta_n (1 - \lambda) E_{\text{av}}]^{1/2}/g_c\} - (U/g_c)}{(2g_c \eta_n E_{\text{av}}{}^{1/2} - U)/g_c} \tag{4.31}$$

which, substituting for λ_{opt} from Eq. 4.30, yields

$$F_R = \frac{(g_c \eta_0 \eta_t E_{\text{av}}/U) + (\eta_n/\eta_0 \eta_t) - U}{(2g_c \eta_n E_{\text{av}})^{1/2} - U} \tag{4.32}$$

This is shown plotted in Fig. 4.25 for the two forward speeds for which it is applicable. The propeller, even with an overall efficiency of 0.7, provides as great a thrust ratio as does a high bypass turbofan at $M = 0.9$ because it yields a high propulsive efficiency. The reasons for the virtual disappearance of the propeller, particularly for high forward speeds, are due to other reasons than propulsive thrust, namely, size, weight, and the effect on aircraft design.

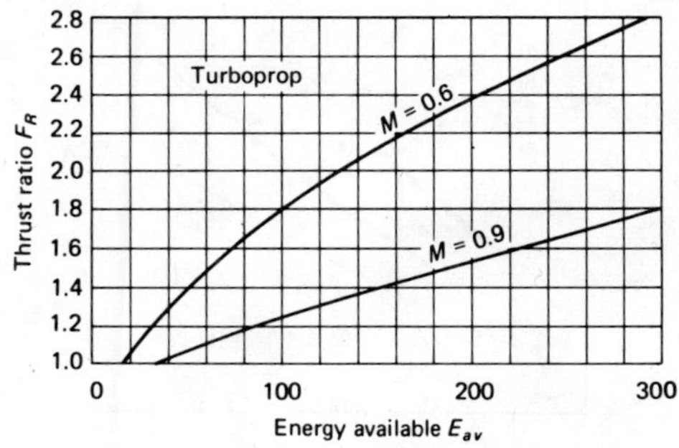

Fig. 4.25 Thrust ratio for turboprop engine.

4.7 Afterburning

The gas discharged from the compressor-turbine is still at a pressure sensibly higher than that of the atmosphere and can be reheated by further combustion of fuel to give additional expansion for a turbojet energy. As the gas does not have to pass through a turbine but flow only through a nozzle where there are no moving parts, the temperature of reheat can be much higher than the normal cycle temperature. In this way the thrust of the turbojet can be increased considerably. The process is known as *afterburning* (sometimes as reheat) and is shown as an ideal process in Fig. 4.26.

The additional work possible is shown thermodynamically by the area 4567 and this can be considerable, some 30–50% in actual operation with losses. However, the efficiency is very poor, so that SFC is high, because the afterburning cycle has a low pressure ratio with large heat input, which a glance back at Figs. 4.12 to 4.15 will demonstrate. Thus it is generally used only in military aircraft where large thrust increments for very short periods can be very advantageous at takeoff, climb, and maneuvering under combat conditions.

There are practical disadvantages as well. The combustion of the fuel occurs at a high initial temperature, a high velocity, and at a low absolute pressure level. The former is beneficial and the last two deleterious, as discussed in Chapter 3. The velocity must be kept high in the exhaust duct as the density is low, and although the pressure is not that of the ambient atmosphere, at high altitude it may still be subatmospheric relative to ground conditions. Baffles have to be provided to stabilize the flame in the exhaust duct and the effect of high velocity and low pressure may make the stability range small, although the high initial temperature does compensate to some extent. In fact it is only that this temperature is close to the self-ignition temperature of hydrocarbon-air mixtures which makes the process viable.

The high temperature and hence very low density of the gas requires that the nozzle be much larger and in fact a variable-area nozzle is required because the afterburning is only occasional. Thus there is a weight penalty due to the combustion equipment, to the larger, variable nozzle and to the apparatus required to operate it. Furthermore, the presence of the combustion baffles in the exhaust duct

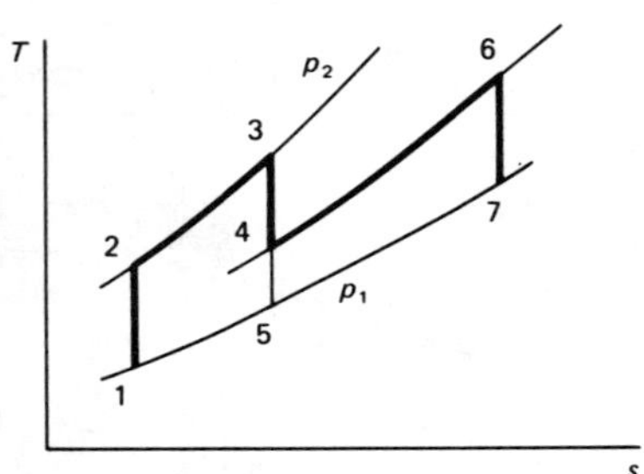

Fig. 4.26 Afterburning process.

at all times leads to a pressure loss that is not negligible and hence reduces the normal performance. Thus, afterburning for turbojets is used only for military aircraft or for civil aircraft when certain takeoff requirements at full load can be met only in this fashion.

The turbofan may have "afterburning" in a sense, that is, combustion after the fan compression, followed by direct expansion through the nozzle. This is used only in supersonic aircraft when there is a large pressure rise due to ram, with the fan making only a small contribution. In this case, the unit acts more as a ramjet and at $M \approx 3$ the process is quite efficient.

4.8 The Ramjet

The ramjet operates on the same Joule–Brayton cycle as does the turbojet but relies on ram diffusion from forward speed instead of a mechanical compressor, and this requires no turbine. This has two major consequences: (1) a ramjet is barely operational until a high enough Mach number is reached to give a useful pressure ratio and (2) a ramjet can utilize higher combustion temperatures because there are no highly stressed moving parts in contact with the hot gas stream. If, for example, a diffusion efficiency (as the stagnation pressure ratio) is used which varies with Mach number according to the A.I.A. specification of Fig. 3.12, then pressure ratios are generated with forward speed as follows:

M		P_r
2	$\approx$	7
3	$\approx$	26
4	$\approx$	72
5	$\approx$	105

Thus the ramjet is barely operative below a Mach number of 2 and the subsonic ramjet is almost ruled out except when the utmost simplicity and lowest weight are required, with fuel consumption a very low priority. The accent is therefore on supersonic ramjets and here a difficulty arises because of the high inlet air temperature to the combustor. In Chapter 3, the effect of dissociation was discussed and in the ramjet this becomes acute because the difference between the inlet air temperature and the maximum combustion temperature coupled with dissociation severely limits the amount of fuel which can be burnt, that is, the energy input and performance are limited by this factor. Figure 4.27 shows the inlet air temperature and the maximum combustion temperature possible plotted against Mach number M_∞ using kerosene as fuel. The decreasing slope of the combustion temperature at the higher Mach numbers represents the effect of dissociation. At a Mach number of about 8, the allowable fuel addition becomes negligible.

One possible solution to this problem is to use a fuel that forms stable combustion products, i.e. those which do not dissociate. Boron fuels (as borohydrides) have good combustion characteristics in this and other respects but a large part of the heating value lies in the heat of condensation of the gases and as these do not condense in the nozzle, then a large part of the advantage is lost. Another possible fuel is powdered metal, for example magnesium or aluminum, mixed with a liquid

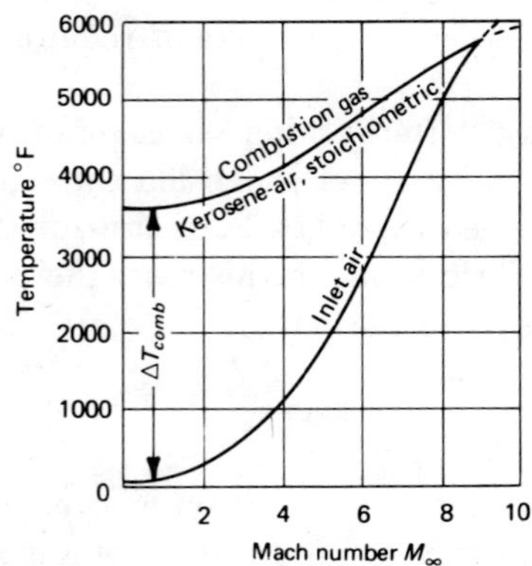

Fig. 4.27 Air and combustion temperatures vs Mach number.

hydrocarbon to form a pumpable and sprayable slurry. Here, although the combustion process is improved, the effect of solid particles in the exhaust can be a drawback (see Sec. 3.12). In any case, the use of such highly specialized fuels is not only expensive but also is limited to special applications at a limited number of locations. As a long-term solution, exotic fuels are not a satisfactory answer. Let us look at the ramjet process in more detail to see how it operates.

A ramjet at supersonic speed would conventionally have a system of oblique-shock diffusion, followed by a normal shock at low supersonic Mach number and then diffusion to a low subsonic Mach number. The process can be shown on a T–s diagram, as in Fig. 4.28(a). Air at P_1 is raised to P_A by oblique shock, to P_B by the normal shock, and to P_2 by subsonic diffusion. Fuel is then injected and burned, and supposing it to occur in a constant-area combustor, the process can be idealized as with negligible friction loss and hence occurs along a Rayleigh line (Sec. 3.9). Maximum energy addition would take place if heating occurred until the flow choked at state x in the figure. The heat added is represented by the area under the curve from state 2 to state x. Expansion then occurs to state 4 through a nozzle. The lower the Mach number at state 2, the more energy can be added, so there is incentive to accomplish the utmost diffusion. However, at high flight Mach number, Fig. 4.27 shows that the static temperature at 2 is so high that dissociation occurs and limits temperature rise. Up to perhaps Mach 3 or 4, this type of process is adequate but at hypersonic speeds ($M \gtrsim 5$), then a different solution may be beneficial.

For the hypersonic regime, it has been proposed that the initial diffusion should be limited and that combustion should occur at supersonic velocities. If then heat is added when $M > 1$ the process occurs along the lower branch of a Rayleigh line and may again take place until the choking point is reached. This is shown in Fig. 4.28(b). The Rayleigh lines are different ones, the specific mass flow $G = \dot{m}/A = \rho V$ being greater for case (b) although the heat added is less and the efficiency lower. It is possible that shock would occur at some point during the heating process, $p - q$, for example, but as shock is an adiabatic process and at constant area, all states must be on the initial Rayleigh line, and the total amount of heat added is not changed.

The balance between a conventional ramjet (CRJ) and the (s)upersonic (c)ombustion ramjet (SCRJ or "scramjet") lies in the efficiencies of the processes (e.g., diffusion) and the initial temperature level (i.e., Mach number). Also at these high temperatures, the degree of frozen or equilibrium flow can have a considerable effect and this is dependent on the fuel. It should be noted that subsonic combustion necessitates a decrease of static pressure (for acceleration of the flow) and that supersonic combustion implies an increase of static pressure (deceleration of the flow) but both are accompanied by a loss of stagnation pressure. The stagnation pressure and temperature at the beginning of expansion then indicate the output available. As an indication of performance, Fig. 4.29 shows a comparison of a CRJ and a SCRJ at a particular condition, with the former shown for two values of diffusion efficiency to emphasize the importance of this in any comparison. All data are for a constant value of inlet dynamic pressure $\rho_\infty V_\infty^2/2 \equiv kp_\infty M_\infty^2/2$ of 350 lb/ft^2 and for the optimum fuel-air ratio at each Mach number (curves adapted from Dugger[3]). It is apparent that diffusion efficiency is controlling for these cases (in this example η_d for the subsonic cases was the kinetic energy efficiency η_k and the value of 0.90 might be considered conservative and the value of 0.94 optimistic).

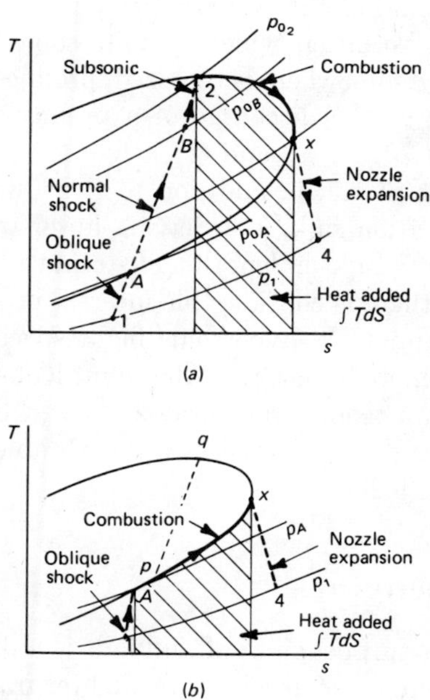

Fig. 4.28 Ramjet processes: (*a*) subsonic combustion; (*b*) supersonic combustion.

[3] G. L. Dugger, "Comparison of Hypersonic Ramjet Engines," Combustion and Propulsion, 4th Agard Colloquium, "High Mach Number Airbreathing Engines," Pergamon Press, 1961.

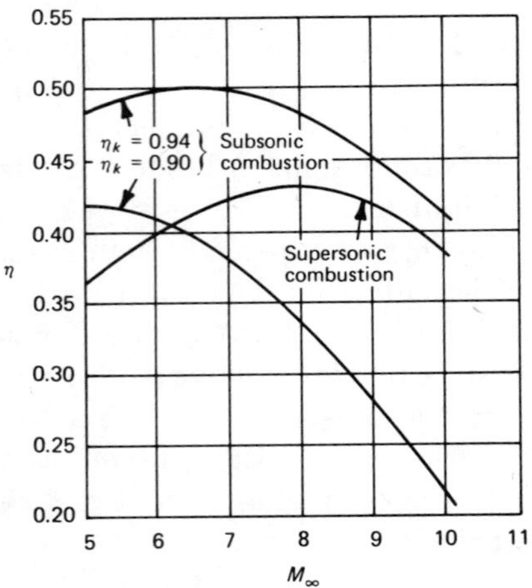

Fig. 4.29 CRJ and SCRJ performance (from Dugger).

The big question for the scramjet is, can the combustion be carried out satisfactorily? The difficulty of successful burning at high velocity was discussed in Chapter 3 and here the situation is an order of magnitude or more severe. The temperature of course is very much higher, which promotes very fast combustion. Pressures too are high, again helpful. Nevertheless, a fuel like hydrogen is almost a necessity and so is a flow situation which permits some stability. It would seem that a standing shock wave somewhere is required, produced by wedges on the duct walls or along its axis or at the discharge of a premixed fuel. These are shown diagrammatically in Fig. 4.30(a), (b) and (c).

The advantage of the SCRJ, in addition to the lower initial or temperature and absence of the dissociation sink, is in having lower pressures and no subsonic diffusion, hence a saving of weight in length and strength.

At the present time, there is considerable interest in supersonic combustion as this is the key to the scramjet. As hypersonic flight speeds are deemed necessary, so will the ramjet become more important. Already it looks advantageous combined with a turbojet for the supersonic transport, $2 < M < 3$, and eventually speeds will go higher. It is also being considered as a recoverable booster for rockets and as part of a hybrid satellite vehicle.

4.9 Fuels for Turboengines

The most common fuel for turboengines of all kinds is a distilled petroleum hydrocarbon approximating a kerosene but with some lighter fractions of a gasoline type. In the United States these fuels are known as JP-fuels, with a number following the letters, of which the most widely quoted is JP-4.

The major desirable characteristics of a fuel are cheapness and availability in quantity, high heating value, low freezing point, high density, storeability and ease of handling, and high flash point but ready ignition under controlled conditions. The JP fuels are reasonably satisfactory on these counts although their safety aspect in the event of an accident could be improved. A composition tending more to a kerosene rather than a gasoline would improve this, but limiting the distillation range decreases the availability and hence increases the cost.

Some constraints on the use of normal hydrocarbon fuels have been discussed in connection with ramjets, with respect to dissociation at high temperatures. Special fuels such as those containing boron or powdered metal slurries are palliatives but not satisfactory answers. The advent of the supersonic transport aircraft (SST) raises some problems of fuel temperature and cooling of engine and aircraft parts, as at a Mach number of 2.7 at 60,000 ft the stagnation temperature is about 500°F. A new fuel that seems to offer several advantages for high-speed aircraft is liquid methane as liquefied natural gas (LNG), which is largely methane.

Some of the advantages of LNG over the JP-fuels are given as follows[4]:

1. 13% higher heating value than JP fuel.
2. Higher heat of vaporization (80%), specific heat (75% as liquid), and low storage temperature (boiling point at 1 atm −259°F) give it a heat sink capacity six to seven times that of JP-fuel.
3. Higher thermal stability—kerosene fuels start to decompose at temperatures corresponding to Mach 3.
4. Clean burning, with no waxes or gums to clog handling and injection apparatus.

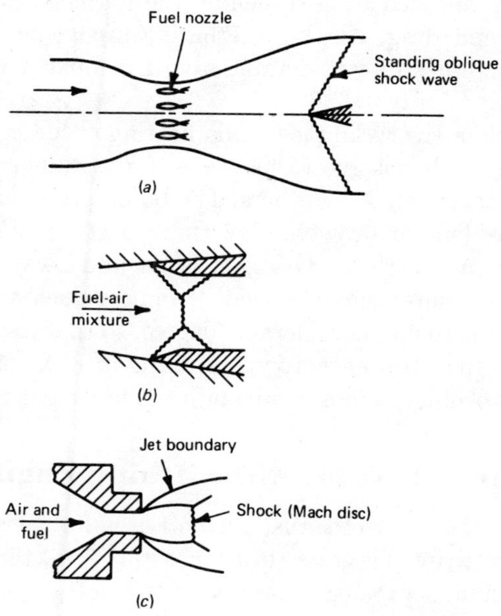

Fig. 4.30 Possible modes of supersonic combustion.

[4] J. B. Whitlaw, Jr., J. D. Eisenberg, and M. D. Shavlin, "Potential of Liquid-Methane Fuel for Mach 3 Commercial Supersonic Transports," NASA TN D-3471, 1966.

5. Price and availability in many parts of the world appear attractive.

Disadvantages are:

1. Density is half that of JP-fuel.
2. Its cryogenic qualities require more careful and special handling and lead to loss through boil-off as pressure is reduced.

The cooling capacity of LNG is most important for turbine blade cooling and may well allow turbine inlet temperatures to rise to 2800°F, yielding a considerable increase in specific thrust and hence decrease of engine weight with consequent improved payload. The LNG is used to cool air for the turbine blades and there is ample capacity available for cabin air conditioning and oil coolers. There is no available cooling capacity for turbine blades with JP-fuels.

Boil-off is a problem that cannot be entirely eliminated by insulation as pressure reduction during climb to cruise altitude is a major cause. Evaporation can be controlled by pressurization and subcooling and actual loss can be minimized by using a vapor pump and selecting the climb schedule carefully so that fuel vapor can be combusted directly. However, a penalty is paid in extra weight and complexity.

The low density of LNG (S.G. $\approx$ 0.42) requires considerably more tankage space, which is usually in the aircraft wings. Individual design determines the degree of any penalty for this, as usually some extra volume is available and presents no great problem to arrange if included in the initial design. However any further space required and obtained by lengthening the fuselage, for example, does incur additional weight and drag. As the alternative cryogenic fuel to LNG appears to be hydrogen, there is still considerable saving compared with this, as hydrogen has a specific gravity of only 0.07.

Figure 4.31[5] plots the weight per unit heating value against the volume per unit heating value, both relative to kerosene, for a number of the possible fuels discussed here together with some others. The boron group looks very good on this basis but has considerable drawbacks with respect to other combustion characteristics as mentioned earlier. Hydrogen is far and away the best on a weight basis but the space requirement is crucial. If improvements in engine and aircraft performance are made to any considerable degree, so that less total fuel is required, then it is possible that the necessary fuel as hydrogen may be accommodated without significant penalty, and a quantum jump in weight reduction would occur.

4.10 Maximum Thrust for Airbreathing Engines

We might calculate the order of thrust per unit area for thermal accelerators using a perfect gas. If we write the gross thrust F as $\dot{m}V_{je}/g_c$, then this is largest for a given unit when complete expansion occurs, i.e., when the discharge pressure is zero. But for this to occur, the nozzle area ratio A/A^* must be infinite. From the gas tables for air, the value of thrust F at any Mach number M to the thrust F^* where

[5] Adapted from L. A. Dawson, and J. B. Holliday, "Propulsion," *Aeron. Jour., R. Aero. Soc.*, 72 (Sept. 1968), 739–747.

$M = 1$ (throat) is 1.43 for $p/p_0 = 0$, complete expansion. Hence maximum F/A occurs at the throat and is F^*/A^*. Now we can write

$$\frac{F^*}{A^*} = \frac{\dot{m}V^*}{g_c} + \frac{p^*A^*}{A^*} \quad \text{assuming exhaust to a vacuum}$$

With

$$\dot{m} = \rho^*A^*V^* = p^*A^*V^*/RT^*,$$

$$F^*/A^* = (p^*A^*V^{*2}/g_cRT^*) + p^* = (k+1)p^* \quad \text{as} \quad V^{*2}/g_cRT^* = kM^{*2} = k$$

Now

$$p^* = \frac{p_0}{[(k+1)/2]^{k/(k-1)}}$$

$$\therefore F^*/A^* = (F/A)_{\max} = \frac{k+1}{[(k+1)/2]^{k/(k-1)}}\, p_0$$

Values of the function of k run from about 1.23 for $k = 1.1$ to 1.3 for $k = 1.67$, so that the maximum thrust per unit area is about 1.25 times the stagnation pressure at the nozzle throat. This allows an order of value to be compared with other propulsion means later on. Note that this is gross thrust.

Appendix 4.1

Referring to Fig. 4.3,

$$E_{\text{av}} = c_{p_n}(T_{04'} - T_5) = c_{p_n}T_{0'4}[1 - (p_1/p_{04})^{\epsilon_n}] \tag{4.A1}$$

where subscript n refers to an average value for the conditions of this expansion.

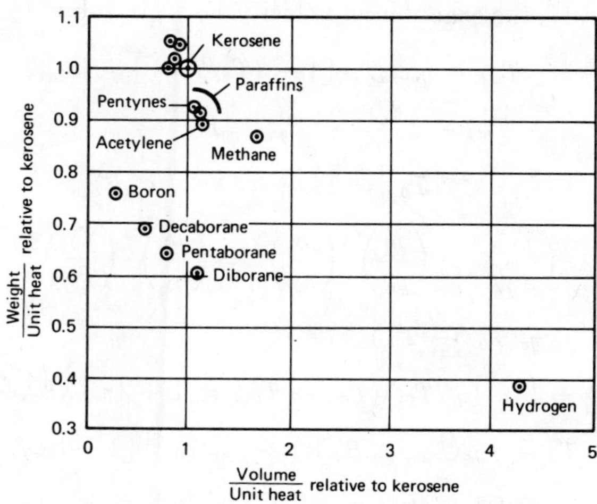

Fig. 4.31 Relative weight and volume of possible fuels (from Dawson and Holliday).

For p_{0_4}, we have

$$p_{0_4} = p_{0_3}(T_{0_3}/T_{0_4})^{1/\epsilon_g} \qquad \text{and} \qquad p_{0_3} = p_{0_2} - \Delta p_c$$

Using $p_{r_c} = p_{0_2}/p_{0_1}$, the compressor pressure ratio, usually called simply the cycle pressure ratio, then

$$p_{0_3} = p_{r_c}p_{0_1}[1 - (\Delta p_c/p_{0_2})] = p_{r_c}p_{0_1}K_{\text{comb}}$$

with K_{comb} being a combustion pressure loss coefficient.

Now equating compressor and turbine work as before,

$$\eta_t c_{p_g}(T_{0_3} - T_{0_4}) = c_{p_a}(T_{0_2} - T_{0_1})/\eta_c \equiv c_{p_a}T_{0_1}[p_{r_c}{}^{\epsilon_a/\eta_s} - 1]$$

$$\therefore T_{0_3} - T_{0_4} = (c_{p_a}/c_{p_g})(T_{0_1}/\eta_t)[p_{r_c}{}^{\epsilon_a/\eta_s} - 1]$$

and

$$T_{0_4}/T_{0_3} = 1 - (c_{p_a}/c_{p_g})(T_{0_1}/T_{0_3})[p_{r_c}{}^{\epsilon_a/\eta_s} - 1]/\eta_t$$

Also

$$c_{p_g}(T_{0_3} - T_{0_4'}) = c_{p_a}T_{0_1}[p_{r_c}{}^{\epsilon_a/\eta_s} - 1]$$

Substituting in Eq. 4.A1 for E_{av},

$$E_{\text{av}} = c_{p_n}T_{0_1}[(T_{0_3}/T_{0_1}) - (c_{p_a}/c_{p_g})(p_{r_c}{}^{\epsilon_a/\eta_s} - 1)$$
$$\times \left\{1 - \left[\frac{p_1/p_{0_1}}{K_{\text{comb}}p_{r_c}[1 - (c_{p_a}/c_{p_g})(T_{0_1}/T_{0_3})(p_{r_c}{}^{\epsilon_a/\eta_s} - 1)/\eta_t]^{1/\epsilon_g}}\right]^{\epsilon_n}\right\} \tag{4.9}$$

Appendix 4.2

Fanjet Velocity V_f

Referring to Fig. 4.17, the exact fanjet velocity V_f is given by

$$V_f^2/2g_c = \eta_n c_{pa}(T_{0'_f} - T_x) = \eta_n c_{pa}T_{0'_f}[1 - (T_x/T_{0'_f})] = \eta_n c_{pa}T_{0'_f}[1 - (p_1/p_{0_f})^{\epsilon_a}]$$

$$T_{0'_f} - T_{0_1} = \eta_t\lambda E_{\text{av}}/c_{pa}B$$

$$\therefore T_{0'_f} = T_{0_1} + (\eta_t\lambda E_{\text{av}}/c_{pa}B)$$

$$\left(\frac{p_1}{p_{0_f}}\right)^{\epsilon_a} = \left(\frac{p_1}{p_{0_1}}\right)^{\epsilon_a}\left(\frac{p_{0_1}}{p_{0_f}}\right)^{\epsilon_a} = \left(\frac{p_1}{p_{0_1}}\right)^{\epsilon_a}\left(\frac{T_{0_1}}{T_{0_f}}\right)$$

$$T_{0_f} - T_{0_1} = \eta_c(T_{0'_f} - T_{0_1})$$

$$\therefore T_{0_f}/T_{0_1} = 1 + (\eta_c/T_{0_1})(T_{0'_f} - T_{0_1}) = 1 + (\eta_c\eta_t\lambda E_{\text{av}}/c_{pa}BT_{0_1})$$

$$\therefore V_f^2/2g_c = \eta_n c_{pa}[T_{0_1} + (\eta_t\lambda E_{\text{av}}/c_{pa}B)]\{1 - (p_1/p_{0_1})^{\epsilon_a}[1 + (\eta_c\eta_t\lambda E_{\text{av}}/c_{pa}BT_{0_1})]^{-1}\}$$

where $(p_1/p_{0_1})^{\epsilon_a} = \{r_0{}^{\epsilon_a}[1 + (k-1)M^2/2]\}^{-1}$ and r_0 is the stagnation ratio given by Eq. 3.8.

Appendix 4.3

Fan pressure ratio p_{rf}

Referring to Fig. 4.17 and equating the total energy given to the fan by the extraction turbine to the total energy given to the bypass air by the fan,

$$\eta_t \lambda E_{av} = c_{pa} B (T_{0'_f} - T_{0_1})$$

and with

$$T_{0'_f} - T_{0_1} = \frac{T_{0_f} - T_{0_1}}{\eta_c} = \frac{T_{0_1}}{\eta_c} (p_{rf}^{\epsilon_a} - 1)$$

$$(c_{pa} B T_{0_1}/\eta_c)(p_{rf}^{\epsilon_a} - 1) = \eta_t \lambda E_{av}$$

and hence

$$p_{rf} = [1 + (\eta_c \eta_t \lambda E_{av}/c_{pa} B T_{0_1})]^{1/\epsilon_a}$$

Substituting the optimum fraction of energy λ_{opt} from Eq. 4.21

$$\lambda_{opt} = \frac{\eta_c \eta_t B - (U^2 B/2g_c \eta_c \eta_t E_{av})}{1 + \eta_c \eta_t B}$$

then the fan pressure ratio p_{rf} may be given as

$$p_{rf} = \left[1 + \frac{E_{av}}{c_{pa} T_{0_1}} \left\{ \frac{(\eta_c \eta_t)^2 - (U^2/2g_c E_{av})}{1 + \eta_c \eta_t B} \right\}\right]^{1/\epsilon_a} \tag{4.25}$$

Problems

4.1 Show for a turbojet cycle with constant specific heat that at the takeoff condition where $M \ll 1$, the maximum energy available for a given temperature ratio and component efficiencies occurs at a pressure ratio given approximately by the expression

$$P_r \approx [\eta_c \eta_t (T_{0_3}/T_{0_1})]^{k/2(k-1)}$$

4.2 For a turbojet cycle with constant specific heats and a diffuser efficiency of unity, show that the minimum temperature ratio to obtain a finite energy available E_{av} is given by

$$\frac{T_{0_3}}{T_1} \geq \frac{\phi^2 (p_r^{\epsilon} - 1)}{\eta_c \eta_t (\phi - 1/p_r^{\epsilon})}$$

where

T_{0_3} = combustion stagnation temperature
T_1 = ambient atmospheric temperature
$\phi = 1 + (k - 1)M^2/2$
M = flight Mach number
η_c = compressor stagnation efficiency
η_t = turbine stagnation efficiency
p_r = compression stagnation pressure ratio
$\epsilon = (k - 1)/k$

Assume combustor pressure loss is included in η_c.

4.3 Show that the efficiency of the Brayton cycle expressed as E_{av}/Q_{in} is given by

$$\eta = \frac{(p_r^{\epsilon-1})[\eta_t(T_r/p_r^{\epsilon}) - (1/\eta_c)]}{(T_r - 1) - (\eta_c)^{-1}(p_r^{\epsilon} - 1)}$$

where, referring to the simple diagram Fig. 4.2,

T_r = temperature ratio = T_3/T_1
p_r = pressure ratio = p_2/p_1
η_c = compression efficiency
η_t = expansion efficiency

Assume constant specific heat and that the only losses are those of compression and expansion.

4.4 A turbojet aircraft is traveling at 40,000 ft (p = 2.73 psia, t = −69°F) at a Mach number of 2. There is a normal shock at inlet (no spillage), followed by diffusion at an isentropic efficiency of 0.7. The turbojet has a mass flow of 180 lb/sec, a compressor pressure ratio of 6.3, a compressor efficiency of 0.83, a combustor loss of 7.4 psi, a turbine inlet temperature of 1700°F, a turbine efficiency 0.88, a propulsion nozzle isentropic efficiency of 0.94. Assume air with c_p = 0.24 and k = 1.4 throughout.

(a) Find the net thrust for a convergent nozzle.
(b) Find the propulsion efficiency.
(c) If, without change of the turbojet performance, an afterburner is used following the turbine to reheat the gases to 2200°F, find the percentage change in nozzle throat area required.

4.5 A turbojet with a convergent nozzle is designed to give complete expansion at sea level, stationary conditions. At an altitude of 53,000 ft, where the atmospheric pressure is 10% of that at sea level, the aircraft speed is 50% of the jet velocity and the stagnation pressure at turbine discharge is the same as at sea level. The nozzle is choked in both instances and the engine operates similarly at both conditions, to the extent that it has the same pressure ratio and the same combustion temperature. Take c_p = 0.265 Btu/lb °R throughout.

(a) Show that the specific thrust at altitude is about 16.7% more than at sea level stationary.
(b) Assuming a diffuser stagnation pressure ratio r_0 effectiveness of 75%, find the flight Mach number at altitude.

4.6 The diagram represents a ramjet traveling in level flight at an altitude of 55,000 ft with a constant speed of 1200 mph. The intake conditions have been simplified to represent the formation of a shock wave immediately at the inlet. It may also be assumed that the pressure external to the duct is everywhere that of the ambient atmosphere. The inlet process, 2–3, is one of isentropic diffusion, and the combustion process, 3–4, occurs through the addition of fuel at constant area in a frictionless duct, the stagnation temperature at 4 being 1840°F.

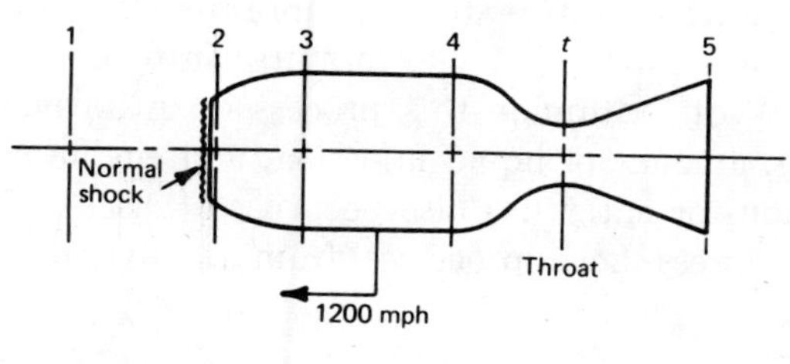

55,000 ft	$A_2 = 1\ \text{ft}^2$
$p_1 = 1.331$ psia	$A_3 = A_r = 2\ \text{ft}^2$
$t_1 = 390°\text{R}$	

Process 4-5 is one of isentropic expansion. Assume that the air-fuel ratio is 40/1 for (g). Take the working fluid to be air with $k = 1.4$ for all processes.

(a) Find the mass flow rate.
(b) Find the throat area at t.
(c) Find the thrust developed by the ramjet if the nozzle expands the gas down to ambient pressure.
(d) Repeat (c) for a convergent nozzle.
(e) Find the propulsion efficiency for (c).
(f) Find the propulsion efficiency for (d).
(g) Find the specific fuel consumption for (c).

4.7 Compare the thrust and SFC of a turbojet and a ramjet for flight at Mach 2.7 at 70,000 ft altitude using the following data:
Assume complete expansion in each case. Use mean values of c_p in Btu/lb °R as follows: compression 0.24, turbine expansion 0.275, nozzle expansion 0.265 and heat addition of 0.265. Use $R = 53.3$ ft lb/lb °R throughout.
Take the ram (diffuser) isentropic efficiency as 0.68 and the nozzle velocity coefficient as 0.95 in both cases.

Turbojet

$p_r = 3.8$
$\eta_c = 0.83$
$\eta_t = 0.88$
$\eta_{comb} = 0.98$
$T_{comb} = 2000°\text{F}$
$\Delta p_{comb} = .06\ p_{max}$
$C_v(\text{nozzle}) = 0.95$
H.V. fuel = 18,000 Btu/lb

Ramjet

Diffusion to Mach 0.25 followed by combustion of fuel at 94% combustion efficiency to a temperature (stagnation) of 2400°F. Take a pressure loss of 0.02 p_0 max for friction and turbulence in addition to momentum loss and assume it occurs after combustion.

4.8 How would the addition of a heat exchanger affect the performance of a turbojet engine? A turboprop engine? (The heat exchanger would use some of the thermal energy of the gas in the exhaust pipe before discharge to heat the air from the compressor before combustion.) Consult the literature for any actual engines or tests made and discuss the results.

4.9 Reducing the inlet temperature to a compressor improves its performance. Spraying water (or a water-methanol mixture) into the intake has been used to do this for takeoff. Examine this process quantitatively with respect to degree of cooling, amount of liquid injection, and engine performance. Water injection at combustor entry has also been used. Discuss the effect of this in similar fashion. Assess both processes from an overall performance point of view.

CHAPTER 5

Mission Analysis I—Central Force Field

Rocket missions can be treated under two main headings, those taking place in a central force field and those in field-free space. Conveniently, if not entirely logically, the former can be analyzed as missions of relatively short duration with launch from the surface of the earth or similar body and as missions of longer duration of an interplanetary nature with launch from an orbit around the earth or planet. This chapter discusses such missions in a central force field.

5.1 Law of Gravitation

The basic fact is expressed by Newton's law of universal gravitation, namely

$$F_{12} = -(GM_1M_2/r_{12}^2) \tag{5.1}$$

where F_{12} = the gravitational force of attraction between two bodies of mass M_1 and M_2, and r_{12} is the distance between the centers of the two bodies. G is the universal gravitational constant, having a value of 3.32×10^{-11} (lbf) (ft²)/(lbm²). For space propulsion, the mass M_2 of the vehicle is very much less than the mass M_1 of the central body (e.g., the earth or sun) and hence the center of the force field is considered as the center of M_1. Thus the force on the vehicle, which is what we are interested in, can be written as

$$F = -(GM_k)M/r^2 = -kM/r^2 \tag{5.2}$$

where subscript k indicates the central body (earth or sun) and the subscript on the vehicle mass is dropped as understood to be that of the vehicle. It is also sometimes convenient to write GM_k as the gravitational parameter k, as this may likely have only the one value for a particular problem. For the earth, $k = 4.37 \times 10^{14}$ (lbf) (ft²)/(lbm) and for the sun, 1.453×10^{20} (lbf) (ft²)/(lbm). The minus sign indicates that the force is directed oppositely to the distance r, i.e. toward the central body and is used when directional sense is needed, but not when only numerical value is in question.

It is customary to express a local gravitational force in terms of local gravitational acceleration g. Thus in our units,

$$F = Mg/g_c \tag{5.3}$$

where g is the local acceleration of gravity in ft/sec² and g_c is the dimensional constant 32.174 (lbm) (ft)/(lbf) (sec²). Equating these two expressions for gravitational force, without regard to sign,

$$k/r^2 = GM_k/r^2 = g/g_c \tag{5.4}$$

From this we get $g \propto 1/r^2$ and hence

$$g/g_0 = R_0^2/r^2 \qquad \text{and} \qquad g = g_0R_0^2/r^2 = g_0R_0^2/(R_0 + h)^2 \tag{5.5}$$

where subscript 0 refers to the value at some standard location such as the surface of the earth and h is the altitude above the earth's surface. Taking $g_0 = 32.174$ ft/sec² and R_0 as 3960 miles, then $g \approx 30.6$ ft/sec² at an altitude of 100 miles, a reduction of nearly 5% from the surface value. Corresponding figures for an altitude of 200 miles are 29.2 ft/sec² and about 9%. Thus, in most instances of rocket missions, even those involving near-earth satellite vehicles, it will be necessary to take a varying g into account. Sometimes this renders a calculation too complex for the purposes of general analysis and it is possible to use an estimated constant mean value.

5.2 Velocity Increment and Mass Ratio

Consider the vehicle to be launched from or near the earth's surface, in general with an initial velocity U_0. The line of action of the thrust is inclined at angle θ with the vertical. Using Eq. 2.7 and c as the equivalent jet velocity for rockets, i.e. $\dot{m}_pc/g_c$ represents the thrust due to momentum and pressure, then we have

$$\frac{\dot{m}_pc}{g_c} - \frac{Mg}{g_c}\cos\theta - D = \frac{M}{g_c}\frac{dU}{dt} \tag{5.6}$$

The drag force D can be expressed as $C_D\rho U^2A/2g_c$, where C_D is the drag coefficient for the vehicle and A is the area on which this coefficient is based. It is next to impossible to use this expression in analysis, because the vehicle has a constantly varying value of C_D as the Mach number and Reynolds number change with altitude and vehicle speed, and the ambient density changes continuously. Either a step-by-step evaluation is necessary, as indeed must be done for real, exact calculations, or an overall empirical mean value can be used.[1] It turns out that for large rockets, the drag has a relatively small value, a few percent of the thrust, and therefore can be omitted for an initial general analysis. For small rockets, such as those used for sounding the upper atmosphere and the like, the drag is an appreciable fraction of the thrust. Here we shall omit the drag factor and hence results will be for an ideal performance in this respect. Likewise the angle θ can be varying continuously, but here we shall assume a constant, mean value of $\cos\theta$, written as $\overline{\cos\theta}$. The value of g will likewise vary with altitude, but as thrust is usually only given to a vehicle in the early part of its ascent, the remainder of the journey being in free-flight or *ballistic*, then it too will be considered to have a constant value and we can write $\overline{g\cos\theta}$ as a mean value of the two.

Introducing a total *burning time* t_b, we can write $\dot{m}_p = M_p/t_b$, where M_p is the initial propellant mass. This assumes a constant rate of propellant consumption, which is largely justified as few rockets as yet have the thrust capable of being varied. Thus in liquid-propellant chemical rockets, propellant and oxidizer are

[1] P. Dergarabedian and R. D. Ten Dyke, "Estimating Performance Capabilities of Boost Rockets," *in* "Ballistic Missile and Space Vehicle Systems" (Seifert and Brown, Eds.) Wiley, New York, 1961.

supplied at a fixed rate and in solid-propellant rockets, the surface is most generally so arranged geometrically as to burn at an even rate. The thrust will not be quite constant for a rocket launched from the earth's surface because of the effect of decreasing back pressure as the vehicle increases altitude, but will again be so assumed.

Equation 5.6 can then be simplified and rearranged to give

$$\frac{M_p c}{t_b} = M\overline{g \cos \theta} + M \frac{dU}{dt} \tag{5.7}$$

The instantaneous mass of the vehicle M is given by

$$M = M_0 - \dot{m}_p t = M_0 - (M_p/t_b)t$$

where M_0 is the total initial (launch) mass of the vehicle. Inserting this into Eq. 5.7, dividing through by M_0, and writing $M_p/M_0 = \lambda$, the *propellant fraction*, then after rearrangement,

$$dU = \frac{\lambda c/t_b}{1 - \lambda t/t_b}\, dt - \overline{g \cos \theta}\, dt$$

Integrating between the interval $t = 0$ when $U = U_0$ and $t = t_b$ when $U = U_b$,

$$\int_{U_0}^{U_b} dU = \frac{\lambda c}{t_b} \int_0^{t_b} \frac{dt}{1 - \lambda t/t_b} - \overline{g \cos \theta} \int_0^{t_b} dt$$

$$\therefore U_b - U_0 = \frac{\lambda c}{t_b} \left[-\frac{t_b}{\lambda} \ln \left(1 - \frac{\lambda t}{t_b} \right) \right]_0^{t_b} - \overline{g \cos \theta}\, t_b \tag{5.8a}$$

$$= -c \ln (1 - \lambda) - \overline{g \cos \theta}\, t_b$$

or

$$U_b - U_0 = \Delta U = c \ln \frac{1}{1 - \lambda} - \overline{g \cos \theta}\, t_b \tag{5.8b}$$

Now $\lambda = M_p/M_0$, hence

$$\frac{1}{1 - \lambda} = \frac{1}{1 - M_p/M_0} = \frac{M_0}{M_0 - M_p} = \frac{M_{\text{initial}}}{M_{\text{final}}} = M_r$$

and Eq. 5.8 becomes

$$\Delta U = c \ln M_r - \overline{g \cos \theta}\, t_b \tag{5.9}$$

with M_r being called the *mass ratio*.

Equation 5.9 is the *basic equation of rocket mechanics*, valid for flight in a gravitational field with a constant value of g and with drag omitted. If the gravitational term is negligible or nonexistent (as in field-free space), then the rocket equation becomes very simply

$$\Delta U = c \ln M_r \tag{5.10a}$$

or

$$M_r = e^{\Delta U/c} \tag{5.10b}$$

The jet velocity c may be replaced by the specific impulse $g_c I_s$ from Eq. 2.16, hence

$$\Delta U = g_c I_s \ln M_r \qquad (5.10c)$$

M_r is a most important parameter, as the velocity increment ΔU varies directly with its logarithm. For large ΔU, the equivalent jet velocity c or specific impulse I_s must be as high as possible and the propellant fraction λ large (large M_r).

Figures 5.1(*a*) and (*b*) show the effect of this logarithmic function on the velocity increment. These figures are for ideal, no-drag performance, with a constant

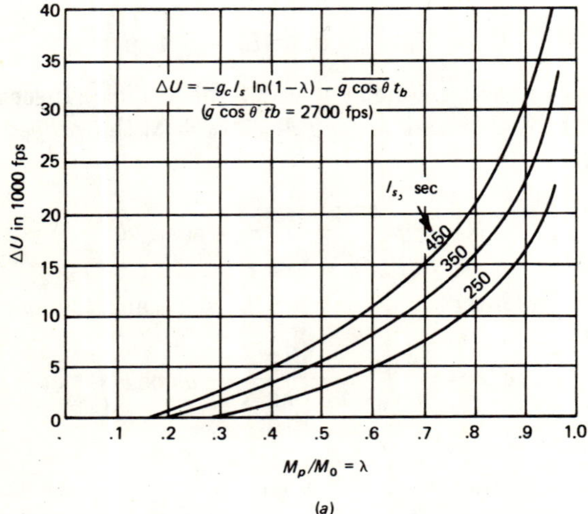

Fig. 5.1 (*a*) Velocity increment vs propellant fraction.

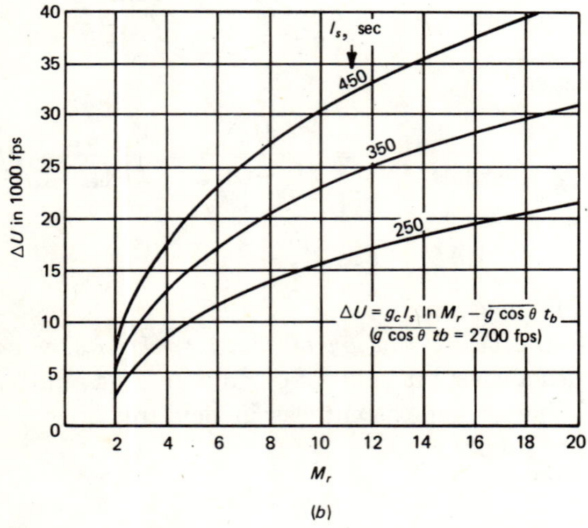

Fig. 5.1 (*b*) Velocity increment vs mass ratio.

value of $\overline{g \cos \theta}\, t_b = 2700$ ft/sec ($g \approx 32$, $\overline{\cos \theta} \approx .7$, $t_b = 120$). It will be seen presently that the velocity required to place a vehicle in orbit round the earth at about 100 miles altitude is approximately 25,700 fps and hence with a representative specific impulse of 350 sec for the propellants, the mass ratio must be 13.6 or the initial propellant fraction must be about 0.935. This leaves only $6\frac{1}{2}\%$ of the takeoff mass for structure and payload. The velocity to escape completely from the earth's gravitational effect is about 36,200 fps and in this case the mass ratio is about 32 and the propellant fraction about 0.97. This illustrates very clearly one of the major difficulties in rocket propulsion from the earth's surface and the great importance of high specific impulse and low structure weight. Hence the seemingly tiny payloads taken into space by monster rockets and the colossal launching rockets required by manned missions.

Equation 5.10 may be changed to show directly the structure and payload fractions or coefficients. These have alternative definitions and are used variously, so the context must be examined when referred to in the literature. If these fractions are defined similarly to λ by dividing by the initial mass, we have

$$\text{Structure fraction} \quad s = M_s/M_0 \tag{5.11a}$$

$$\text{Payload fraction} \quad l = M_L/M_0 \tag{5.11b}$$

and hence, using the simple rocket Eq. 5.10a for illustration,

$$\Delta U = c \ln M_0/(M_s + M_L) = c \ln 1/(s + l) \tag{5.12}$$

Alternatively, the definitions are

$$\text{Structure coefficient} \quad \sigma = \frac{M_s}{M_0 - M_L} = \frac{M_s}{M_s + M_p} \tag{5.13a}$$

$$\text{Payload coefficient} \quad \mu = \frac{M_L}{M_s + M_p} \tag{5.13b}$$

and hence

$$\Delta U = c \ln (1 + \mu)/(\sigma + \mu) \tag{5.14}$$

Again, a mixture sometimes appears, using Eqs. 5.11b for payload fraction and 5.13a for structure coefficient in which case

$$\Delta U = \ln [\sigma(1 - l) + l]^{-1} \tag{5.15}$$

It will be seen from Eq. 5.9 that the burning time should be as short as possible for maximum velocity increment. Thus, in the example used previously the value of $\overline{g \cos \theta} t_b$ was about 10% of the orbiting velocity. The basic reason for this effect is that the shorter the burning time, the less energy is used in lifting propellant against gravity, as this gain in potential energy of the propellant is lost as the propellant is used up. This is the first indication that in a central force field, the thrust should be *impulsive*, i.e. ideally instantaneous, rather than continuous. This feature of propulsion will be encountered again.

It can be shown that the burning time must always be less than the specific impulse in seconds. Starting with the expression for thrust, $F = \dot{m}_p c/g_c = M_p c/g_c t_b$, we get

$$t_b = \frac{M_p I_s}{F} = \frac{M_p}{M_0}\frac{M_0}{F} I_s = \frac{\lambda}{a_0/g_c} I_s$$

where $a_0/g_c = F/M_0$, the initial acceleration. Now a_0/g_c must be greater than unity for any acceleration at all from the earth's surface and λ must be less than unity for any structure mass whatsoever, hence with $a_0/g_c > 1$, and $\lambda < 1$, $t_b < I_s$. In the limit t_b may approach I_s. With $\lambda \approx 0.9$ and $a_0/g_c \approx 2$, then $t_b \approx I_s/2$ or 125–200 sec.

5.3 Burnout Distance

The distance S covered from initial fire to burnout is given by

$$S_b = \int_0^{t_b} U\, dt$$

Using Eq. 5.8a but with the limit of integration any time $t = t$ rather than $t = t_b$, we can substitute

$$U - U_0 = -c \ln [1 - (\lambda t/t_b)] - \overline{g \cos \theta}\, t$$

Hence

$$S_b = U_0 \int_0^{t_b} dt - c \int_0^{t_b} \ln [1 - (\lambda t/t_b)]\, dt - \overline{g \cos \theta} \int_0^{t_b} t\, dt$$

Integration and simplification gives

$$S_b = U_0 t_b + c t_b \{[(1 - \lambda)/\lambda] \ln (1 - \lambda) + 1\} - \tfrac{1}{2}\overline{g \cos \theta}\, t_b^2 \qquad (5.16a)$$

or alternatively,

$$S_b = U_0 t_b + c t_b [(M_r - 1)^{-1} \ln (M_r)^{-1} + 1] - \tfrac{1}{2}\overline{g \cos \theta}\, t_b^2 \qquad (5.16b)$$

The effect of burning time on the gravity factor should again be noted, this time proportional to t_b^2. From these relationships for velocity increment and distance, it is concluded that in the absence of drag, U_b and S_b are independent of rocket size and very dependent on specific impulse I_s and the mass ratio M_r or propellant fraction λ.

5.4 Trajectory and Gravity Turn

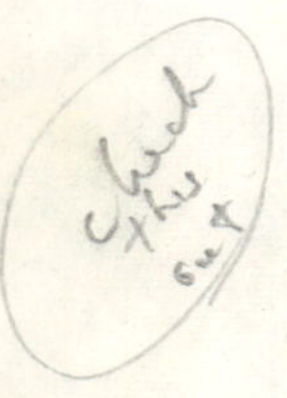

The expressions as given include the gravity term with a constant value of $\cos \theta$ and thus are accurate only for relatively short intervals unless the flight is vertical, i.e. $\cos \theta = 1$. Usually a vehicle has no lift surfaces and the thrust and velocity are colinear. If θ is greater than zero, then the vehicle undergoes a *gravity turn*, that is, its path is curved. In an initial interval of time Δt_1, the vehicle has an overall velocity increment ΔU_1 which is the vector sum of ΔU_{F_1} due to thrust and ΔU_{g_1}, due to

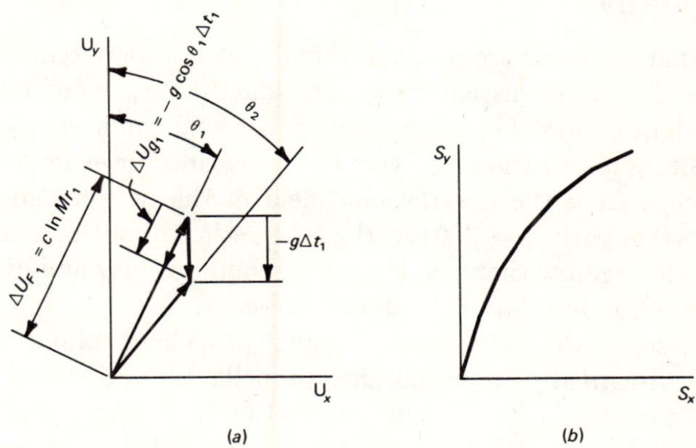

Fig. 5.2 Gravity turn

gravity. This is shown in Fig. 5.2(*a*). In the next interval of time Δt_2, the vehicle may be considered to be traveling at angle θ_2 and has components of velocity ΔU_{F_2} and ΔU_{g_2}. This can be carried on for any number of time intervals Δt, which need not necessarily be the same if more accuracy is required over parts of the trajectory, say where $\cos\theta$ is changing very rapidly. The components of overall velocity increments in the vertical and horizontal directions when multiplied by the respective Δt's give the height and range distances, which may then be plotted as in Fig. 5.2(*b*). With sufficiently small time intervals, the trajectory can be given by fairing the linear segments to give a smooth curve.

5.5 Coasting Height

For vertical flight, as might be nearly the case for a sounding rocket, the coasting height after burnout is considerable, as the burnout velocity is high and drag is very small because the atmospheric density is very low above the burnout height. In the absence of drag, the coast height h_c may be readily calculated by equating the kinetic energy of the vehicle at burnout to the increase in potential energy due to gain in altitude, i.e.,

$$MU_b^2/2g_c = \int_0^{h_c} M(g/g_c)\,dh$$

Equation 5.5 for variable g may be substituted if accuracy is required, yielding

$$U_b^2/2g_0 = h_cR_0/(h_c + R_0)$$

and

$$h_c = R_0U_b^2/(2g_0R_0 - U_b^2) \tag{5.17a}$$

When a constant mean value of gravity $\bar{g}$ is assumed, then

$$h_c = U_b^2/2\bar{g} \tag{5.17b}$$

For a burnout velocity of 20,000 fps, which can be the order of velocity for a small rocket, then the coasting height is over 1000 miles.

5.6 Multistaging

It is apparent that payloads are minimal if high velocities are required. The curves of Fig. 5.1 show that some missions are impossible if the specific impulse is limited, as structure weight cannot be reduced below a certain minimum, i.e., there will always be a limiting mass ratio even with zero payload. For example, the necessary velocity for escape from the gravitational field of the earth is about 36,500 fps. Figure 5.1 shows that with $I_s = 350$ sec, the mass ratio is about 24 and the propellant fraction about 0.96, even without taking into account gravity and drag. Some relief from such a restriction is obtained by *multistaging*.

It can be seen intuitively that if empty propellant tanks and supporting structure can be discarded at intervals, then propellant is not wasted in accelerating these masses to a final velocity. The principle of multistaging is that instead of one large rocket, several smaller rockets are fastened in tandem, with each empty unit being jettisoned when its propellant is burned. Two- and three-stage rockets are common, with four stages as a maximum. It might be wondered if the gain is worthwhile if the structure fraction is small anyway. The point is that toward the end of the burning time, the overall mass is small, i.e. structure plus payload only, and therefore an apparently small reduction in structure mass makes a considerable change in the mass ratio, to which the overall velocity increment is very sensitive. To gain some idea of the improvement, consider a one-stage rocket of initial mass 25,000 lb, structure mass of 2000 lb and payload of 500 lb. The mass ratio is then $25{,}000/2500 = 10$ and, for a specific impulse of 300 sec, the velocity increment from Eq. 5.10c is 22,250 fps. This is short of orbiting velocity.

Now suppose that a two-stage rocket is used, with the same total initial mass, structure mass and payload as before, with the specific impulse of both stages the same as the original. This is shown diagrammatically in Fig. 5.3.

Let the payload fraction $l = M_L/M_0$ and the structure fraction $s = M_s/M_0$ be the same for each stage. Using the notation shown in the figure

$$l_1 = l_2 = M_{L_1}/M_{0_1} = M_{L_2}/M_{0_2}$$

With $M_{L_1} = M_{0_2}$ and $M_{L_2} = M_L'$, then

$$M_{0_2} = [(M_{0_1})(M_L')]^{1/2} = [(25{,}000)(500)]^{1/2} = 3540 \text{ lb}$$

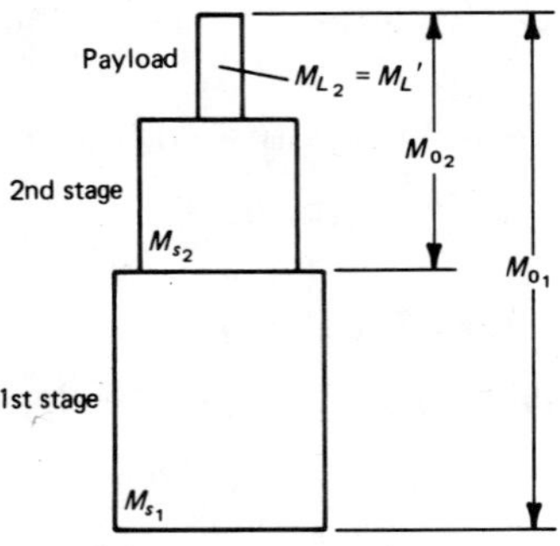

Fig. 5.3 Two-stage rocket—schematic diagram

Again,

$$s_1 = s_2 = M_{s_1}/M_{0_1} = M_{s_2}/M_{0_2}$$

Now

$$M_{s_1} + M_{s_2} = M_s' = 2000$$

$$\therefore (2000 - M_{s_2})/25{,}000 = M_{s_2}/3540$$

$$\text{yielding } M_{s_2} = 248 \text{ and } M_{s_1} = 1752$$

Now the stage mass ratios can be calculated.

$$M_{r_1} = \frac{M_{0_1}}{M_{s_1} + M_{L_1}} = \frac{M_{0_1}}{M_{s_1} + M_{0_2}} = \frac{25{,}000}{1752 + 3540} = 4.72$$

$$M_{r_2} = \frac{M_{0_2}}{M_{s_2} + M_{L_2}} = \frac{M_{0_2}}{M_{s_2} + M_L'} = \frac{3540}{248 + 500} = 4.72$$

Hence both velocity increments are

$$\Delta U_1 = 9660 \ln 4.72 = 15{,}000 \text{ fps}$$

giving a final velocity of twice this, or 30,000 fps, which is well over the satellite orbiting velocity.

This is an arbitrary example and in general it is necessary to make a stage-by-stage calculation using the appropriate data for each stage. However, it is useful to generalize the problem to get an idea as to how many stages are desirable. In order to get such a general result it is necessary to assume that the specific impulse and structure and payload fractions are the same for all stages.

For any stage n of N stages, we have from Eq. 5.12

$$\Delta U_n = c \ln (s_n + l_n)^{-1}$$

where s_n and l_n are the stage fractions.

Now using an overall payload fraction l_N as the ratio of final payload of the Nth stage to the initial total mass M_0', then using a three-stage rocket as example,

$$l_n = M_L'/M_{0_3} = M_{0_3}/M_{0_2} = M_{0_2}/M_{0_1}$$

and

$$l_N = \frac{M_L'}{M_0'} = \frac{M_L'}{M_{0_3}} \cdot \frac{M_{0_3}}{M_{0_2}} \cdot \frac{M_{0_2}}{M_0'} = l_n^3$$

or, in general, $l_N = (l_n)^N$ and $l_n = (l_N)^{1/N}$.

With equal mass ratios and specific impulses, the velocity increments ΔU_n of all N stages are similar, hence the overall velocity increment ΔU_N is

$$\Delta U_N = cN \ln \left(\frac{1}{s_n + (l_N)^{1/N}} \right) \tag{5.18}$$

This is plotted in Fig. 5.4, which shows that two stages give about 50% increase in ΔU and that more than four stages is not worthwhile.

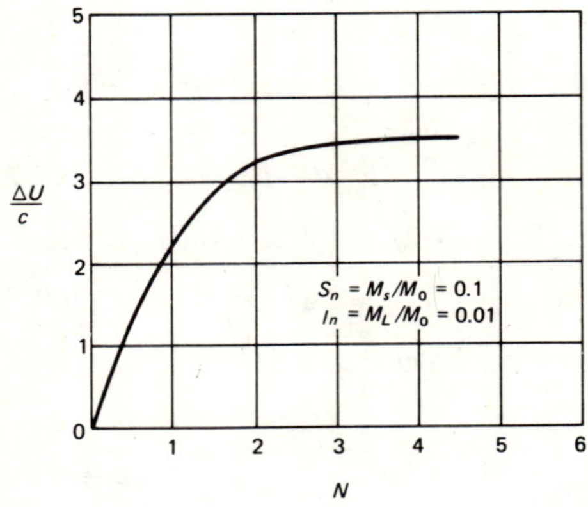

Fig. 5.4 Effect of multistaging on velocity increment.

5.7 Satellite and Escape Velocity

Very simple analysis can give the necessary minimum velocity required for a body to have a circular trajectory around the earth (satellite orbit) or to escape entirely from the earth's gravitational field.

For the former, it is reasoned that the gravitational force on the body is balanced by the centripetal acceleration, i.e.

$$M(g/g_c) = (M/g_c)(U_s^2/r)$$

where U_s is the linear velocity at radius r from the center of the earth

Thus

$$U_s = (gr)^{1/2} \tag{5.19}$$

Acceleration of gravity g is the local value at radius r and using Eq. 5.5 and putting $r = R_0 + h$,

$$U_s = \{g_0[R_0/(R_0 + h)]^2(R_0 + h)\}^{1/2}$$

$$= R_0[g_0/(R_0 + h)]^{1/2} \tag{5.20}$$

For $h = 100$ miles, satellite velocity, $U_s \approx 25{,}600$ fps.

The orbital period τ is the orbital circumference divided by the satellite velocity, thus

$$\tau = \frac{2\pi r}{U_s} = \frac{2\pi(R_0 + h)}{R_0[g_0/(R_0 + h)]^{1/2}} = \frac{2\pi}{R_0(g_0)^{1/2}}(R_0 + h)^{3/2} \tag{5.21}$$

For $h = 100$ miles, the period $\tau \approx 87.7$ mins. For an orbiting body to remain fixed relative to a point on the earth's surface, the period would have to be 24 hours and the corresponding altitude from Eq. 5.21 is about 22,400 miles.

The escape velocity U_e may be found by equating the kinetic energy to the work required to move the body from radius r to infinity against the gravitational force, hence

$$MU_e^2/2g_c = (M/g_c)\int_r^\infty g\,dr$$

and

$$U_e^2 = 2\int_r^\infty g\,dr$$

$$= 2g_0R_0^2\int_r^\infty \frac{dr}{r^2} = 2g_0R_0^2\left[-\frac{1}{r}\right]_r^\infty = \frac{2g_0R_0^2}{r} = \frac{2g_0R_0^2}{R_0+h}$$

$$\therefore U_e = R_0[2g_0/(R_0+h)]^{1/2} \tag{5.22}$$

It should be noted that the escape velocity U_e is $(2)^{1/2}$ times the satellite velocity U_s or the kinetic energy for escape is twice that for orbital motion. Escape velocity from 100 miles altitude is then 36,200 fps.

For a body in satellite orbit, the change in velocity from orbit to escape is $(2)^{1/2} - 1 = 0.414$ of the orbiting velocity. The necessary applied velocity increment is exactly this if the thrust is directed circumferentially [Fig. 5.5(*a*)]. It might be thought that a radially oriented thrust would be more direct, but Fig. 5.5(*b*) shows that the increment if applied perpendicular to the motion would have to be equal to the orbiting velocity.

5.8 Orbital Mechanics

Turning to the more general question of the performance of vehicles in orbital motion, it is necessary to examine the mechanics of bodies in a central force field. The discussion will be simple, with a view to understanding the development rather than presenting a rigorous or involved mathematical analysis.

It was Kepler in the early seventeenth century who first evolved the planetary laws relating to orbits. Briefly these may be stated as

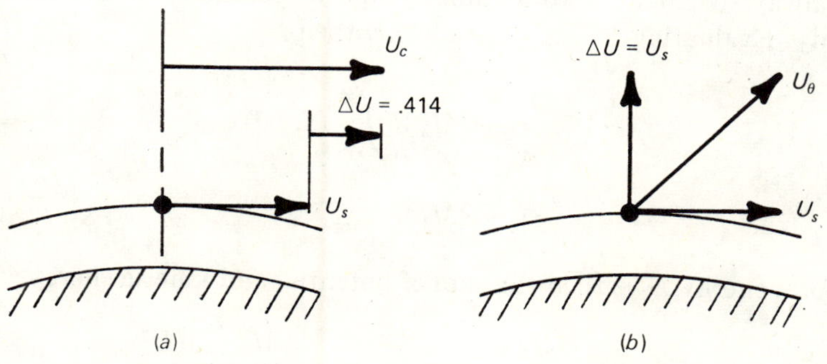

Fig. 5.5 Velocity diagrams for (*a*) circumferential thrust addition and (*b*) radial thrust addition.

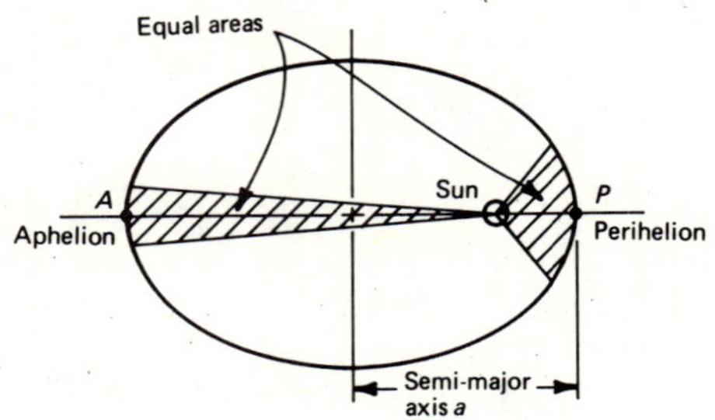

Fig. 5.6 Planetary orbit.

(1) The orbit of planets is an ellipse with the sun at one focus.
(2) A straight line from the planet to the sun sweeps out equal areas in equal times.
(3) The square of the orbital period is proportional to the cube of the semi-major axis, i.e. $\tau \propto a^{3/2}$.

Figure 5.6 shows these elements of planetary motion.

It is seen that the orbital velocity is greatest at point P, the *perihelion* and least at point A, the *aphelion*. (For motion round the earth, the corresponding minimum and maximum radial distances are called the *perigee* and *apogee*, respectively. In general, the terms are *pericenter* and *apocenter*.) Newton explained Kepler's laws some three-quarters of a century later by his law of universal gravitation which, exclusive of relativistic mechanics, laid the foundation of mechanical analysis.

From our knowledge of the laws of elementary mechanics, we can state from a consideration of a body in orbit as shown in Fig. 5.6, that the total energy is the sum of a kinetic energy due to motion and a potential energy due to the gravitational field. The kinetic energy (KE) is simply $MU^2/2g_c$ where U is the linear velocity of the body with respect to the central body. The potential energy (PE) is obtained from a consideration of the gravitational force and the radial distance r between the body and the central body. The potential energy is defined as zero at infinity and the force is given by Eq. 5.2, $F = -GM_kM/r^2$. As a result, potential energy is always negative, zero at infinity and a maximum at the surface of the central body. Evaluating PE at any radius r, then

$$\mathrm{PE} = -GM_kM \int_r^{\infty} (dr/r^2)$$

$$= -GM_kM/r \tag{5.23}$$

The total energy E_t is then the sum of potential and kinetic energy

$$E_t = -\frac{GM_kM}{r} + \frac{MU^2}{2g_c} = -\frac{kM}{r} + \frac{MU^2}{2g_c} \tag{5.24}$$

Also from a knowledge of physical laws, we can state that the angular momentum of the body is constant, because the only force acting is radial. In symbols, angular momentum $= M_\omega = Mr^2\omega/g_c = MUr/g_c =$ constant, where ω is the angular velocity $= U/r$.

However it is necessary to delve a little more deeply into the mechanics of the motion in order to obtain quantitative expressions and the conservation of energy and momentum will be shown to be constant in the course of this.

Figure 5.7 shows the polar coordinate system for a body of mass M moving in the gravitational field of body M_k.

From Newton's law $\mathbf{F} = Md\mathbf{U}/dt$, we have

$$\mathbf{F} = (M/g_c)(d/dt)(\mathbf{u}_r + \mathbf{u}_\theta)$$

Using unit vectors $\mathbf{e}_r$ and $\mathbf{e}_\theta$ and substituting $F = -GM_kM/r^2$, then

$$-\frac{GM_kM}{r^2}\mathbf{e}_r = \frac{M}{g_c}\frac{d}{dt}(u_r\mathbf{e}_r + u_\theta\mathbf{e}_\theta)$$

and

$$-\frac{g_cGM_k}{r^2}\mathbf{e}_r = u_r\frac{d\mathbf{e}_r}{dt} + \mathbf{e}_r\frac{du_r}{dt} + u_\theta\frac{d\mathbf{e}_\theta}{dt} + \mathbf{e}_\theta\frac{du_\theta}{dt}$$

Now $u_r = dr/dt$ and $u_\theta = rd\theta/dt$. Also $d\mathbf{e}_r/dt = \mathbf{e}_\theta d\theta/dt$ and $d\mathbf{e}_\theta/dt = -\mathbf{e}_r d\theta/dt$. Making these substitutions, we have after reduction, and using the notation $dx/dt = \dot{x}$, etc.,

$$-(g_cGM_k/r^2)\mathbf{e}_r = \mathbf{e}_r(\ddot{r} - r\dot{\theta}^2) + \mathbf{e}_\theta(r\ddot{\theta} + 2\dot{r}\theta)$$

The component scalar equations may then be stated thus: For $\mathbf{e}_\theta$: $0 = r\ddot{\theta} + 2\dot{r}\dot{\theta}$, which may be written as

$$0 = (d/dt)(r^2\dot{\theta}) \tag{5.25}$$

For $\mathbf{e}_r$:

$$-(g_cGM_k/r^2) = \ddot{r} - r\dot{\theta}^2 \tag{5.26}$$

It is seen that $r^2\dot{\theta} = r^2\omega$ represents the angular momentum per unit mass and therefore Eq. 5.25 shows that this momentum is constant as was assumed earlier. Using the previous symbols, $r^2\dot{\theta} = g_cM_\omega/M$, angular momentum per unit mass.

Turning to Eq. 5.26, no immediate conclusion can be made and further analysis

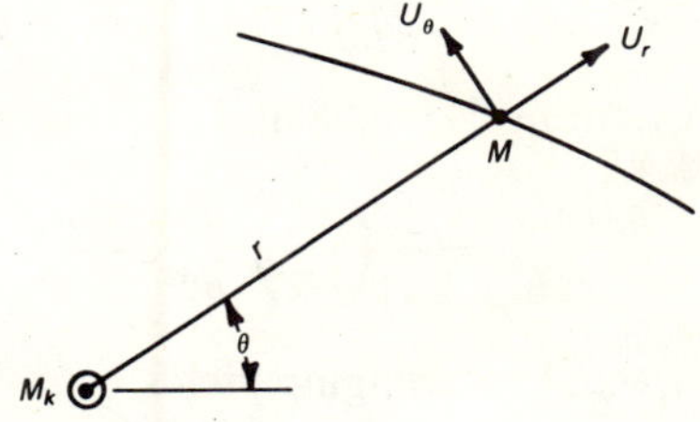

Fig. 5.7 Coordinate system for orbit analysis.

is necessary. Putting $GM_k = k$ and multiplying through by $\dot{r}$,

$$-g_c k\dot{r}/r^2 = \dot{r}\ddot{r} - \dot{r}\dot{\theta}^2 r$$

Substituting from the angular momentum relationship, $\dot{\theta} = g_c M_\omega/Mr^2 = m_\omega/r$ where $m_\omega = g_c M_\omega/M$, momentum per unit mass,

$$-g_c k\dot{r}/r^2 = \dot{r}\ddot{r} - m_\omega{}^2(\dot{r}/r^3)$$

This can be written as

$$g_c k\left[\frac{d}{dt}\left(\frac{1}{r}\right)\right] = \frac{1}{2}\frac{d}{dt}\left[\dot{r}^2 + \left(\frac{m_\omega}{r}\right)^2\right]$$

Direct integration gives

$$g_c k/r = \tfrac{1}{2}[\dot{r}^2 + (m_\omega/r)^2] + \text{constant}$$

Calling this constant $g_c E$ and rearranging,

$$(2g_c)^{-1}[\dot{r}^2 + (m_\omega/r)^2] - (k/r) = E \tag{5.27}$$

Now $\dot{r}^2$ is the radial velocity component squared and substituting $r^2\dot{\theta}$ for m_ω, the second term in parentheses becomes $(r\dot{\theta})^2 = (r\omega)^2$, which is the tangential velocity component. Thus the complete first term on the left-hand side represents the kinetic energy per unit mass, $U^2/2g_c$. The second complete term is the potential energy per unit mass. The sum of these two terms is then the total energy per unit mass and is equal to the constant of integration E. Thus the total energy per unit mass has been shown to be constant, again as assumed previously in the simple analysis.

We would like to use these equations to obtain r as a function only of θ, eliminating t, that is to find the shape of the orbit. To do this, we make substitutions as follows.

$$\frac{d}{dt}\left(\frac{1}{r}\right) = \frac{1}{r^2}\frac{dr}{dt} \quad \text{and} \quad \frac{dr}{dt} = -r^2\frac{d}{dt}\left(\frac{1}{r}\right) = -r^2\frac{d\theta}{dt}\frac{d}{d\theta}\left(\frac{1}{r}\right)$$

Substituting $r^2 d\theta/dt = m_\omega$ and letting $s = 1/r$,

$$dr/dt = m_\omega(ds/d\theta)$$

and

$$\frac{d^2r}{dt^2} = \ddot{r} = -m_\omega\frac{d^2s}{d\theta^2}\frac{d\theta}{dt} = -\left(\frac{m_\omega}{r}\right)^2\frac{d^2s}{d\theta^2} \tag{5.28}$$

Returning to Eq. 5.26, substituting Eq. 5.28 and $\dot{\theta} = m_\omega/r^2$, then

$$-\frac{g_c GM_k}{r^2} = -\left(\frac{m_\omega}{r}\right)^2\frac{d^2s}{d\theta^2} - \frac{m_\omega{}^2}{r^3}$$

which, substituting $k = GM_k$ and rearranging yields

$$(d^2s/d\theta^2) + s = g_c k/m_\omega{}^2$$

The solution of this differential equation is straightforward,

$$s = C \cos (\theta - \theta_0) + (g_c k/m_\omega^2) \tag{5.29}$$

Of the two constants of integration, θ_0 can be eliminated if the initial condition is $\theta = 0$. Since we are concerned only with the shape of the orbit and not any particular location, this can be done. For constant C, we use the constant total energy condition, writing Eq. 5.27 as

$$E = \text{KE} + \text{PE} = (U^2/2g_c) - (k/r) \tag{5.30}$$

and

$$U^2 = u_r^2 + u_\theta^2 = \dot{r}^2 + r^2\dot{\theta}^2$$

From the foregoing analysis $\dot{r} = m_\omega ds/d\theta$ and $\dot{\theta} = m_\omega/r^2$, hence

$$U^2 = m_\omega^2(ds/d\theta)^2 + (m_\omega^2/r)^2 = m_\omega^2[(ds/d\theta)^2 + s^2]$$

Differentiating Eq. 5.29,

$$ds/d\theta = -C \sin \theta.$$

$$\therefore U^2 = m_\omega^2\{C^2 \sin^2 \theta + [C \cos \theta + (g_c k/m_\omega^2)]^2\}$$

Substituting for total energy, Eq. 5.30, with $1/r = s$ from Eq. 5.29,

$$E = (m_\omega^2/2g_c)[C^2 \sin^2 \theta + C^2 \cos^2 \theta + (2g_c k/m_\omega^2)C \cos \theta + (g_c k/m_\omega^2)^2]$$
$$- (g_c k^2/m_\omega^2) - kC \cos \theta = (C^2 m_\omega^2/2g_c) - (g_c k^2/2m_\omega^2)$$

and

$$C = (g_c k/m_\omega^2)[1 + (2Em_\omega^2/g_c k^2)]^{1/2}$$

Substituting back in Eq. 5.29 and using $r = 1/s$

$$r = \frac{m_\omega^2/g_c k}{1 + [1 + (2Em_\omega^2/g_c k^2)]^{1/2} \cos \theta} \tag{5.31}$$

Now Eq. 5.31 is of the same form as the general equation for a conic section,

$$r = p/(1 + \epsilon \cos \theta) \tag{5.32}$$

where ϵ is the *eccentricity* and p is the "parameter." The four conic sections are the circle, the ellipse, the parabola, and the hyperbola. Equation 5.32 then shows that all bodies in a central force field must follow one of these orbits. The two most common orbits in propulsion are the ellipse and the hyperbola, with the circle regarded as a special case of the ellipse. The parabola is a boundary orbit separating the closed elliptical orbit and the open hyperbolic orbit.

Comparing Eqs. 5.31 and 5.32, it is seen that $\epsilon = [1 + 2Em_\omega/g_c k^2]^{1/2}$. m_ω and k must always be positive, and then E must be either positive or negative depending on the value of ϵ. For the ellipse, $\epsilon < 1$ and hence E must be negative, implying that the kinetic energy is less than the potential energy, which is that energy necessary for the body to escape the force field and travel to infinity. This means that the body is captured in the force field and the orbit is closed.

For the parabola, $\epsilon = 1$ and hence $E = 0$. Then from Eq. 5.30,

$$U^2/2g_c = k/r = GM_k/r$$

and

$$U = (2GM_kg_c/r)^{1/2}$$

From Eq. 5.4, we can then transform this to

$$U = (2GM_kg_c/r)^{1/2} = (2gr^2/r)^{1/2} = (2gr)^{1/2} = R_0[2g_0/(R_0 + h)]^{1/2} \quad (5.33)$$

which is, of course, the same as that obtained before (Eq. 5.22). The parabola is the boundary case, giving the minimum velocity for escape. If $\epsilon > 1$, then E is positive and there is kinetic energy in excess of that for escape.

We would like to define a satellite, i.e. elliptic, orbit more exactly yet by finding a value for the energy E in terms of simple orbit characteristics, in particular the eccentricity as defined by pericenter and apocenter distances.[2] Figure 5.8 shows these distances, denoted by R_p and R_A, respectively, with corresponding velocities U_p and U_A.

Starting with the energy equation, we have

$$E = -(k/r) + (U^2/2g_c) \quad (5.34)$$

From the conservation of energy condition, $E_p = E_A$, hence

$$-\frac{g_ck}{R_p} + \frac{U_p^2}{2} = -\frac{g_ck}{R_A} + \frac{U_A^2}{2} \quad (5.35)$$

and

$$g_ck\left(\frac{1}{R_A} - \frac{1}{R_p}\right) = \tfrac{1}{2}(U_A^2 - U_p^2)$$

From the conservation of momentum condition, $U_AR_A = U_pR_p$ and substituting for U_A and rearranging,

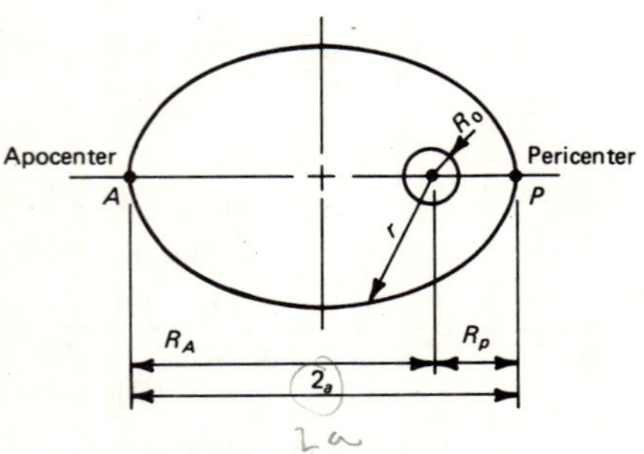

Fig. 5.8 Orbit symbols.

[2] This analysis follows that of M. Camac, "Reduction of Flight Time and Propellant Requirements of Satellites with Electric Propulsion by the Use of Stored Electrical Energy," AVCO Res. Lab., Res. Rep. 36, 1958.

$$g_c k\left(\frac{R_p - R_A}{R_A - R_p}\right) = \frac{U_p^2}{2}\left(\frac{R_p^2 - R_A^2}{R_A^2}\right) = \frac{U_p^2}{2}\frac{(R_p - R_A)(R_p + R_A)}{R_A^2}$$

$$\therefore \quad \frac{g_c k}{R_p + R_A} = \frac{U_p^2}{2}\frac{R_p}{R_A} \tag{5.36}$$

and similarly

$$\frac{g_c k}{R_p + R_A} = \frac{U_A^2}{2}\frac{R_A}{R_p}$$

Using the values of E_A and E_p given in Eq. 5.35, we have (since $E_p = E_A$),

$$E = \frac{E_p + E_A}{2} = -\frac{g_c k}{2}\left(\frac{1}{R_A} + \frac{1}{R_p}\right) + \frac{1}{2}\left(\frac{U_p^2 + U_A^2}{2}\right)$$

From Eq. 5.36,

$$\frac{U_p^2}{2} = \frac{g_c k}{R_A + R_p}\frac{R_A}{R_p} \qquad \text{and} \qquad \frac{U_A^2}{2} = \frac{g_c k}{R_A + R_p}\frac{R_p}{R_A}$$

$$\therefore E = \frac{g_c k}{2}\left(\frac{R_A + R_p}{R_A R_p}\right) + \frac{1}{2}\frac{g_c k}{R_A + R_p}\left(\frac{R_A}{R_p} + \frac{R_p}{R_A}\right)$$

which reduces to

$$E = -g_c k/(R_A + R_p) \tag{5.37a}$$

Substituting this in Eq. 5.34 gives

$$-g_c k/(R_p + R_A) = -(g_c k/r) + (U^2/2) \tag{5.37b}$$

and since $R_p + R_A = 2a$, the major axis of the ellipse, there is finally obtained,

$$E = -g_c k/2a = -(g_c k/r) + (U^2/2) \tag{5.38a}$$

In an alternative form written explicitly for the velocity,

$$U^2 = 2g_c k[(1/r) - (1/2a)] \tag{5.38b}$$

Equations 5.38a and b are forms of the "vis viva" equation, so called in the time of Newton before the idea of the conservation of energy was formulated explicitly.

The velocities at pericenter and apocenter are

$$U_p = [2g_c k/(R_A + R_p)]^{1/2}(R_A/R_p)^{1/2} \tag{5.39a}$$

and

$$U_A = [2g_c k/(R_A + R_p)]^{1/2}(R_p/R_A)^{1/2} \tag{5.39b}$$

It will be noted that for a circular orbit, $R_A = R_p = R_0 + h$ and hence

$$U = [2g_c k/2(R_0 + h)]^{1/2} = R_0[g_0/(R_0 + h)]^{1/2}$$

which once again is the satellite velocity U_s of Eq. 5.20.

Equation 5.38 is an important and useful one. It shows that for a body launched in a central force field with a certain velocity at a certain radius, then $2a$ is fixed, that is, the ellipse is defined (or in general the orbit whatever the conic section). For a vehicle given the exact satellite velocity U_s at an altitude relatively close to the earth's surface, then the angle at which it is injected is critical. If exactly tangential, the orbit will be circular. If at an angle to the tangential, then various ellipses are possible, but all with a semimajor axis of a. This is shown in Fig. 5.9 and it will be seen that some of the orbits are impossible, i.e., they intersect the earth or part may be so near the earth that the atmosphere exerts a braking effect. On the other hand, injection at the correct angle but with a velocity greater or less than the designed value will lead to a similar situation. Of course if the velocity is much higher, it can lead to escape.

Equation 5.38 can be used to show how the initial KE is utilized and how small differences in initial velocity lead to large variations in distance. For simplicity, assume vertical travel from the earth's surface. The total energy of the body will be $E = -k/R_0 + U^2/2g_c$. If $U < U_{\text{escape}}$, then the body will eventually have zero velocity and would start to fall back to the earth. At the point of maximum altitude, $E = -k/r + 0$. Equating the two values of energy, we can get

$$U^2 = 2g_c k\left(\frac{1}{R_0} - \frac{1}{r}\right) = \frac{2g_c k}{R_0}\left(1 - \frac{R_0}{r}\right) = U_e^2\left(1 - \frac{R_0}{r}\right)$$

where U_e = escape velocity = $(2g_c k/R_0)^{1/2}$ from Sec. 5.1 and Eq. 5.22. Hence

$$U/U_e = [1 - (R_0/r)]^{1/2} \tag{5.40}$$

If $r \gg R_0$, then $(1 - R_0/r)^{1/2} \approx 1 - R_0/2r$. If $r = 50R_0 \approx 200{,}000$ miles, then the velocity must be 99% of the escape velocity. Note that the moon distance is about 240,000 miles and thus that a journey to the moon requires very nearly escape velocity. In general, most of the initial kinetic energy is used in the first part of the journey, although the remaining energy may result in a considerable increase of range. Again suppose an initial velocity U_1 resulted in a distance r_1, and an initial velocity U_2 resulted in distance r_2. From Eq. 5.40,

$$\frac{U_2}{U_1} = \left[\frac{1 - (R_0/r_2)}{1 - (R_0/r_1)}\right]^{1/2} = \left[\frac{r_1\,(r_2 - R_0)}{r_2\,(r_1 - R_0)}\right]^{1/2} \tag{5.41}$$

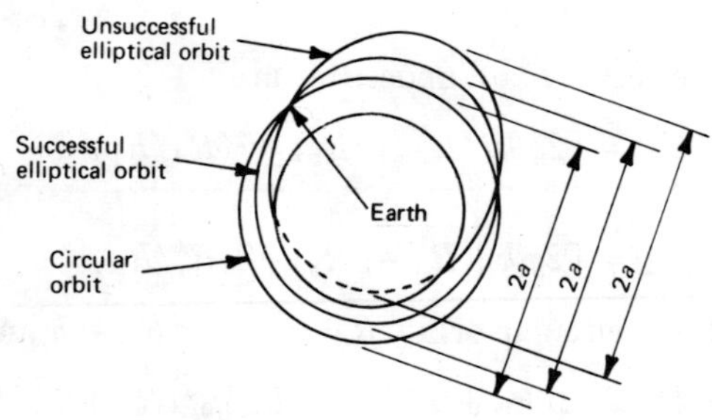

Fig. 5.9 Covelocity orbits.

Now if r_1 and r_2 are both very much greater than R_0, the terms in parentheses reduce almost to r_1 and r_2 and hence the right-hand side is almost unity. Thus, provided that $R_0 \ll r_1$ or r_2, r_1 and r_2 can differ significantly between themselves but there is very little difference between U_2 and U_1. This leads to the conclusion that a small error in velocity can lead to a large difference in range. Using the moon distance again as example with $R_0 = 4000$ miles and $r_1 = 240{,}000$ miles, a value of actual velocity U_2 only 1% less than required velocity U_1 leads to the actual range r_2 being only 122,500 miles, or just halfway to the moon. (This is only an example, as the simple example of vertical trajectory ending in zero velocity is not very realistic and, of course, a trajectory to the moon is not linear, and the example does not take into account the gravitational effect of the moon.)

5.9 Impulsive versus Continuous Thrust

It was seen previously that due to gravity, a larger vehicle velocity increment is obtained when the burning time is short and in the limit, the maximum value is obtained when the thrust is impulsive. This might be formalized by stating that because gravitational fields are conservative energy fields, a smaller ΔU is required for any maneuver if it is added impulsively. We might also note that a given ΔU added to a fast-moving vehicle increases the vehicle KE more than the same ΔU added when the vehicle is moving more slowly. Thus

$$U_2 = U_1 + \Delta U$$

and

$$U_2^2 = (U_1 + \Delta U)^2$$

$$\Delta \text{KE} \propto U_2^2 - U_1^2 = (U_1 + \Delta U)^2 - U_1^2 = 2U_1\Delta U + \Delta U^2$$

As $\Delta U \ll U_1$, the energy increment is directly proportional to the original velocity U_1.

It was also seen that because the vehicle acceleration must be greater than the gravitational acceleration for the vehicle to move out from the earth, then the burning time had to be less than the specific impulse. For chemical rockets, which for the foreseeable future appear to be the only kind capable of giving the very large thrust needed, this means that the time is of the order of 200 sec. If the total orbital time is large compared with this, then we can assume that the thrust is impulsive and the orbit is ballistic.

5.10 Orbital Transfer Missions

A common type of mission in gravitational fields is that between two orbits, either both centered on the earth, for example, or from an earth orbit to an orbit around another planet. From the idea that impulsive thrust is more economical of energy than continuous thrust, it seems reasonable that an initial boost from the first orbit giving a path to the location required, followed by a second boost placing the body in the correct orbit at that location would be suitable. This is the basis of the *Hohmann* transfer plan and it can be shown rigorously that in most cases it provides a minimum energy orbit.

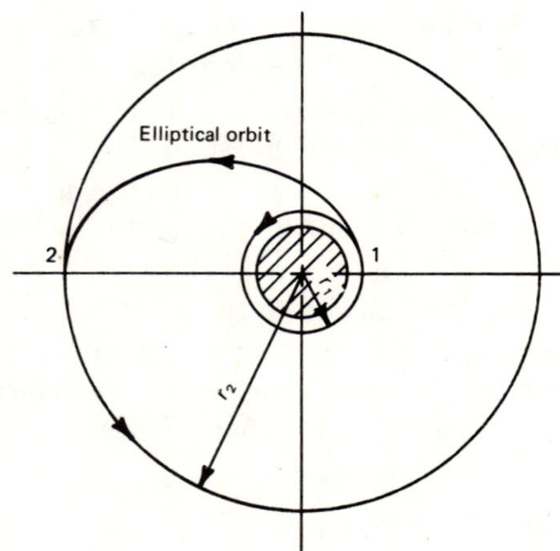

Fig. 5.10 Hohmann-type transfer orbit.

A Hohmann-type transfer is shown in Fig. 5.10, with the starting point at 1, and the body in a circular orbit of radius r_1 around a central body. An impulse at 1 gives an elliptic orbit out to point 2, which is at the radius r_2 required for the new orbit. An impulse at 2 sufficient to transfer the body into a circular orbit of radius r_2 then completes the mission. The same maneuvers can be carried out for the return trip, first an impulse from the circular orbit to an elliptical orbit back to radius r_1, then an impulse to circularize the orbit again.

The relationships necessary to calculate the velocity increments are available from the previous analyses. Figure 5.11 shows the geometry of a Hohmann transfer. The body is in circular orbit at an altitude h_1 above the surface of the central body which has a radius R_0. Thus $R_0 + h_1 = R_p$, where R_p is the pericenter of the transfer ellipse. Then the velocity U_{c_1} in initial circular orbit is $(g_ck/R_p)^{1/2}$. The velocity U_p at pericenter to give the elliptical orbit required is, from Eq. 5.39a,

$$U_p = [2g_ck/(R_A + R_p)]^{1/2}(R_A/R_p)^{1/2} \tag{5.39a}$$

The velocity increment is the difference between U_{c_1} and U_p, and with $U_{c_1} = (g_ck/R_{p_1})^{1/2}$,

$$\Delta U_p = U_p - U_{c_1} = (g_ck/R_p)^{1/2}\{[2R_A/(R_A + R_p)^{1/2}] - 1\} \tag{5.42}$$

The required new circular orbit is at radius R_A, the apocenter of the transfer ellipse. The velocity U_A at apocenter is from Eq. 5.39b,

$$U_A = [2g_ck/(R_A + R_p)]^{1/2}(R_p/R_A)^{1/2} \tag{5.39b}$$

The velocity U_{c_2} in circular orbit at radius R_A is $(g_ck/R_A)^{1/2}$ and hence the velocity increment at apocenter is

$$\Delta U_A = U_A - U_{c_2} = (g_ck/R_A)^{1/2}\{[2R_p/(R_A + R_p)]^{1/2} - 1\} \tag{5.43}$$

As an example, Fig. 5.12 (following Camac) shows these velocities and velocity increments for an initial pericenter radius at 100 n mi above the earth's surface, for various altitudes h_2 at apogee. U_p and U_a are the velocities at perigee and apogee for the elliptical orbit. U_{c_1} is the satellite (circular) velocity at 100 n mi altitude above the earth and U_{c2} is the satellite (circular) velocity at altitude h_2 above the earth's surface, i.e., at radius R_a.

The velocity increments ΔU_p and ΔU_a are then clearly shown as the distances between U_p and U_{c_1}, and between U_a and U_{c2}, respectively. Also the line representing the sum of the velocity increments ΔU_p and ΔU_A is shown, since this is the total increment necessary for going from circular orbit at U_{c_1} to circular orbit at U_{c2}

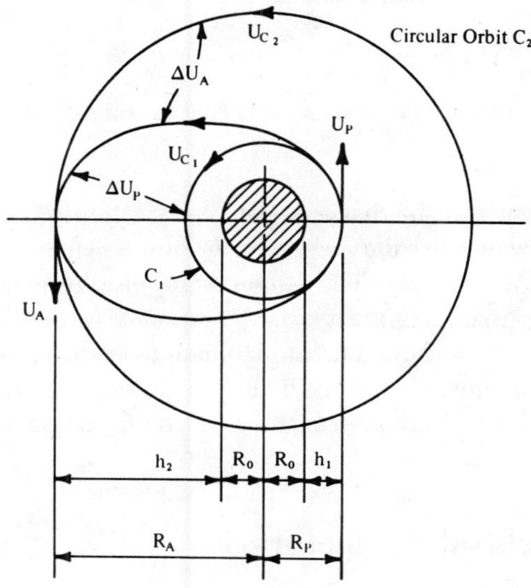

Fig. 5.11 Geometry of a Hohmann orbit.

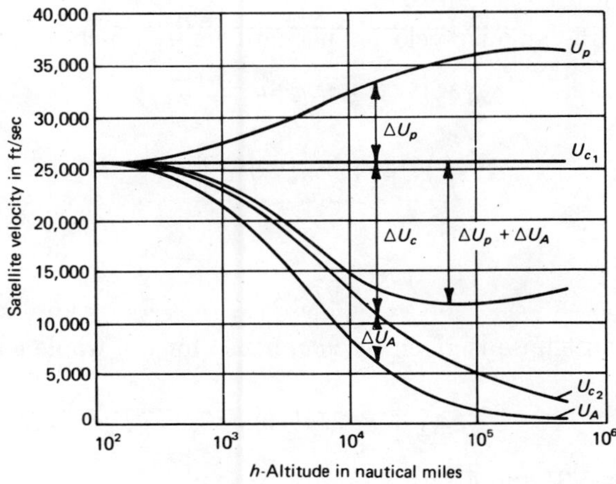

Fig. 5.12 Satellite velocity for Hohmann transfer (after Camac.)

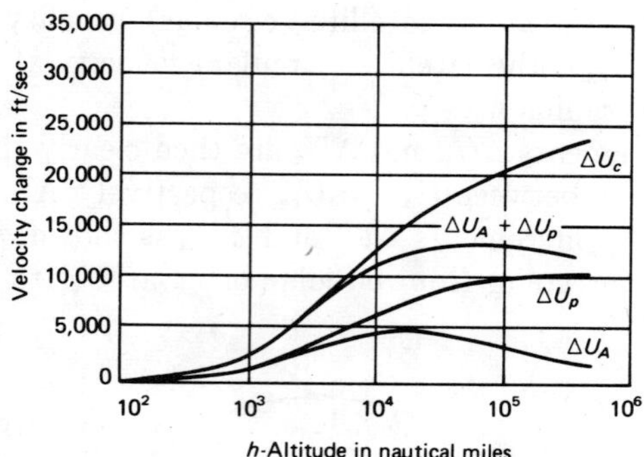

Fig. 5.13 Velocity increments for Hohmann transfer (after Camac.)

via two impulsive thrusts. The distance between U_{c_1} and U_{c_2} represents the difference of velocity between circular orbits at h_1, an altitude of 100 miles, and any altitude h_2. $\Delta U_c = U_{c_1} - U_{c_2}$ is then the total impulse for a spiral orbit between h_1 and h_2 and it is seen that it is always larger than $\Delta U_p + \Delta U_A$, thus showing the minimum energy characteristic of the Hohmann transfer ellipse. The velocity increments are shown more definitely in Fig. 5.13 and the difference between the spiral transfer and the Hohmann transfer is seen to be considerable at the larger distances.

5.11 Time of Orbital Trajectory

The time of a transfer mission is important and this can be obtained simply for an elliptical trajectory. From Fig. 5.14, a differential area dA is seen to be

$$dA = r(rd\theta)/2 = \tfrac{1}{2}r^2\, d\theta$$

But $r^2 d\theta/dt = m_\omega$, the angular velocity per unit mass, which is constant.

$$\therefore dA/dt = \tfrac{1}{2}r^2(d\theta/dt) = m_\omega/2$$

$$\therefore \int dt = (2/m_\omega) \int dA$$

and

$$t = (2/m_\omega)A$$

The total area of an ellipse is given by πab, hence for the whole ellipse, the period τ is given by

$$\tau = 2\pi ab/m_\omega$$

We have from Eqs. 5.31 and 5.32,

$$r = (m_\omega^2/g_c k)/(1 + \epsilon \cos\theta)$$

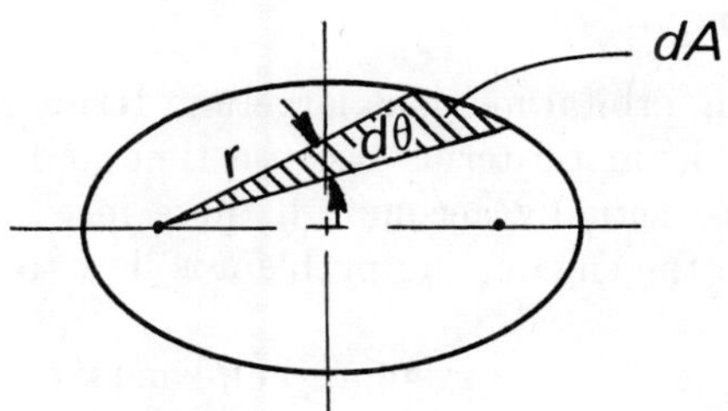

Fig. 5.14 Elliptical orbit—differential area.

Referring back to Fig. 5.8,

$$\text{At } P, \theta = 0 \quad \text{and hence} \quad R_p = (m_\omega{}^2/g_c k)/(1 + \epsilon)$$

$$\text{At } A, \theta = \pi \quad \text{and hence} \quad R_A = (m_\omega{}^2/g_c k)/(1 - \epsilon)$$

$$a = \frac{R_p + R_A}{2} = \frac{m_\omega{}^2}{2g_c k}\left(\frac{1}{1+\epsilon} + \frac{1}{1-\epsilon}\right) = \frac{m_\omega{}^2}{g_c k(1-\epsilon^2)}$$

or

$$(1 - \epsilon^2)^{1/2} = m_\omega/(g_c k a)^{1/2}$$

From analytic geometry we have the relationship between the eccentricity and the major and minor semi-axes, a and b, thus

$$b^2 = a^2(1 - \epsilon)$$

and

$$b = a(1 - \epsilon)^{1/2}$$

$$\therefore ab = a^2(1 - \epsilon^2)^{1/2} = a^2[m_\omega/(g_c k a)^{1/2}]$$

Substituting in the expression for the period,

$$\tau = \frac{2\pi ab}{m_\omega} = \frac{2\pi a^2}{m_\omega}\frac{m_\omega}{(g_c k a)^{1/2}} = \frac{2\pi a^{3/2}}{(g_c k)^{1/2}} = 2\pi\left(\frac{a^3}{g_c k}\right)^{1/2}$$

(Note that is a form of Kepler's third law, Sec. 5.8.) Thus the time from one circular orbit to another along a Hohmann semi-elliptical trajectory is

$$t_{Hoh} = \tau/2 = \pi(a^3/g_c k)^{1/2} \tag{5.44}$$

We might note that the mean angular velocity ω for an elliptical orbit is

$$\omega = 2\pi/\tau = (g_c k/a^3)^{1/2} \tag{5.45}$$

Another geometrical relationship from mathematics is obtained from the two relationships $R_p = a(1 - e)$ and $R_A = a(1 + e)$. Their ratio is

$$R_p/R_A = (1 - e)/(1 + e) \tag{5.46}$$

This is useful for relating perigee and apogee to eccentricity and hence the general shape of the orbit.

5.12 Mission Energy

We have been discussing orbital missions largely in terms of velocity and velocity increment, that is, in kinematic terms because time and distance are governing parameters. However to actually accomplish these missions, we must deal with energy quantities, since the engineering problem is how to physically supply such energy increments.

Thinking in terms of the solar system, the planets, of which the earth is one, have slightly elliptical orbits around the sun as focus. The planetary orbits are not exactly coplanar, but sufficiently so for us to make the assumption here, together with that of circular orbits, without great error. Table 5.1 gives some data on the planets in terms of the earth as a basis. There are two planets, Mercury and Venus, inside the earth's orbit and six outside. It will be seen from consideration of the distances why Venus and Mars are the two planets on which attention has been fixed in these early days of space propulsion.

The vehicle must escape from earth and as it travels out from the earth its velocity relative to the earth is decreasing. Suppose it is given just the correct escape velocity at the earth's surface, say at a radius of 4000 miles. Its total energy is then zero at all times (see Sec. 5.8) and from the energy equation we have

$$0 = -(k_E/r) + (U^2/2g_c)$$

and

$$U = (2g_c k_E/r)^{1/2} = 2.31 \times 10^6/(r_{\text{mi}})^{1/2} \tag{5.47}$$

where r_{mi} is in miles. When the velocity has decreased to about 1000 fps, r is about 5,350,000 miles. A velocity of 1000 fps is only about 1% of the earth's orbital velocity around the sun and the value of r, which superficially seems to be a long way, is less than 6% of the earth's distance from the sun. The time to get there is the order of 7 months.

Thus to all intents and purposes, after 7 months the body is still in earth orbit around the sun and must be given a velocity increment to change its heliocentric orbit. This may be called "hyperbolic excess velocity,"[3] ΔU_{exc}. However, the total kinetic energy per unit mass of a body given this excess velocity is not given by $(U_{\text{esc}} + \Delta U_{\text{exc}})^2$ but by

$$E_{\text{tot}} = E_{\text{esc}} + \Delta E_{\text{exc}}$$

or, for unit mass,

$$U_{\text{tot}}^2 = U_{\text{esc}}^2 + \Delta U_{\text{exc}}^2 \tag{5.48a}$$

and

$$U_{\text{tot}} = (U_{\text{esc}}^2 + \Delta U_{\text{exc}}^2)^{1/2} \tag{5.48b}$$

This principle of adding energies rather than velocities when evaluating propulsion requirements must be used at all times. Some velocity increments are actually decelerations and ΔU is negative. However, it takes just as much energy to slow down by the amount ΔU as it does to speed it up by ΔU and, therefore, ΔU's are always additive when considered as energy increments.

[3] R. F. Porter, "Trajectories of Unmanned Spacecraft," *Battelle Technical Review* (1966).

TABLE 5.1

Earth: Mean radius = 3959 miles = 3438 n mi
Mass = 1.3173×10^{25} lb
Distance from sun = 1 A.U. = 92.96×10^6 miles = 80.7×10^6 n mi
$k_E = GM_E$ = 4.375×10^{14} (lbf) (ft^2)/lbm
g_0 = 32.174 ft/sec^2

Gravitational constant G = 3.32×10^{-11} (lbf) (ft^2)/lbm

Sun: Mean radius (E = 1) = 109
Mass (E = 1) = 332,950

*Planetary Data**

	Mercury	Venus	Earth	Mars	Jupiter	Saturn	Uranus	Neptune	Pluto
Distance from sun (mean)	0.387	0.723	1.000	1.524	5.203	9.54	19.19	30.07	39.46
Diameter	0.39	0.972	1.000	0.535	11.2	9.47	3.69	3.5	1.1 (?)
Mass	0.055	0.815	1.000	0.108	318.4	95.2	14.6	17.3	0.87 (?)
Surface Gravity	0.36	0.87	1.000	0.38	2.64	1.13	1.07	1.41	0.5 (?)
Orbital Velocity	1.605	1.175	1.000	0.81	0.44	0.325	0.278	0.18	0.158
Period of revolution	0.241	0.615	1.000	1.88	11.86	29.46	84.01	164.8	247.7

Conversion Data: 1 n mi = 1.1515 miles (6080 feet)
1 mile = 0.8684 n mi
1 knot = 1 n mi/hr
1 km = 0.6214 mile
1 mile = 1.61 km
1 m = 3.28 ft

* Based on Earth = 1. Data vary with source and have been averaged or rounded off in some cases.

As an example, consider a mission to Mars. From Fig. 5.10 we see that we need ΔU_p from the earth orbit around the sun and escape from the earth.

Using Eq. 5.39a, with $k = k_s$ (for the sun) $= 1.41 \times 10^{20}$ (lbf ft^2)/lbm and $R_A + R_p = R_E + R_M$ (Mars-sun radius) $= [92.9 + 1.524(92.9)]\ 10^6$ miles, we find $U_p = 106{,}800$ fps. But since the body already has the velocity of the earth as a satellite of the sun, $U_E = 97{,}600$ fps, then $\Delta U_p = U_p - U_E = 106{,}800 - 97{,}600 = 9200$ fps.

The earth escape velocity $U_e = 36{,}700$ fps and therefore we might say $U_{\text{initial}} = U_e + \Delta U_p = 36{,}700 + 9200 = 45{,}900$ fps. However, we should add energies and instead write

$$U_{\text{initial}}^2 = U_e^2 + \Delta U_p^2$$

or

$$U_i = (U_e^2 + U_p^2)^{1/2} = [(36{,}700)^2 + (9{,}200)^2]^{1/2} = 37{,}800 \text{ fps},$$

which is a somewhat different value from that obtained from simple addition of velocities. In terms of Eqs. 5.48a and b, $U_{\text{initial}} = U_{\text{tot}}$ and $\Delta U_p = \Delta U_{\text{exc}}$ for the same answer.

Now a mission to a Jupiter orbit from an earth orbit will be demonstrated. Figure 5.15 shows the necessary symbols:

U_e (earth)	$= 35{,}800$	
U_s (earth)	$= 25{,}300$	
ΔU_{se} (earth)		$= 10{,}500$
U_p from Eq. 5.39a	$= 126{,}500$	
U_E	$= 97{,}600$	
ΔU_p		$= 28{,}900$
U_A from Eq. 5.39b	$= 24{,}300$	
U_J	$= 42{,}750$	
ΔU_A		$= -18{,}450$

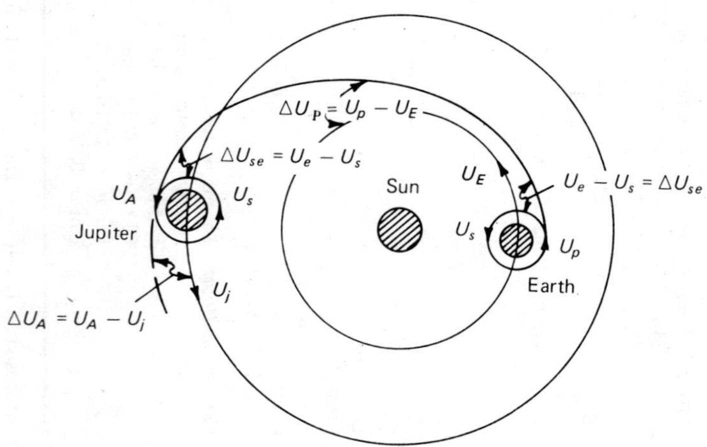

Fig. 5.15 Earth-Jupiter transfer orbit.

$$U_e \text{ (Jupiter)} = 198{,}000$$
$$U_s \text{ (Jupiter)} = 140{,}000$$
$$\Delta U_{es} = -58{,}000$$

Direct addition of absolute values of velocity increments gives $\Sigma\Delta_{\text{vel}} =$ $10{,}500 + 28{,}900 + 18{,}450 + 58{,}000 = 115{,}850$ fps.

In terms of energy, we can combine the first two steps in the earth environment,

$$U_1 = (\Delta U_{se}^2 + \Delta U_p^2)^{1/2} = 30{,}700$$

The two Jupiter maneuvers can be combined, the ΔU_A to get it into the Jupiter orbit and the "kick" to put it into Jupiter satellite orbit, i.e.,

$$\Delta U_2 = [(\Delta U_A)^2 + (\Delta U_{es})^2]^{1/2} = 60{,}800 \text{ fps}$$

The sum is therefore

$$\Sigma\Delta U_{\text{energy}} = 30{,}700 + 60{,}800 = 91{,}500 \text{ fps}$$

The considerable difference should be noted between velocity increments for a planetary journey starting from an earth satellite orbit and from the earth's surface. Thus for the Mars example, the initial velocity required for escape and orbit transfer was 37,800 fps. For the Jupiter case in spite of the much larger ΔU_p for orbit transfer, it was only 30,700 fps. Thus the task for the transfer mission propulsion arrangement is much lighter. The vehicle must still be lifted from the surface to satellite orbit, but this could be the task of a separate rocket system that could be discarded before the real journey is started.

Now it turns out that although the Hohmann orbits require the minimum energy, they also require the longest time and the times to the planets are very great. Thus for the three "nearby" planets, Mercury, Venus and Mars, the times are respectively 15 weeks, 21 weeks, and 37 weeks. These seem possible, but for the other planets the times are measured in years. Thus for Jupiter it is 2¾ years, for Saturn 6 years, while for the outermost one, Pluto, it is 45 years. For these planets ballistic trajectories of greater initial velocity and higher energy will be required or, alternatively, thrust must be supplied from an engine. Camac suggests the use of

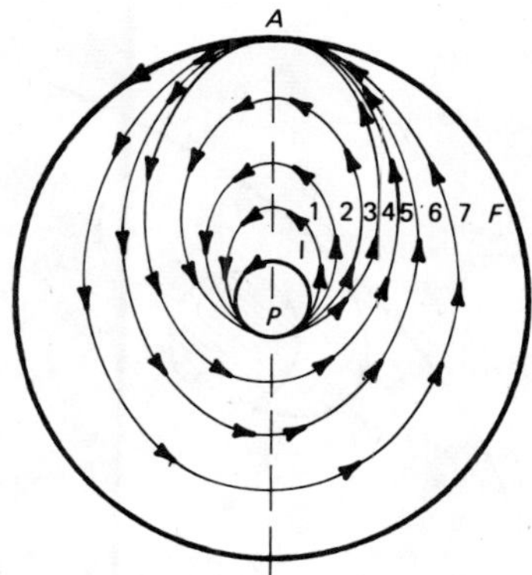

Fig. 5.16 Periodic impulsive thrusts using stored energy (after Camac)

periodic impulsive thrusts if the provision of one initial thrust is difficult or impossible with a given technology. This is shown in Fig. 5.16, whereby successive fractional thrusts are given at perigee, gradually increasing the major axis of the elliptical orbit until the apogee of the rendezvous planet is reached. Then fractional thrusts are given at apogee to circularize that orbit. A similar idea could be used on the return journey. The interval between velocity increments could be used for generating and storing a supply of energy for the next boost, utilizing solar energy to recharge batteries, for example.

5.13 Reentry

The problem of bringing a vehicle back from space through the earth's atmosphere to a landing on the earth's surface is known as the *reentry* problem and has been one of the well-known aspects of space travel. If the vehicle is to achieve a "soft" landing, then the problem becomes one of recovery as well. Usually it is only missiles for which recovery is not desired and this makes the problem more difficult in respect of velocity control but possibly somewhat easier with respect to the exact position of touchdown. It is also well-known that the earth's atmosphere is used as the decelerating agent so that it is not necessary to have a motor of equal thrust to that of launching. The problem is to control the deceleration in view of the tremendous heating caused by air friction on the vehicle at high velocities.

Considering first the reentry of a satellite, it has to be initially decelerated by a *retrorocket* to reduce its velocity and allow it to "fall" towards the earth, or more accurately, to put it into an elliptic orbit which intersects the earth's atmosphere (Fig. 5.17). Only a little knowledge of heat transfer is required to realize that the exact trajectory is crucial, so that a great deal of deceleration may be accomplished as far out as possible where the atmosphere is less dense. On the other hand, time may be important, that is, how long the vehicle is exposed to intense heating.

The vehicle is subject to gravitational force accelerating it and the atmospheric drag force decelerating it. Fortunately we find that over most of the trajectory, the gravitational effect is very small compared to that of the atmosphere and we can neglect it. Hence the problem is simply one of a single force, the drag, acting on a

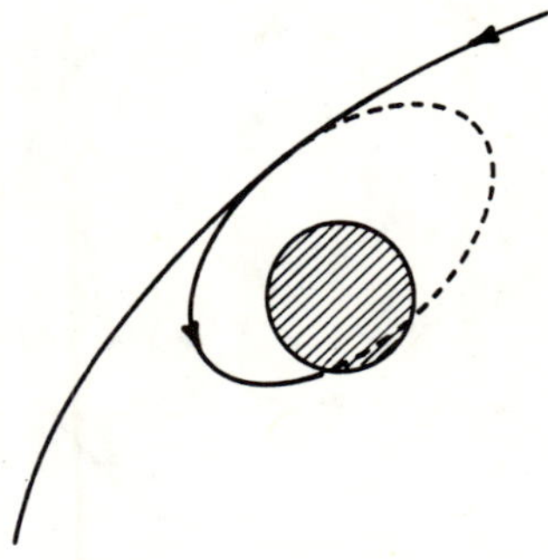

Fig. 5.17 Reentry orbit.

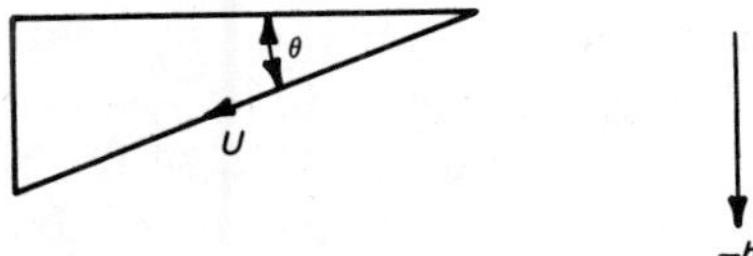

Fig. 5.18 Reentry velocity—height relationship

mass with an initially high velocity. Thus we can write

$$M(dU/dt) = -C_D\rho(U^2/2)A \tag{5.49}$$

Now only M and A are constants, but for many purposes and certainly here for demonstration purposes, we may also take C_D as a constant. It is customary to form the relationship M/C_DA as a parameter that is very useful in characterizing reentry performance. We are dealing mostly with the high velocity region, where the Mach number effect on C_D is relatively invariant and where change of Reynolds number does not have a marked effect. Thus with M/C_DA constant we need only a relationship for the density and confining ourselves to the upper atmosphere where most of the "reentry" occurs, then an analytical expression for density in terms of altitude can be given fairly simply. There are several expressions available, mostly in empirical form, and using one of the simplest we have

$$\rho = \rho_s e^{-\beta h} \tag{5.50}$$

where ρ_s is a reference value $= 0.11$ lb/ft^3, $\beta = 4.5 \times 10^{-5}$/ft and h is the altitude in feet. This expression is reasonably valid above about 50,000 ft. Thus we have

$$dU/dt = -(C_DA/M)\rho_s e^{-\beta h}(U^2/2) \tag{5.51}$$

Now we would like U as a function of altitude and, as we can neglect gravity, the velocity in the reentry region is directly proportional to altitude and is directed at a constant angle θ with respect to the tangential direction (Fig. 5.18). Hence

$$dh/dt = -U \sin\theta \tag{5.52}$$

From Eqs. 5.51 and 5.52 we get

$$\frac{dU}{dh} = \frac{\rho_s e^{-\beta h} U}{2(M/C_DA)\sin\theta} \tag{5.53}$$

The only variable on the right-hand side is h so that this equation can be integrated directly. The limits of integration are the initial state where $U = U_E$, the entry velocity at infinite altitude, and any state where $U = U$ at $h = h$. Thus we get

$$\frac{U}{U_E} = \exp\left(-\frac{\rho_s e^{-\beta h}}{2\beta(M/C_DA)\sin\theta}\right) \tag{5.54}$$

For any given value of M/C_DA and U_E, the altitude at which a specified velocity U is reached is fixed and given by Eq. 5.54. Regarding M/C_DA as a design variable, then we can express this alternatively by saying that the parameter M/C_DA

determines the altitude at which a certain value of reentry (as U/U_E) is achieved. For example, using $\theta = 2°$ (a value shortly to be justified), a value of U/U_E of 0.5 at a height of 140,000 ft requires a value of M/C_DA of about 87 lb/ft^2. Halving this value of M/C_DA gives the same velocity reduction at about 155,000 ft. Noting that $\rho_s e^{-\beta h} = \rho$ in Eq. 5.54, it is seen that $\rho \propto M/C_DA$ and hence the density where $U/U_E = 0.5$ is halved when M/C_DA is halved. Although the altitude difference seems small, the corresponding density variation is large. Values of M/C_DA of the order of 100 lb/ft^2 are possible. Thus, taking a reentry vehicle as having the characteristics of a flat disc, a diameter of 12 ft (Apollo size) with a drag coefficient of unity and a mass of 12,000 lb (Apollo), yields $M/C_DA \approx 106$ lb/ft^2.

From Eq. 5.54 we obtain

$$\rho = -2\beta(M/C_DA) \sin\theta \ln(U/U_E) = \rho_s e^{-\beta h}$$

and if this is substituted back in Eq. 5.51, it is found that

$$\begin{aligned} dU/dt = a &= \beta \sin\theta U^2 \ln(U/U_E) \\ &= \beta U_E^2 \sin\theta (U/U_E)^2 \ln(U/U_E) \end{aligned} \tag{5.55}$$

Thus at any given reentry velocity fraction U/U_E, the deceleration is independent of M/C_DA and hence also is the time to achieve that velocity. To find the maximum deceleration, Eq. 5.55 is differentiated with respect to U and set equal to zero, with the result that the maximizing condition is found to be

$$\ln(U/U_E) = -\tfrac{1}{2} \qquad \text{or} \qquad U/U_E = e^{-1/2} = 0.61 \tag{5.56}$$

Thus the velocity at which maximum deceleration occurs depends only on the reentry velocity. Substituting this result, Eq. 5.56, back into Eq. 5.55 we get

$$a_{\max} = (\beta \sin\theta U_E^2)/2e \tag{5.57}$$

For an entry velocity of 25,000 fps for the previous example, the maximum deceleration from Eq. 5.57 is about 182 ft/sec^2 or 5.6 g's, and this occurs when the velocity is 15,250 fps.

The vehicle must be retarded by a retrorocket to enter the atmosphere. From the simple rocket equation $\Delta U = c \ln M_r$, we see that ΔU must be very small in order to keep down the size of the necessary retrorocket. Thus for a velocity decrement of 500 fps, and assuming a rocket exhaust velocity of 5500 fps, the mass ratio is 1.095. For a vehicle mass of say 5 tons, then the propellant mass is 950 lb, a mass which would have had to be raised from the earth initially. This velocity decrement of 500 fps gives an entry angle θ of just over 2° at an altitude of 300,000 ft and larger angles require ever-increasing retrorocket masses. Hence from this point of view alone, entry angles must be small. Equation 5.57 shows that small θ keeps the deceleration at a low value, which is also desirable.

The heating of the vehicle surface can be a complex problem but whether the flow is laminar or turbulent, the heat rates are proportional to powers of velocity and density. It is thus advantageous to decelerate in the upper reaches of the atmosphere and this requires a low value of M/C_DA. The amount of heat to be dissipated is enormous, corresponding to the loss of kinetic energy in decelerating, and the solution of the problem may take various forms. Heat is of course dissipated by radiation but this is usually insufficient at any practical material temperature. Cooling by forced convection is a possible solution, but the additional mass and complexity practically rules it out. Because the heating is of relatively short duration, a heat sink can be helpful but again this is insufficient by itself. The answer for extreme cases has been found in *ablation*, that is, the provision of a protective covering that melts and evaporates, thus dissipating heat rapidly by both change of phase and convection, and which at the same time protects the structural material by having a low thermal conductivity.

The ideal answer to reentry then appears to lie in having low values of M/C_DA for the vehicle and entering the atmosphere at very small angles. However two points must be borne in mind. First, deceleration at high altitudes implies that the velocity is small at altitudes which are in the domain of "weather." Thus for the last part of the descent, wind can affect the final landing point to a considerable degree. Secondly, a small reentry angle or very flat trajectory means that the reentry program must be started at a long range (distance along the earth's surface) and that slight deviations in velocity decrement and in vehicle orientation can lead to very large errors in landing range. All in all, although the reentry and recovery technique has been proved successful many times, it requires exact calculation and pinpoint accuracy in the values of quantities involved. Some degree of maneuvering ability, such as by being able to vary the drag coefficient or area, or by providing lift would be most helpful.

The reentry problem for vehicles returning from interplanetary missions is even more difficult, because the velocities, unless considerably modified by retrorockets, are near escape speed. At these speeds, a low entry angle can be insufficient to allow "capture" of the vehicle as it traverses a section of atmosphere so that it exits again with an elliptic orbit which may require considerable time before return. This angle is about 4°. On the other hand, higher angles mean excessive deceleration, with high heating rates, so that only a narrow band of reentry angles is possible, between 5° and 6°. This so-called "corridor" is then only a few miles wide, with error leading either to possible overheating or to a longer journey in space.

5.14 Modification of Specific Impulse for Actual Conditions

It was stated at the beginning of this chapter that the drag force would be neglected and elsewhere it has usually been assumed that the specific impulse would be constant. The effect of gravity has been taken into account as $\overline{g \cos \theta} t_b$, with an average value $\overline{g \cos \theta}$. Such simplifying assumptions enable concepts to be examined and many useful results to be obtained but eventually an actual mission must be dealt with under actual conditions. This may involve step-by-step calculations but again a need arises for some analyses to be made without having to go to this extreme and this becomes pressing when comparing different propulsion means, such as turbojets, ramjets and rockets for some kinds of mission. It is then highly desirable to be able to carry out analyses with a value of specific impulse which is representative of actual flight conditions without each time having to specifically include the modifying factors. One such parameter is the use of an "equivalent, effective specific impulse" as developed by Frank.[4] It is denoted by I^* and is tantamount to the usual I_s compensated for drag/gravity losses (hence "effective") and representing a mean value over the entire flight (hence "equivalent").

Using the gross thrust as F_n in place of $\dot{m}_p c/g_c$, which in turn represents the sum of the momentum term $m_p V_j/g_c$ and the pressure term $A_j(p_j - p_a)$, we can write Eq. 5.6 as

$$F_n - D - (Mg/g_c)\cos\theta = (M/g_c)(dU/dt) \tag{5.58}$$

From the definition of I_s as $F_n/\dot{m}_p$, then with $\dot{m}_p = -dM/dt$,

$$dM/dt = -(F_n/I_s) \tag{5.59}$$

Substituting this into Eq. 5.58, writing U as U_f to indicate *flight velocity* (i.e. with drag, etc. as opposed to an ideal velocity U_i) and rearranging, we get

$$dU_f = -g_c I_s(dM/M) - [(g_c D/M) + g\cos\theta]\,dt \tag{5.60}$$

This integrates to

$$\Delta U_f = -g_c \int_{M_1}^{M_2} I_s(dM/M) - \int_{t_1}^{t_2} [(g_c D/M) + g\cos\theta]\,dt \tag{5.61}$$

If I_s is constant, then the first term on the right-hand side becomes $g_c I_s \ln(M_1/M_2)$ as used previously. If drag and gravity are neglected and using constant I_s, we have

$$\Delta U_i = g_c I_s \ln(M_1/M_2)$$

where ΔU_i is the *ideal velocity* increment.

If I_s is not constant, then without drag and gravity, from Eq. 5.60,

$$\frac{1}{g_c}\int_{U_{i_1}}^{U_{i_2}} \frac{dU_i}{I_s} = \int_{M_1}^{M_2} \frac{dM}{M} = \ln\frac{M_2}{M_1}$$

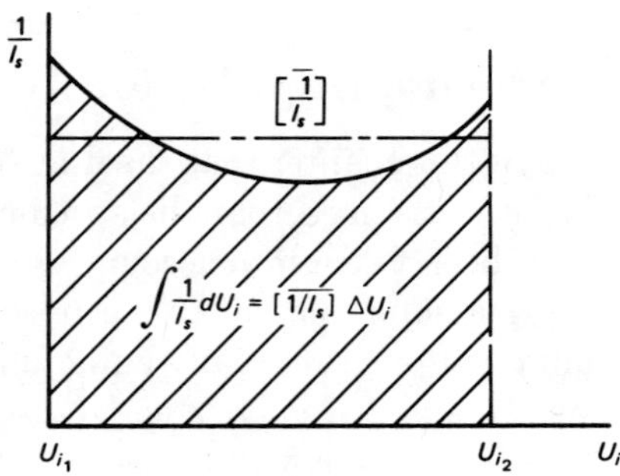

Fig. 5.19 Evaluation of $[\overline{1/I_s}]\Delta U_i$

[4] H. Frank, supporting analysis for "Composite (Rocket/Airbreathing) Engines: Key to the Advanced (Non-staged) Space Transport Vehicle," W. J. D. Escher, presented at Univ. of Tennessee Space Institute, August 1969.

The integral on the left-hand side is equivalent to the area under the curve of $1/I_s$ vs U_i, as in Fig. 5.19. If an average value of $1/I_s$ can be obtained then

$$\int_{U_{i_1}}^{U_{i2}} (dU_i/I_s) = [\overline{1/I_s}]\Delta U_i$$

and thence

$$\Delta U_i = g_c \frac{1}{[\overline{1/I_s}]} \ln \frac{M_1}{M_2} \tag{5.62}$$

It should be noted that an average value of I_s should not be used, as $\overline{I_s} \neq 1/(\overline{1/I_s})$. An alternative formulation from Eqs. 5.59 and 5.60 yields

$$\frac{dM}{M} = -\frac{dU_f}{g_c I_s[1 - (D/F_n) - (Mg/F_n g_c) \cos\theta]} \tag{5.63}$$

The denominator on the right-hand side suggests modifying I_s by the drag and gravity terms to give an effective specific impulse I_{eff}, i.e.

$$dM/M = -(dU_f/g_c I_{\text{eff}}) \tag{5.64}$$

where $I_{\text{eff}} = I_s(1 - D/F_n - M_g \cos\theta/F_n g_c)$. Integrating Eq. 5.64,

$$\int_{M_2}^{M_2} \frac{dM}{M} = -\frac{1}{g_c}\int_{U_{f_1}}^{U_{f2}} \frac{dU_f}{I_{\text{eff}}}$$

and following the previous analysis,

$$\ln (M_1/M_2) = (1/g_c)(\overline{1/I_{\text{eff}}})\Delta U_f$$

and

$$\Delta U_f = g_c \frac{1}{(\overline{1/I_{\text{eff}}})} \ln \frac{M_1}{M_2} \tag{5.65}$$

Now let $1/(\overline{1/I_{\text{eff}}}) = I^*$, the equivalent effective specific impulse, as it embodies both an average value and a modification for drag and gravity. Hence

$$\Delta U_f = g_c I^* \ln M_1/M_2 \tag{5.66}$$

which is the basic simple form of Eq. 5.10c but with an actual flight velocity increment ΔU_f and a modified actual I_s.

To evaluate I^*, two methods are possible. From the above analysis we can write

$$dM/M = dU_i/g_c I_s = dU_f/g_c I_{\text{eff}}$$

whence

$$I_{\text{eff}} = I_s(dU_f/dU_i)$$

Thus I_{eff} can be evaluated from plots of U_f vs U_i. Finally

$$I^* = U_f \bigg/ \int (dU_f/I_{\text{eff}}) \tag{5.67}$$

Alternatively by definition above

$$I_{\text{eff}} = I_s[1 - D/T - (Mg/F_n g_c) \cos \theta]$$

and I_{eff} can be found from a curve of U_f vs the term in brackets.

Problems

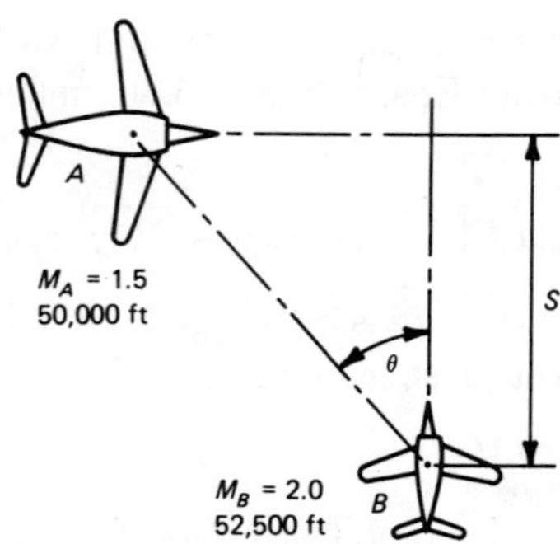

5.1 Planes A and B are flying on mutually perpendicular straight-level courses at constant speed, with B 2500 ft above A. Plane B carries a rocket of mass ratio 2.0, total initial mass of 500 lb, yielding a thrust of 5000 lb with a propellant of specific impulse 200 lb sec/lb.

If the rocket is launched in a horizontal direction directly along the line of flight of plane B, find the required distance S and the angle θ which it must have at the moment of firing in order for it to hit plane A. Assume zero drag, constant rocket thrust and propellant consumption.

5.2 Find the ratio of the velocity increments of two vehicles, one powered by a liquid chemical rocket and the other by a solid chemical rocket, when they are used for acceleration of a 10,000 lb payload. Both vehicles have a total initial mass of 510,000 lb. The liquid-propellant rocket has 60% greater specific impulse and 30% greater structure mass. The solid-propellant rocket has a structure coefficient $\sigma = 0.08$. Neglect the effects of gravity and drag.

5.3 A rocket has an initial mass of 10 tons, propellant fraction of 0.7, exhaust velocity of 8640 fps and a propellant flow rate of 137.7 lb/sec.

(a) Assuming a vertical trajectory and an average value of $g \cos \theta$ of 30 ft/sec^2 and neglecting drag, find: burnout velocity, burnout height, total height.

(b) If 0.1 of the total initial mass is a second-stage rocket of mass ratio 10 yielding a thrust of 18,000 lb which fires at the first-stage burnout altitude when the first stage vehicle is jettisoned, find the necessary second-stage specific impulse to give it escape speed.

5.4 Show that using structure coefficient σ and payload fraction l, the overall velocity increment for N stages of a multistage rocket with equal mass ratios and specific impulses is expressed by

$$\Delta U_N = -cN \ln [\sigma(1 - l_N^{1/N}) + l_N^{1/N}]$$

and that as $N \to \infty$, $\Delta U_N \to -c(1 - \sigma) \ln l_N$

5.5 A vehicle having a chemical rocket is to be used for escape of a payload of 100,000 lb from the earth. The vehicle is to have two stages, each with a structure fraction s of 0.06 and a specific impulse of 400 sec. Assume as an approximation that the value of $\overline{g \cos \theta}$ is 30 fps and that the total burning time is 200 sec. Estimate the total mass of the vehicle.

5.6 Show for a one-stage rocket fired vertically from the earth that the maximum height h_{max} from its coast may be computed from the expression

$$\frac{h_{max}}{R_0} = \frac{1}{1 - (U_i/U_e)^2} - 1$$

where R_0 = radius of the earth
U_i = velocity of injection at R_0
U_e = escape velocity from R_0

Hence show that if the injection velocity is the satellite (orbiting) velocity, then the height is equal to the radius of the earth. (It is assumed that gravity varies in the standard manner.)

5.7 Show that a minimum launch velocity of about 55,000 fps from the surface of the earth is required for a solar escape trajectory (the "absolute escape velocity").

5.8 Find the total velocity increment required from an initial orbit of altitude 100 nautical miles to a circular orbit at the moon distance (taken as 200,000 nautical miles) and return to the 100-mile orbit by (a) continuous low thrust with a spiral path and (b) impulse thrust method.

5.9 A small, solid-fuel rocket called the ARCAS (Atlantic Research, a division of The Susquehanna Corporation) is used as an atmospheric probe, taking a 12.5-lb payload to an altitude of about 200,000 ft. The brochure information includes the following data:

Chamber pressure	1020 psi	Total initial weight	77 lb
Average thrust	336 lb	Burnout weight	36 lb
Time to burnout	29 sec	Burnout velocity	3650 fps
Time to max. altitude	128 sec	Burnout altitude	47,300 ft
Nozzle throat area	0.196 in.^{2}	Max altitude	210,000 ft
Nozzle exit area	2.55 in.^{2}	Launch velocity	150 fps

Assume a value of $k = 1.25$ for the propellant gases and that the trajectory is vertical. The initial launch velocity at $t = 0$ is developed by momentarily confining the exhaust gases in a closed breech.

(a) Calculate the average effective specific impulse, the thrust coefficient and the characteristic velocity c^*.

(b) Estimate the value of the actual exhaust (jet) velocity.

(c) Calculate the ideal (no-drag) burnout velocity, burnout height and maximum altitude for this rocket.

(d) Discuss the ideal versus actual performance of the ARCAS rocket.

5.10 The Astrobee† 1500 Sounding Rocket of the Space General Company, a Division of Aerojet-General Corporation, is a two-stage, solid-propellant rocket suitable for wide applications in space research. It will carry payloads of 50 to 300 lbs to altitudes of 1600 to 730 n.mi, respectively. It is launched from a boom-type launcher (a kind of portable derrick and jib arrangement), hence it can be considered for missions anywhere.

First-stage propulsion is provided by an Aerojet Junior Motor augmented by two Thiokol Recruit boosters that supply extra off-the-launcher thrust. The second stage is an Aerojet Alcor 1-B Motor. At takeoff, the vehicle has four stabilizer fins at the rear. Overall, it is 31 in. in diameter and about 32 ft long.

Nominal vehicle weight (less payload) is 11,541 lb at launch and 139.7 lb at second-stage burnout. A nominal flight sequence with a 100-lb payload with an initial "Quadrant Elevation" (angle) of 10° with the vertical is thus: the first-stage Aerojet Junior Motor is ignited and a motion-sensing switch ignites two side-mounted Recruit motors almost simultaneously (these boosters increase initial acceleration and produce velocity necessary for stability and low wind sensitivity). The Aerojet Junior Motor boosts the vehicle to 122,200 ft with a burnout velocity of 6127 fps. After first-stage burnout, the vehicle coasts for 15 sec. The second stage is ignited at 200,000 ft, ignition causing separation of the second stage and payload from the rest of the vehicle. The motor fires for 28 sec and velocity increases to 18,618 fps at a burnout altitude of 520,000 ft. The expended second-stage motor and payload coast to a summit altitude of 1227 nautical miles and a range of 678 nautical miles in 965 sec flight time. Available data:[5]

	First Stage	Booster	Second Stage
Total impulse, lb sec	1,681,678	59,000	257,000
Average I_s, sec	226.9	220.5	281.5
Burning time, t_b, sec	40	2.4	28
A_j/A^* (nozzle)	6.46	7.10	
Propellant weight, M_p, lb	7413	268	925
Inert parts weight, lb	1404	103.8	85
Max. acceleration, g's	11		41

Weight of payload structure and despin system, 21.6 lb

For a payload of 100 lb, you are asked to:

(1) Compare the actual performance with the ideal no-drag performance as to velocities and altitudes. For each phase (first-stage burn, coast, second-stage burn, coast) take the initial conditions as those actually given, i.e., compare actual and ideal performance phase-by-phase.

† Registered trademark.

[5] Adapted from the brochure, "Astrobee 1500," of the Space General Company.

(2) Find the value of the drag thrust ratio D/F for the actual figures at burnout of the first stage to be achieved. If you can, from outside reading, calculate or estimate the effect of drag as velocity reduction or a value for the drag for a rocket of this nature, it would be helpful.

(3) Assuming the nozzle discharge diameter for the first stage is 30 in., estimate the first-stage nozzle discharge Mach number, velocity and pressure, and hence the fractions of thrust from momentum and pressure at the burnout condition.

(4) Compare calculated and quoted values of maximum acceleration in g's for both first- and second-stage phases.

You will have to make a number of assumptions for parts of the analysis and perhaps do a little extra reading for information. Differences between the overall weight as given and the sum of the component weights are presumably vehicle structure (casing, fins, etc.) and must be allocated as indicated to each phase of the flight.

5.11 "The Nitehawk 9 is a fully developed data-gathering rocket system which is versatile, inexpensive, and capable of carrying relatively large instruments to extremely high altitudes."

The above is a quotation from a report issued in 1966 by the Sandia Corporation. Relevant data relating to the rocket engine and its performance are given below, being abstracted from the report. You are asked to analyze these data to provide the information requested.

Available information:

The Nitehawk 9 is a two-stage rocket with an M-5 Nike motor as first stage (booster) and a Thiokol Tomahawk motor as second stage. (Both are solid-propellant motors.) The unit overall is about 30 ft long, with an initial mass of 1972 lb including payload. The second stage is 207 in. long and weighs 664 lb at second-stage fire and 267 lb at burnout. The payload is 69 in. long and weighs 125 lb.

First Stage (Nike motor)			*Second Stage* (Tomahawk motor)		
Casing	430 lb	motor	Casing	88 lb	motor
Propellant	750 lb	1180	Propellant	397 lb	485
Fin assembly	91 lb		Fin assembly	54 lb	
Coupling	37		Payload	125 lb	
	1308 lb			664 lb	
Burning time	3.5 sec		Burning time	9.0 sec	
Thrust	44,500 lb		Thrust	10,367 lb	
Specific impulse	198 sec		Specific impulse	242 sec	
			Pressure (average)	876 psia	
			Pressure (ax)	1010 psia	

Thrust coefficient (meas)	1.42	
" " (theor)	1.595	
Characteristic velocity	5521	fps
Density, propellant	108.8	lb/ft^3
$k = 1.16$		
Throat area (initial)	8.50	in.^{2}
Exit area (initial)	57.162	in.^{2}
Nozzle half-angle	15°	

Actual performance

Time (sec)	Altitude (ft)	Velocity (fps)	Mach No.	
0	0	0	0	
3.50	4604	2570	2.34	1st-stage burnout
12.00	22,042	1731	1.68	2nd-stage fire
20.94	63,656	8202	8.47	2nd-stage burnout
276	1,025,966	1894	2.24	
525	103,437	7690	7.62	
540.96	208	3768	3.38	

A typical flight profile is given thus:

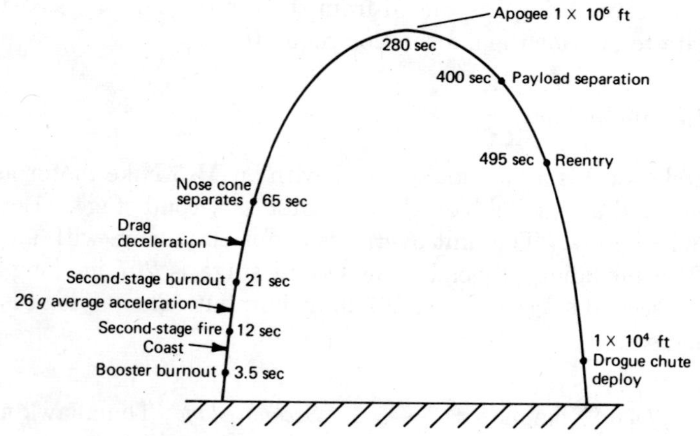

The following information is required:

(a) Calculation of no-drag performance; velocities and altitude at fire and burnout stations, final altitude.
(b) Calculation of no-drag performance; velocities and altitude at fire and burnout stations, final altitude, but using actual initial velocities and altitude for each phase.
(c) Analyze and discuss the ideal versus actual values.
(d) Estimate the average nozzle exit pressure and Mach number of the Tomahawk unit.
(e) Assume a reasonable value of combustion temperature and calculate the thrust of the Tomahawk. Compare your value with that quoted.

CHAPTER 6

Mission Analysis II—Field-Free Space

6.1 Basic Definitions

We now turn to consideration of missions outside any gravitational field and requiring thrust. It is implied that the mission times are large, that is, they are not missions which could be accomplished reasonably with a ballistic trajectory. Thrusts are low, with fractional accelerations, and are provided by a propellant that is quite distinct from the fuel of a power plant which is the energy source for the propellant. Such a propulsive means is then quite different from the chemical rocket and it is known as a "separately powered" rocket.

The simple relationships of power, thrust and exhaust velocity show one of the major characteristics of performance of the separately powered rocket. This is, that for a given thrust, high exhaust velocity with low mass flow rate entails a higher power level than the converse. Thus

$$F_n = \dot{m}_p c/g_c \quad \text{and} \quad P = \dot{m}_p c^2/2g_c$$

$$\therefore P = F_n c/2 \quad \text{or} \quad F_n^2/(2\dot{m}_p/g_c)$$

For a given thrust then $P \propto c$ and $\propto 1/\dot{m}_p$. Some simple examples are helpful in emphasizing this by considering the power requirements with a constant thrust F_n of 1 lb from various devices. (The power figures are rounded off to provide a ready comparison.)

	F_n lb	c fps	I_s sec	$\dot{m}_p$ lb/sec	P hp
Garden hose	1.0	32.2	1.0	1.0	1/30
Chemical rocket	1.0	9660	300	0.0033	9
Separately powered (thermal)	1.0	32,200	1000	0.001	30
Separately powered (ion)	1.0	161,000	5000	0.0002	150

The power level is seen to vary from that of a hairdryer motor to that of a medium-sized car.

The higher the specific impulse, then the bigger must be the power plant for a given thrust but also, of course, for a given mission time, the mass of propellant will be smaller. However, the power increases at a faster rate than the propellant rate and this is a considerable problem. We shall see that it is possible to design for optimum values for a given mission.

The arrangement of the separately powered rocket is shown diagrammatically in Fig. 6.1. The total *initial* mass M_0 consists first of a *payload* M_L, which for a

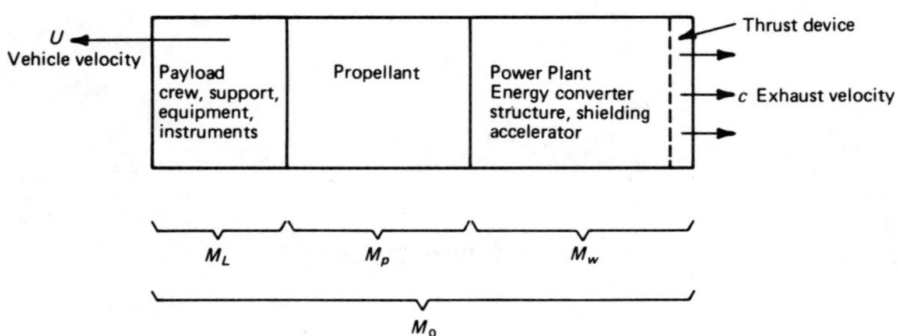

Fig. 6.1 Vehicle mass fractions.

manned mission includes the personnel, their living quarters, their support apparatus, and so forth. Then there is a *power plant* of mass M_w which supplies power in some form, probably electrical, to energize the propellant. The fuel for the power plant is included in M_w, because it has to be a very small amount in any case and is most likely to be nuclear, in which case the mass is practically constant and a relatively small fraction of the reactor and power plant mass. The *propellant* mass is M_p and the propellant itself can be chosen with considerable latitude, as it does not have to undergo any self-energizing chemical reaction. Finally there is the thrust-producing device itself, which may take any of a wide variety of forms. Its mass is constant and is most likely to be but a fraction of the power plant mass and is therefore assumed to be included in it. Likewise any other structural elements are assumed to be included in M_w. For a nuclear-powered rocket, considerable shielding may be necessary, particularly for a manned vehicle. This too must be included in the power plant mass. Thus we have

$$M_0 = M_L + M_w + M_p \tag{6.1}$$

where in effect M_w includes any mass which is not clearly payload nor propellant. In more detailed analyses, it is desirable to break down M_w more closely, as various parts scale differently with output but this is not necessary here.

It is very useful to analyze the performance in dimensionless parameters, expressing the component masses in terms of the initial gross mass. Thus we have a payload fraction M_L/M_0, a power plant fraction M_w/M_0, and a propellant fraction M_p/M_0. Obviously

$$\frac{M_L}{M_0} + \frac{M_w}{M_0} + \frac{M_p}{M_0} = 1 \tag{6.2}$$

The thrust $F_n = \dot{m}_p c/g_c$, and the jet or thrust power $P_j = \dot{m}_p c^2/2g_c$, where $\dot{m}_p$ is the propellant flow rate. With an efficiency η_j from power plant to propulsive jet, then the power plant output $P_w = P_j/\eta_j$. This efficiency η_j is important because its value can vary considerably and some otherwise desirable types of thrust device may have low efficiencies and an optimum arrangement must be sought.

Another very useful parameter is the ratio of the mass of the power plant to its output, called the *specific mass* α. Then $\alpha = M_w/P_w$ and is given in lb/hp or kg/kw. (Note that in field-free space, we are concerned only with mass, which is constant, not weight, which is variable. Nevertheless customary usage often still refers to specific *weight* and indeed to weight in general. Here we shall use *mass* as the constant quality involved.) It is possible to use the parameter α as its inverse, namely specific power, which has some advantages, but it seems best to stay with the original usage as specific mass here. Specific mass α is a useful parameter, as it expresses engineering possibility and shows what has to be achieved by engineering design skill to attain certain objectives.

6.2 Constant Exhaust Velocity Performance

We shall first discuss the mission performance with a constant exhaust velocity c, which with constant propellant flow rate $\dot{m}_p$ gives constant thrust and constant power. We can write $\dot{m}_p = M_p/t$, where t is the total "burning" time, or rather, mission time, which is a better expression here.

We return to the basic rocket equation (5.10a) giving velocity increment in terms of c and the mass ratio, with no gravitational term, i.e.

$$\Delta U = c \ln M_r = c \ln [M_0/(M_0 - M_p)] = c \ln (1 - M_p/M_0)^{-1} \tag{6.3a}$$

It is desired to rearrange this relationship introducing the vehicle parameter α and the mission parameter t. From Eq. 6.1, $M_p + M_w = M_0 - M_L$ and therefore

$$M_p\left(1 + \frac{M_w}{M_p}\right) = M_0\left(1 - \frac{M_L}{M_0}\right),$$

from which

$$\frac{M_p}{M_0} = \frac{1 - M_L/M_0}{1 + M_w/M_p}$$

Substituting this value into Eq. 6.3a and rearranging,

$$\Delta U = c \ln \frac{1 + M_w/M_p}{M_L/M_0 + M_w/M_p} \tag{6.3b}$$

Now $M_w = \alpha P_j/\eta_j$ and $P_j = \dot{m}_p c^2/2g_c = M_p c^2/2g_c t$, hence

$$M_w/M_p = \alpha c^2/2g_c\eta_j t \tag{6.4a}$$

It has been found very useful to call the parameter $(2g_c\eta_j t/\alpha)^{1/2}$ a *characteristic velocity,** denoted by V_c. Hence

$$V_c = (2g_c\eta_j t/\alpha)^{1/2}$$

and

$$M_w/M_p = c^2/V_c^2 \tag{6.4b}$$

* The terminology is still in a transitional state. Some authorities refer to ΔU as the "characteristic velocity." Here we shall use "velocity increment" for ΔU and "characteristic velocity" for V_c, which was so called by J. H. Irving who introduced it.

It is seen to be a kind of normalizing function which allows us to form another dimensionless parameter c/V_c. It is also useful to remember that it expresses the ratio of power plant mass to propellant mass, as $c^2/V_c^2 = M_w/M_p$. Substituting in Eq. 6.4, we get

$$\frac{\Delta U}{c} = \ln \frac{1 + c^2/V_c^2}{(M_L/M_0) + c^2/V_c^2} \tag{6.5a}$$

or

$$e^{\Delta U/c} = \frac{1 + c^2/V_c^2}{(M_L/M_0) + c^2/V_c^2} \tag{6.5b}$$

This can be solved explicitly for the payload fraction,

$$\frac{M_L}{M_0} = \frac{1 + c^2/V_c^2}{e^{\Delta U/c}} - \frac{c^2}{V_c^2} \tag{6.6}$$

There are several ways in which this relationship may be demonstrated graphically, each emphasizing a particular piece of information. First, Eq. 6.5a may be plotted as the inverse ratio $c/\Delta U$ against c/V_c with M_L/M_0 as parameter and this is shown in Fig. 6.2. It will be observed that for a reasonable payload, the exhaust velocity c must be a significant fraction of the velocity increment ΔU, at least one-third and preferably one-half. For a given payload and exhaust velocity, there is a limiting value of characteristic velocity V_c beyond which ΔU does not increase appreciably. This implies that there is a limiting maximum value of mission time and minimum value of power plant weight beyond which no significant gain is apparent.

A more significant diagram is obtained if the parameter $\Delta U/V_c = (\Delta U/c)(c/V_c)$ is plotted against c/V_c. This is given in Fig. 6.3, which shows the additional very important fact that for a given payload, mission time and power plant mass (i.e., value of V_c), there is a value of exhaust velocity which maximizes the velocity increment. Or alternatively, by Eq. 6.6 as shown in Fig. 6.4, for a given ΔU and V_c the payload is maximized for a particular value of c; since $c = g_c I_s$, we can say

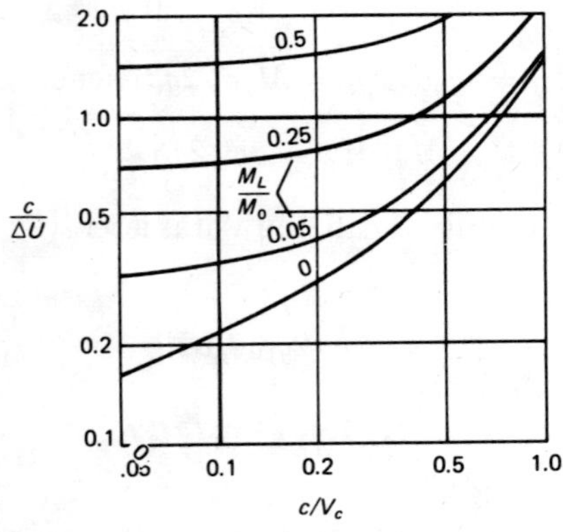

Fig. 6.2 Vehicle velocity increment vs characteristic velocity, payload as parameter.

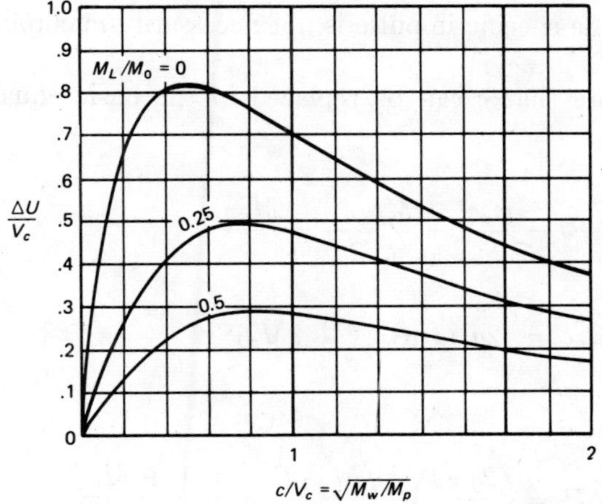

Fig. 6.3 Vehicle velocity increment vs exhaust velocity, payload as parameter.

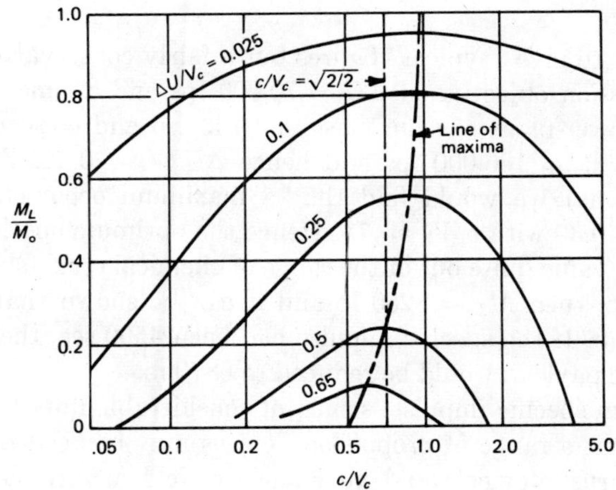

Fig. 6.4 Payload vs exhaust velocity, vehicle velocity increment as parameter.

that there is an optimum value of specific impulse for any mission and that at this optimum, the payload is maximized for a given velocity increment or, that for a given payload, the velocity increment is maximized.

Furthermore, we see that the maximum part of the curve is quite flat for the larger payloads and associated values of $\Delta U/c$. The latter implies lower velocity increments, longer times and heavier power plants, what Stuhlinger[1] calls a "truck" type of operation. For smaller payloads, the maximum is somewhat sharper and the values of ΔU, t and V_c are the inverse of the above, leading to a "sports-car"

[1] E. Stuhlinger, "Ion Propulsion for Space Flight," McGraw-Hill, New York, 1964.

type of operation, fast journey, short time and low payload. For travel in free space, therefore, the specific impulse is not necessarily maximized but depends on the mission.

We might note that c can be replaced in the basic equation (6.3b) in the following manner:

$$P_j = \frac{\dot{m}_p c^2}{2g_c} = \frac{M_p}{t}\frac{c^2}{2g_c} = \frac{M_w \eta_j}{\alpha}$$

$$\therefore c = (2g_c \eta_j t/\alpha)^{1/2}(M_w/M_p)^{1/2}$$

Thus

$$\Delta U = \left(\frac{2g_c\eta_j t}{\alpha}\right)^{1/2}\left(\frac{M_w}{M_p}\right)^{1/2} \ln \frac{1 + M_w/M_p}{M_L/M_0 + M_w/M_p}$$

Hence for given mass fractions and power plant, the velocity increment is proportional to the square root of the mission time and this is a simple and useful design criterion.

It is difficult to give "typical" figures but a fairly conservative example might be to have a mission objective of ΔU = 70,000 fps and a time of 4 months (120 days), with a power-plant specific mass of 10 lb/hp and conversion efficiency of 75%. This gives $V_c \approx$ 166,000 fps and hence $\Delta U/V_c = 0.42$. From the plot (or from the equations) we would find that a maximum occurred with a payload fraction of about 0.33 with $c/V_c \approx .78$. Hence the optimum specific impulse is just about 4000 sec, a value quite out of the range of chemical rockets. If the initial mass M_0 was 25,000 lb, then $M_L \approx$ 8250 lb and it can be shown that $M_p \approx$ 10,600 lb and $M_w \approx$ 6150 lb. If the specific impulse had been 4500 sec, then keeping V_c and ΔU the same, the payload would be reduced to 6870 lb.

The optimum specific impulse is then of considerable importance and leads to the conclusion that a range of propulsion devices may be needed or ideally that a given type of thrust device could be made to yield a variable specific impulse without sacrificing other desirable features.

From Fig. 6.4, it is seen that the optimum values of c/V_c lie in a range from about 0.5 to 1 or, more realistically limiting the possible payload fraction, from 0.5 to 0.8. We shall later use a value of $c/V_c = (2)^{1/2}/2 = .707$ as a general or typical value for a quick approximate calculation. Note that as engineering design still improves and α is reduced, then specific impulse will have to be increased to provide optimum values. In the above example, a reduction of α to 5 lb/hp would increase the optimum I_s to about 6500 sec. However, we should probably take the opportunity to decrease mission time, which can vary directly with power plant mass for the same characteristic velocity. Thus we can trade off a mission parameter with a power plant parameter and vice versa.

We can find optimum values by the usual process of differentiating and setting equal to zero. Applying this to Eq. 6.6, differentiating with respect to c/V_c and holding $\Delta U/V_c$ constant, the optimizing parameter is found to be

$$e^{(\Delta U/V_c)/(c/V_c)} = \frac{\Delta U/V_c}{2c/V_c}\left(1 + \frac{V_c^2}{c^2}\right) + 1 \tag{6.7}$$

(Note that we are keeping the parameters $\Delta U/V_c$ and c/V_c rather than reducing directly to $\Delta U/c$.)

Substituting this back into Eq. 6.6 gives

$$\left.\frac{M_L}{M_0}\right|_{\max} = \frac{1 + c^2/V_c^2}{[(\Delta U/V_c)/(2c/V_c)][1 + (c^2/V_c^2)^{-1}] + 1} - \frac{c^2}{V_c^2} \tag{6.8}$$

The relationship of these factors is shown in Fig. 6.5, with c/V_c as abscissa and either $\Delta U/V_c$ or M_L/M_0 as ordinate. This shows clearly that c/V_c has a cutoff

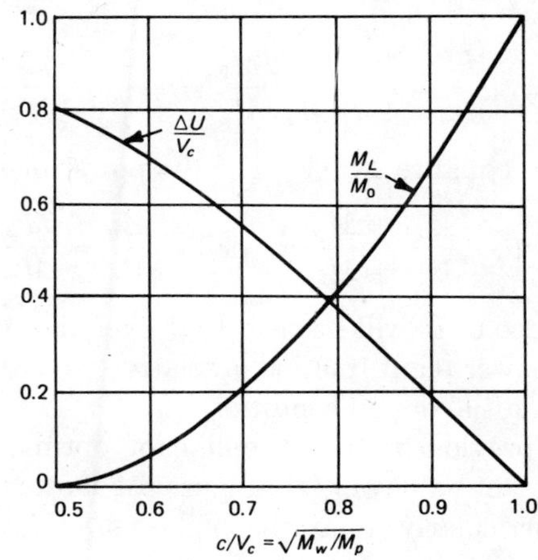

Fig. 6.5 Optimum values of vehicle velocity and payload vs exhaust velocity.

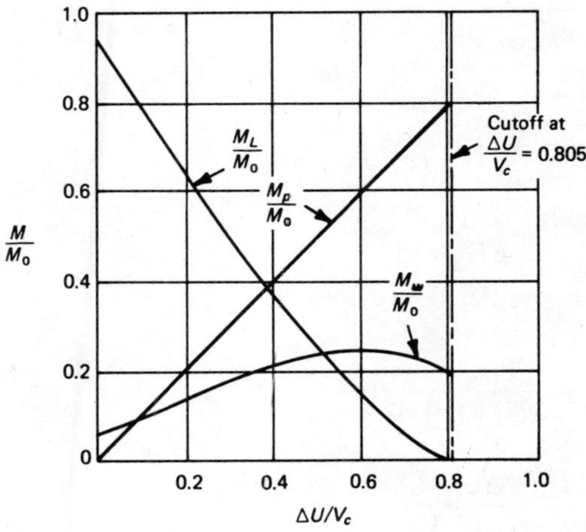

Fig. 6.6 Optimum mass fractions vs vehicle velocity.

point for zero payload with a value of about 0.5 (0.505) and that for payload factors up to 0.5, it rises to about 0.85. Likewise $\Delta U/V_c$ has a cutoff at 0.805 at the zero payload point. From these values we get a limiting, no payload value of 1.593 for $\Delta U/c$ and, using the original equation (6.3), $M_p/M_0 = 0.796$ and $M_w/M_0 = 0.204$. Since $M_r = M_0/(M_0 - M_p)$, then for zero payload again, $M_r = 4.92$ as a limiting value. We should also note that in addition to this being an optimizing procedure for either maximum M_L at fixed ΔU or maximum ΔU at a given M_L, it also represents minimum time for given M_L and ΔU, as we have used the parameter c/V_c, and V_c is a function of time.

Values of the mass fractions, M_p/M_0 and M_w/M_0, corresponding to the optimum values of M_L/M_0 can be found as follows. From Eq. 6.2,

$$1 - \frac{M_L}{M_0} = \frac{M_p}{M_0} + \frac{M_w}{M_0} = \frac{M_p}{M_0}\left(1 + \frac{M_w}{M_p}\right)$$

$$\therefore \frac{M_p}{M_0} = \frac{1 - M_L/M_0}{1 + M_w/M_p} = \frac{1 - M_L/M_0}{1 + (c^2/V_c^2)} \tag{6.9}$$

which it can be shown is equal to $\Delta U/V_c$. For the power plant fraction,

$$\frac{M_w}{M_0} = 1 - \frac{M_L}{M_0} - \frac{M_p}{M_0} \quad \text{or} \quad \frac{M_w}{M_0} = \frac{M_w}{M_p} \cdot \frac{M_p}{M_0} \tag{6.10}$$

These are shown in Fig. 6.6. It will be seen that over the probable range of lower payload fractions, the power plant fraction is relatively constant around $M_w/M_0 \approx 0.25$, which suggests an initial design condition.

It was suggested previously that the line of optima for M_L/M_0 could be approximated by a constant value of c/V_c. Figure 6.4 shows that a value $(2)^{1/2}/2 = 0.707$ is reasonable, particularly as at the higher M_L/M_0 values the curves are flatter and a departure from the optimum c/V_c does not have a great effect. If $c/V_c = (2)^{1/2}/2$ is inserted in Eq. 6.8 for maximum payload, then

$$\left.\frac{M_L}{M_0}\right|_{\max} = \frac{1 + c^2/V_c^2}{\frac{1}{2}(\Delta U/c)[(c^2/V_c^2)^{-1} + 1] + 1} - \frac{c^2}{V_c^2} = \frac{1 + \frac{1}{2}}{\frac{1}{2}(\Delta U/c)(2 + 1) + 1} - \frac{1}{2}$$

$$= \frac{3}{3\Delta U/c + 2} - \frac{1}{2}$$

Now

$$\frac{\Delta U}{c} = \frac{\Delta U}{V_c} \cdot \frac{V_c}{c} = \frac{\Delta U}{V_c}(2)^{1/2}$$

$$\therefore \left.\frac{M_L}{M_0}\right|_{\max} = \frac{3}{3(2)^{1/2}[(\Delta U/V_c)] + 2} - \frac{1}{2}$$

$$= \frac{4 - 3(2)^{1/2}(\Delta U/V_c)}{4 + 6(2)^{1/2}(\Delta U/V_c)} \tag{6.11}$$

This is a useful approximation for initial possible values but should be regarded as only that. It becomes inaccurate at the lowest payload values.

The characteristic velocity V_c has physical meaning in the following ways. From $V_c = (2g_c\eta_j t/\alpha)^{1/2}$, we get

$$\frac{V_c^2}{2g_c} = \frac{\eta_j t}{\alpha} = \frac{\eta_j t}{M_w/P_w} = \frac{P_j t}{M_w} = \frac{\text{jet energy}}{\text{power plant mass}}$$

or jet energy $= M_w V_c^2/2g_c$. If this energy had all been utilized with perfect efficiency in increasing the KE of the power plant alone, then $V_c = \Delta U$. However it was seen that even with zero payload, $\Delta U = .805 V_c$, the remainder of the KE being wasted as residual exhaust energy, i.e., the propulsive efficiency is less than unity. As payload increases, there is less acceleration and ΔU is smaller, i.e., the propulsive efficiency is decreasing. Another way of looking at this is to consider the general relationship of vehicle velocity ΔU to exhaust velocity c for a rocket (Fig. 2.5). With $\Delta U/c \approx U/V_j = r$, then a very low value of $r < 1 (c \gg \Delta U)$ and a very high value of $r > 1 (c \ll \Delta U)$ both give a poor propulsive efficiency. With a high value of c, the power is high (see Sec. 6.1) and hence M_w is excessive. With a low value of c, the propellant mass needed to provide thrust becomes very high.

We can get an expression for distance by using the distance equation (5.16b) without the gravity term and considering $U_0 = 0$, thus

$$S = ct[(M_r - 1)^{-1} \ln (1/M_r) + 1]$$

and substituting $M_r = e^{\Delta U/c}$ from Eq. 5.10b, we get

$$\frac{S}{ct} = 1 - \frac{\Delta U/c}{e^{\Delta U/c} - 1} \tag{6.12}$$

Then using the optimizing condition Eq. 6.7,

$$\frac{S}{ct} = 1 - \left[\frac{\Delta U/V_c}{c/V_c} \Big/ \frac{\Delta U/V_c}{2c/V_c}\left(1 + \frac{1}{c^2/V_c^2}\right)\right]$$

and

$$\frac{S}{ct} = \frac{1 - c^2/V_c^2}{1 + c^2/V_c^2} \tag{6.13}$$

This is shown in Fig. 6.7 together with the corresponding optimum values of payload

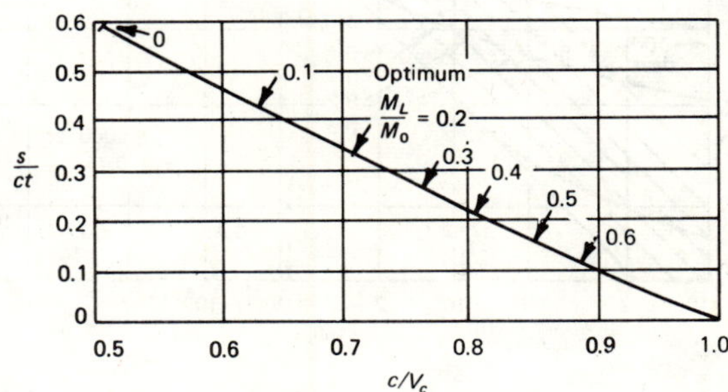

Fig. 6.7 Optimum distance vs exhaust velocity.

fraction. As c/V_c cannot be less than .505 for the optimum condition, even with $M_L/M_0 = 0$, then the abscissa is shown only from .5 to 1.0. The plot is very nearly linear and can be represented very closely by the relationship

$$S/ct = 1.225[1 - 1.02\ (c/V_c)] \tag{6.14}$$

which is another simple approximation for initial design. It is of interest to know the order of mission time and this is shown for a specific case in Fig. 6.8(*a*), which uses Eq. 6.13 to show S vs t with $I_s(=c/g_c)$ as parameter for a specific power plant

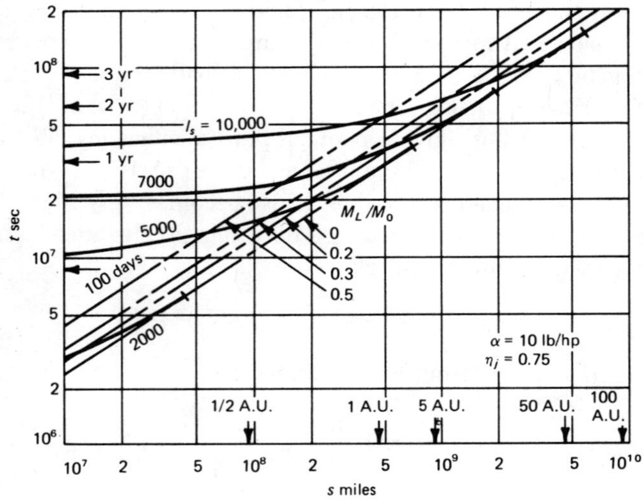

Fig. 6.8 Time distance relationship: (*a*) fixed α, I_s as parameter;

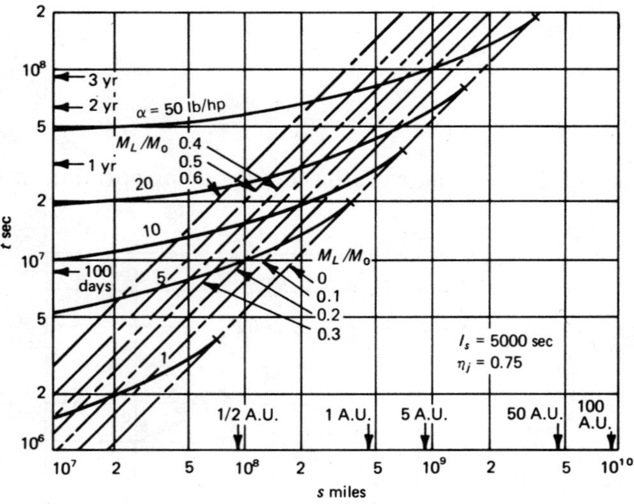

(*b*) fixed I_s, α as parameter.

mass of 10 lb/hp with a conversion efficiency of 0.75. Note that there is a maximum time for each specific impulse, representing the cutoff condition of $c/V_c = 0.505$. Figure 6.8(b) shows time and distance for a fixed value of I_s of 5000 sec, with power plant specific mass α as parameter. Both figures show that at the higher payload factors at small distances, then time varies only slightly with distance.

Values of thrust and hence of acceleration are very small. For the example used previously, with $M_0 = 25{,}000$ lb, propellant mass is 10,600 lb and the time is 120 days. Propellant flow rate is then 1.02×10^{-3} lb/sec and with the exhaust velocity 128,000 fps ($I_s \approx 4000$ sec), the thrust is

$$F = \frac{\dot{m}_p C}{g_c} = \frac{1.02 \times 10^{-3} \times 1.28 \times 10^5}{32.2} \approx 4 \text{ lb}$$

For the initial mass of 25,000 lb, the initial acceleration is

$$a_0 = \frac{F}{M_0/g_c} \approx 5.2 \times 10^{-3} \text{ ft/sec}^2$$

which is only about 0.016% of the value of g at the earth's surface.

Thus we see that the propulsive system for long journeys in field-free space is entirely different from what we associate with flight in a gravity field. For the above example, the power plant has an output of 615 hp but is giving a thrust of only 4 lb. The vehicle would have to be lifted to a satellite orbit by a booster rocket and preferably given an initial mission velocity by another stage.

6.3 System Efficiencies

There are several efficiencies connected with the behavior of the complete propulsion system which we might define. They could be called vehicle, mission or system efficiencies. Several of them might better be termed "effectiveness" rather than efficiency, because their value may be greater than unity or less than zero and it seems desirable to restrict the term efficiency to a ratio of quantities which can be unequivocally and uniquely defined and whose ratio lies always between zero and unity.

One such parameter is defined as the ratio of the rate at which a vehicle gains kinetic energy to the rate of kinetic energy generated by the engine. This is sometimes called the kinetic power efficiency. Denoting it by ϵ_p, then

$$\epsilon_p = \frac{dE_v/dt}{dE_j/dt}$$

where $E_v = MU_t^2/2$ and $dE_j/dt = \dot{m}_p c^2/2$, with U_t the vehicle velocity at any time t. Thus the numerator becomes $MU_t\dot{U}_t - \dot{m}_p U_t^2$, with the minus sign because $\dot{m}_p = -dM/dt$. From the simple rocket equation, $\dot{m}_p c = M\dot{U}_t$, and so making these substitutions,

$$\epsilon_p = \frac{MU_t\dot{U}_t - \dot{m}_pU_t^2/2}{\dot{m}_pc^2/2} = \frac{MU_t\dot{U}_t - M\dot{U}_tU_t^2/2c}{M\dot{U}_tc^2/2c}$$

$$= 2(U_t/c) - (U_t/c)^2 = 2r - r^2 = r(2 - r)$$

From Eq. 5.10a with $U_t = \Delta U$, then $r = U_t/c = \ln M_{rt}$.

$$\therefore \epsilon_p = (2 - \ln M_{rt}) \ln M_{rt} \tag{6.15}$$

where M_{rt} is the mass ratio at any time t and equal to $M_0 - \dot{m}_pt = M_0 - M_ft/\tau$, with M_f the final mass and τ the total propulsion time.

This is shown in Fig. 6.9, from which it is seen that ϵ_p has a maximum value of unity at $r = 1$ ($U_t = c$), values of zero at $r = 0$ and $r = 2$, and is negative for $r > 2$. These results require explanation. In the first place, only for $r = 1$ is the propulsion efficiency unity and for all other values the exhaust jet has residual kinetic energy that has not been given to the vehicle. Secondly, the negative values occur because the kinetic energy of the vehicle is increasing due to increase of velocity but is decreasing due to decrease of mass. Thus from the previous relations we have

$$dE_v/dt = \epsilon_p dE_j/dt = (\dot{m}_pc^2/2)(2r - r^2)$$

$$= \frac{\dot{m}_pc^2}{2}\left(2\frac{U_t}{c} - \frac{U_t^2}{c^2}\right) \tag{6.16a}$$

$$= (\dot{m}_pc)U_t - (\dot{m}_pU_t^2/2) \tag{6.16b}$$

The first term is the rate of increase of energy due to thrust ($F \times U$) and the second term is the loss of energy due to the KE of the propellant. From Eq. 6.16a it is seen that dE_v/dt is negative when $U_t/c > 2$.

Another parameter is the overall energy efficiency or effectiveness, ϵ_E, defined as the ratio of the vehicle KE at the end of propulsion to the total energy generated by the engine during that time τ, thus

$$\epsilon_E = \frac{M_eU^2/2}{(\dot{m}_pc^2/2)\tau} = \frac{M_eU^2}{M_pc^2}$$

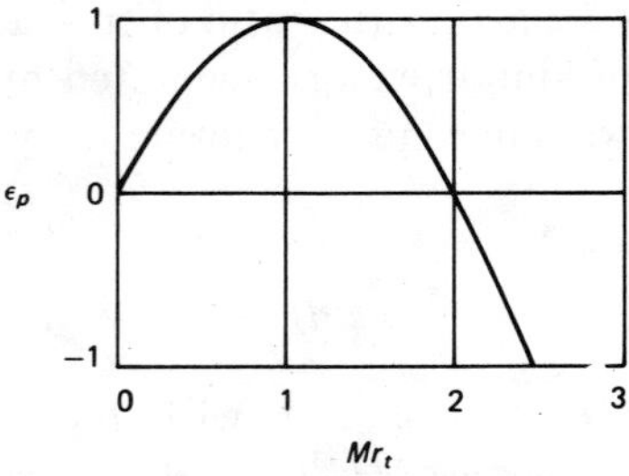

Fig. 6.9 Kinetic power efficiency vs mass ratio.

Substituting from Eq. 5.10b,

$$\epsilon_E = \frac{1}{e^{U/c} - 1}\frac{U^2}{c^2} = \frac{r^2}{e^r - 1} \quad \text{or} \quad \frac{(\ln M_r)^2}{M_r - 1} \tag{6.17}$$

Differentiating, we find a maximum value of 0.65 at $M_r = 4.92$ or $V/c = 1.593$ (note that this is the limiting zero payload condition). Values of ϵ_E are shown in Fig. 6.10, plotted against r and M_r, and it is seen that the curve is fairly flat over a wide range of r or M_r, decreasing slowly from the maximum value.

These mission effectivenesses are very useful in understanding the rocket performance and in giving guidance to selection of variables wherever this is possible. They are, however, seldom used quantitatively in the way that component efficiencies are evaluated and used. They help to evaluate a type of power plant for a given job and show why effort should be made to evolve a different solution of a mission problem.

6.4 Variable Exhaust Velocity

Constant exhaust velocity is the simplest method of operation, implying constant propellant flow, thrust and power required; that is, a fixed plant setting. One must inquire into the possible advantage of a variable velocity and look for an optimum program.

Although c, $\dot{m}_p$ and F can vary, the power cannot exceed a certain maximum and the operation is "power-limited." We have to decide on what quantity we wish to optimize and what conditions are fixed. The most likely case is one in which the time and velocity increment are fixed and the payload is to be maximized, although there are missions in which maximum velocity increment or minimum time can be important. Here we shall analyze the maximum M_L/M_0 mission following Irving,[2] noting that other approaches and relevant references are discussed by Stuhlinger. The analysis seeks to optimize the acceleration program for the given conditions and then to solve for the consequent exhaust velocity program required.

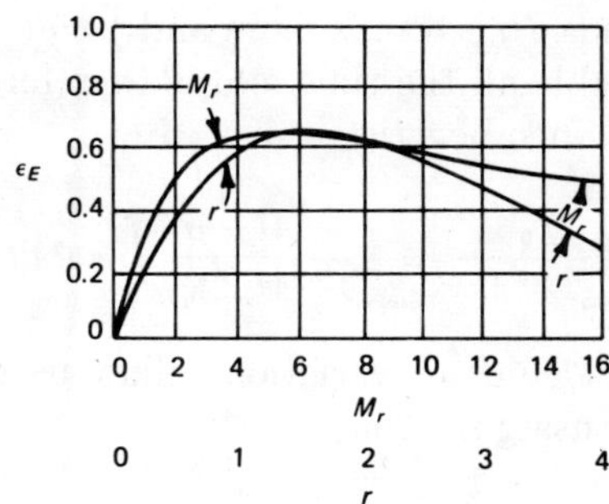

Fig. 6.10 Overall energy efficiency vs mass ratio

[2] J. H. Irving, "Low Thrust Flight: Variable Exhaust Velocity in Gravitational Fields," *in* "Space Technology" (H. S. Seifert, Ed.) Wiley, New York, 1959.

The acceleration is given by

$$a = F/M = \dot{m}_p c/M$$

where all values are instantaneous values at any time $t = t$. The power is given by

$$P_j = \dot{m}_p c^2/2$$

and eliminating c from these two relationships by writing $c^2 = (Ma/\dot{m}_p)^2$ and substituting, then

$$M^2a^2/\dot{m}_p^2 = 2P_j/\dot{m}_p$$

and

$$a^2/2P_j = \dot{m}_p/M^2$$

Now a positive value of $\dot{m}_p$ implies a decrease of instantaneous mass M, that is, $\dot{m}_p = -dM/dt$, hence

$$\frac{\dot{m}_p}{M^2} = -\frac{dM/dt}{M^2} = \frac{d}{dt}\left(\frac{1}{M}\right)$$

$$\therefore \frac{d}{dt}\left(\frac{1}{M}\right) = \frac{a^2}{2P_j}$$

Integrating from $t = 0$, $M = M_0$ to $t = t$, $M = M$,

$$\frac{1}{M} - \frac{1}{M_0} = \frac{1}{2}\int_0^t (a^2/P_j)\, dt \tag{6.18}$$

For the total time T of a given mission, the mass M is $M_0 - M_p = M_L + M_w$, hence

$$\frac{1}{M_L + M_w} = \frac{1}{M_0} + \frac{1}{2}\int_0^T (a^2/P_j)\, dt \tag{6.19}$$

For given values of M_w and M_0, then M_L will be a maximum when the right-hand side is a minimum, i.e., when P_j is a maximum and $\int a^2dt$ is a minimum. Thus the full power should be available all the time and is therefore a constant, with $P_j = \eta_j M_w/\alpha$. Equation 6.19 can then be written

$$\frac{M_0}{M_L + M_w} = 1 + \frac{M_0}{M_w}\frac{\alpha}{2\eta_j}\int_0^T a^2\, dt \tag{6.20}$$

The methods of the calculus of variations allow us to find the provision for minimizing the integral by using the condition*

$$\frac{\partial F}{\partial y} - \frac{d}{dx}\left(\frac{\partial F}{\partial y'}\right) = 0$$

* Consistent with the title of the mathematical topic, this is known variously as the Euler, the Lagrange, or the Euler-Lagrange equation.

where in this case, $F = a^2$, $x = t$ and $y = \Delta U$. So we have

$$y' = dy/dx = d(\Delta U)/dt = a$$

and with

$$\frac{d}{dx}\left(\frac{\partial F}{\partial y'}\right) = \frac{\partial^2 F}{\partial x \partial y'} + y' \frac{\partial^2 F}{\partial y \partial y'} + y'' \frac{\partial^2 F}{\partial y'^2}$$

we obtain the condition $da/dt = 0$. Hence a = constant and we get the simple answer that for M_L/M_0 to be a maximum, we must have constant acceleration. Therefore

$$\frac{M_0}{M_L + M_w} = 1 + \frac{M_0}{M_w} \frac{\alpha a^2 T}{2\eta_j}$$

and, for constant acceleration, $\Delta U = aT$, so that

$$\frac{M_0}{M_L + M_w} = 1 + \frac{M_0}{M_w}\left(\frac{\alpha}{2\eta_j T}\right)(\Delta U)^2 = 1 + \frac{M_0}{M_w}\left(\frac{\Delta U}{V_c}\right)^2$$

or

$$\frac{M_L}{M_0} = \frac{1}{1 + (M_0/M_w)(\Delta U/V_c)^2} - \frac{M_w}{M_0} \tag{6.21}$$

where V_c is the characteristic velocity previously defined.

Differentiating with respect to M_w (with M_0 constant) and setting equal to zero yields

$$\frac{M_w}{M_0} = \frac{\Delta U}{V_c} - \left(\frac{\Delta U}{V_c}\right)^2 = \frac{\Delta U}{V_c}\left(1 - \frac{\Delta U}{V_c}\right) \tag{6.22}$$

and substituting back in Eq. 6.21, we get

$$(M_L/M_0)\big|_{\max} = [1 - (\Delta U/V_c)]^2 \tag{6.23}$$

For the propellant fraction,

$$\frac{M_p}{M_0} = 1 - \frac{M_L}{M_0} - \frac{M_w}{M_0}$$

$$\therefore \frac{M_p}{M_0} = \frac{\Delta U}{V_c} \tag{6.24}$$

Equations 6.22, 6.23 and 6.24 for the mass fractions corresponding to maximum payload are shown in Fig. 6.11. The relative proportions of payload, propellant and power-plant masses are shown more clearly in Fig. 6.12.

We can get a value for the optimum acceleration as

$$a = \frac{\Delta U}{T} = \frac{\Delta U}{V_c} \cdot \frac{V_c}{T}$$

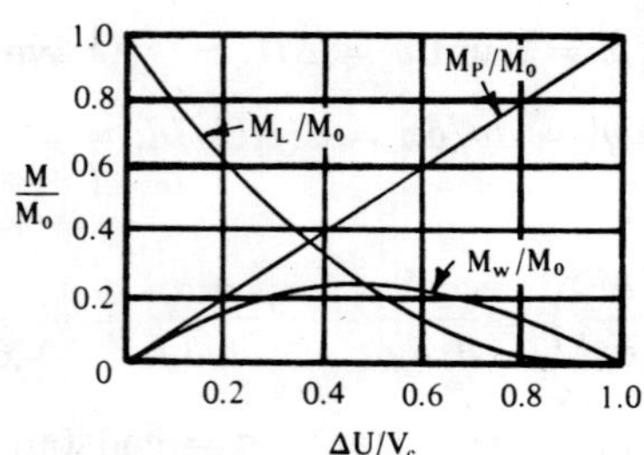

Fig. 6.11 Optimum mass fractions—variable exhaust velocity.

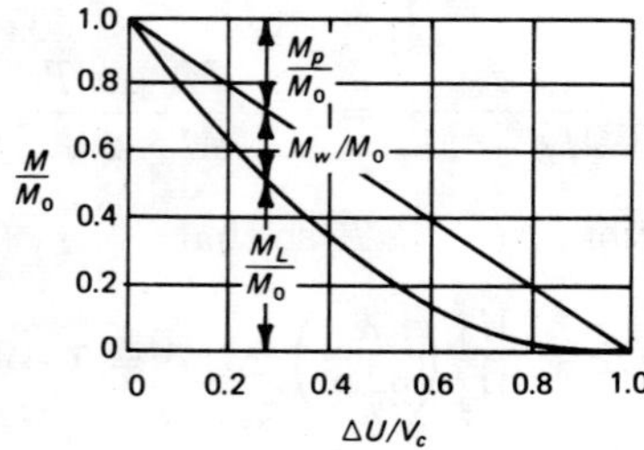

Fig. 6.12 Optimum mass fractions—variable exhaust velocity.

From Eq. 6.23 we have the optimum value

$$\Delta U/V_c = 1 - (M_L/M_0)^{1/2}$$

Substituting this, together with $V_c = (2\eta_j T/\alpha)^{1/2}$, we get

$$a = (2\eta_j/\alpha T)^{1/2}[1 - (M_L/M_0)^{1/2}] \tag{6.25}$$

For the exhaust velocity program we use Eq. 6.18 with constant acceleration and introduce exhaust velocity from the power relationship, i.e. from Eq. 6.18,

$$\frac{1}{M} = \frac{1}{M_0} + \frac{a^2 t}{2P_j} \tag{6.26}$$

and by definition $P_j = (\dot{m}_p c^2)/2$, which with $\dot{m}_p c = M(dU/dt) = Ma$ gives $P_j = Mac/2$. Eliminating M from Eq. 6.26 yields

$$c = (2P_j/aM_0) + at \tag{6.27}$$

Hence at $t = 0$, the exhaust velocity has an initial value

$$c_0 = 2P_j/aM_0 \tag{6.28}$$

Thus $c = c_0 + at = c_0 + \Delta U_t$, where ΔU_t is the vehicle velocity increment at any time t.

From $P_j = Mac/2$ we have at any time,

$$c = 2P_j/Ma$$

and at the end of the mission, $M = M_0 - M_p$.

$$c_T = 2P_j/(M_0 - M_p)a$$

Substituting $P_j = \eta_j M_w/\alpha$

$$c_T = \frac{2\eta_j M_w}{a\alpha(M_0 - M_p)} = \frac{2\eta_j M_w/M_0}{a\alpha(1 - M_p/M_0)}$$

and substituting the optimum value of M_w/M_0, Eq. 6.22, and of M_p/M_0, Eq. 6.24, together with $a = \Delta U/T$, this reduces to

$$c_T = V_c \tag{6.29}$$

From Eqs. 6.27, 6.28, and 6.29 we see that the exhaust velocity for maximum payload

(1) has a particular initial value at the start equal to $2P_j/aM_0$,
(2) increases linearly with time and this increase is equal to the velocity increment of the vehicle during this time, ΔU_t, and
(3) has a final value equal to the characteristic velocity V_c.

As a consequence of the second statement, we see that the absolute velocity of the propellant, which is the vector sum of the relative velocity (exhaust velocity) and the vehicle velocity is constant with respect to the coordinate system at $t = 0$, so that with V for fluid velocity and U for vehicle velocity,

$$\mathbf{V}_{\text{abs}} = \mathbf{V}_r + \mathbf{U}$$

Hence, with initial vehicle velocity U_0, the propellant absolute velocity at any time t is, in scalar values,

$$\begin{aligned} V_{\text{abs}} &= c - (U_0 - \Delta U_t) \\ &= (c_0 + \Delta U_t) - (U_0 + \Delta U_t) = c_0 - U_0 \end{aligned}$$

which is constant. Another way of looking at this behavior is that it represents a better propulsion efficiency, because exhaust velocity increases as vehicle velocity increases.

Figure 6.13 shows the relationship of exhaust and vehicle velocities for the initial and final conditions. Since $c = c_0 + \Delta U_t$ and $c_T = V_c$, then $c_0 = V_c - \Delta U_T$

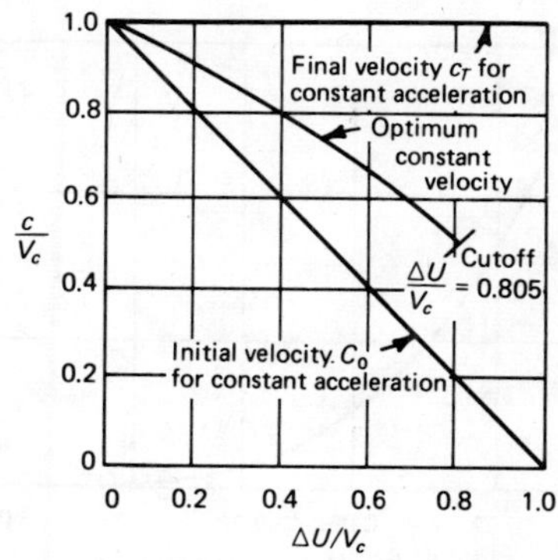

Fig. 6.13 Initial and final exhaust velocities—constant acceleration.

or $c_0/V_c = 1 - \Delta U_T/V_c$. c_0/V_c is then a straight line of slope minus one with respect to $\Delta U/V_c$. Final velocity $c_T/V_c = 1$ and the figure then shows the variation of exhaust velocity c for maximum payload at any given condition $\Delta U/V_c$ dictated by available power plant and mission specification.

For comparison, the optimum value for constant exhaust velocity from Eq. 6.7 and Fig. 6.5 is also shown in Fig. 6.13. For any value of $\Delta U/V_c$, it has a value roughly the mean of the initial and final values for variable exhaust velocity. One significant point is that there is a cutoff at $\Delta U/V_c$ of 0.805, whereas the variable velocity allows ΔU to equal V_c before the payload goes to zero.

A comparison of payloads at constant and variable exhaust velocity is shown in Fig. 6.14. For the lower values of $\Delta U/V_c$, there is scarcely any difference in payload. At about $\Delta U/V_c = 0.5$, it may be said to become important, as the constant velocity payload is approaching zero and becomes so at the cutoff point of 0.805.

Figure 6.15 shows the ratio of payloads for optimum variable velocity and constant velocity as a function of the payload at constant exhaust velocity. This

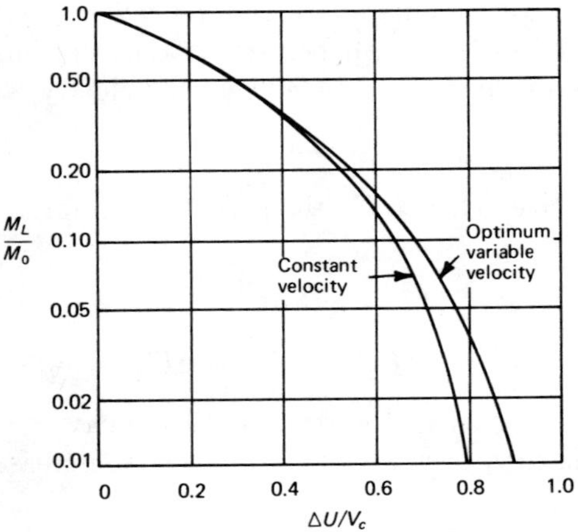

Fig. 6.14 Payload comparison—constant and variable exhaust velocity.

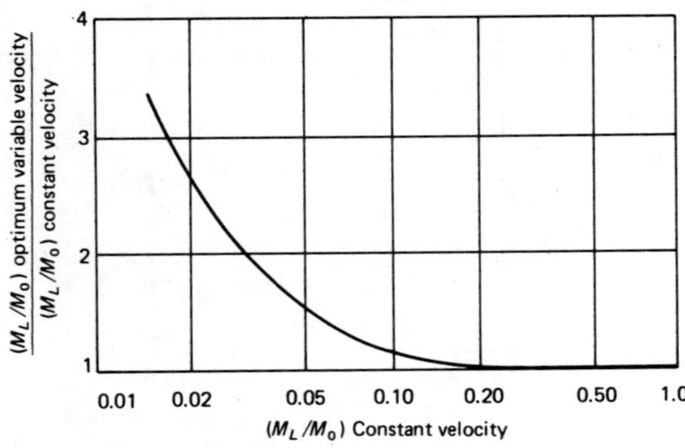

Fig. 6.15 Ratio of payload—constant and variable exhaust velocity.

shows clearly that there is little to be gained by variable velocity if the payload fraction is greater than 0.10. Thus variable velocity can be very beneficial for high velocity increments and low characteristic velocities associated with low payloads (sports car-type vehicle), but has very little effect in the opposite case (truck-type vehicle). It could be important in borderline cases, where the payload is marginal with constant exhaust velocity. This conclusion can have an important effect on engine development, as it is possible that variable velocity types might be very much more complicated and thus have increased risk of failure.

6.5 Low Thrust, Central Force Field Motion

In Chapter 5, we discussed missions in a central force field from the viewpoint of a large thrust, either impulsive as for orbit transfer or for a short time ($t < I_s$) for boost from the earth's surface. In this chapter to this point we have dealt with missions in field-free space, with these missions being carried out with a continuous but low-valued thrust, either constant or variable. There remains the possibility of missions with a continuous, low thrust in a central force field. Such missions must start from a satellite orbit, as the thrust is quite insufficient to place it in such an orbit.

The equations of motion can be set up thus,

$$\ddot{r} = r\dot{\theta}^2 - (k/r^2) + a_r \tag{6.30a}$$

$$\frac{d}{dt}(r^2\dot{\theta}) = ra_\theta \tag{6.30b}$$

where the symbols have the meaning of Sec. 5.8, with a_r and a_θ being the radial and circumferential components of the thrust acceleration of the vehicle. The three terms on the right-hand side of Eq. 6.30a represent, respectively, the centrifugal force, the gravitational force, and the radial propulsive thrust, all per unit mass. Equation 6.30b relates the rate of change of angular momentum to the torque per unit mass (= circumferential acceleration a_θ multiplied by radius r). These are the differential equations governing the motion, and the boundary conditions pertain to departure from a particular point at a particular velocity and to arrival at another specified point at a particular velocity after a specified time interval, at the same time maximizing the payload. The optimizing process for acceleration is a very complex process and few if any closed solutions are possible without making certain assumptions or forgoing certain conditions. It is necessary to make trial solutions by numerical methods, hoping that a useful result as a "near miss" can be obtained, even if a precise value is impossible. It often turns out that the absolute value of the total thrust acceleration remains almost constant, which is consistent with the field-free space mission solution with variable exhaust velocity analyzed in Sec. 6.4.

Any solution will result in the description of a spiral orbit around the central body, with gradually increasing velocity until the escape value is reached, as shown in Fig. 6.16. Only a very few turns are shown, and in actual solutions there will be a very tight spiral with dozens of turns before escape, because the low thrust means a fractional acceleration.

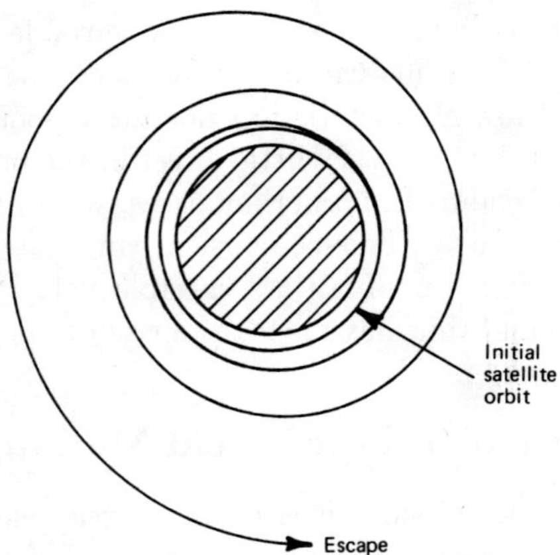

Fig. 6.16 Spiral orbit in a central-force field with low thrust.

The tightness of the spiral allows us to make the simplifying assumption that the velocity at any point is the velocity of a circular orbit at that point and with this an approximate but simple analysis of the effect of initial conditions to be made. Thus, following Stuhlinger, for a circular orbit of radius r and for orbital velocity U, with constant tangential thrust F, and with E_t the total energy of the body,

$$dE_t/dt = FU$$

With $E = E_t/M$, energy per unit mass,

$$dE/dt = aU$$

where a is the thrust acceleration of the vehicle.

Using Eq. 5.34 and substituting from Eqs. 5.4 and 5.5,

$$E = -g_0R_0^2/2r$$

and hence

$$dE/dr = g_0R_0^2/2r^2$$

Combining the two expressions involving derivatives of E,

$$dr/dt = (2r^2/g_0R_0^2)aU = 2aUr^2/g_0R_0^2$$

and with the assumption that $U \simeq U_s = R_0(g_0/r)^{1/2}$, then

$$\frac{dr}{dt} = \frac{2a}{R_0}\left(\frac{r^3}{g_0}\right)^{1/2}$$

The ratio of initial acceleration a_0 to acceleration a at time t is given by

$$a_0/a = (M_0 - \dot{m}t)/M_0$$

and with $\dot{m} = F/c = M_0 a_0/c$, then

$$a = \frac{a_0}{1 - (a_0 t/c)}$$

Substituting back,

$$\frac{dr}{dt} = \frac{2a_0 r^{3/2}}{R_0 (g_0)^{1/2}(1 - a_0 t/c)}$$

which can be integrated between $r = r_0$ at $t = 0$ and $r = r$ at $t = t$, yielding

$$r/r_0 = \{[(c/R_0)(r_0/g_0)^{1/2} \ln (1 - a_0 t/c) + 1]^2\}^{-1} \tag{6.31}$$

If we make the further assumption, justified in all practical cases, that $a_0 t \ll c$, Eq. 6.31 can be simplified to

$$r/r_0 \approx \{[1 - (a_0 t/R_0)(r_0/g_0)^{1/2}]^2\}^{-1} \tag{6.32}$$

as $\ln (1 - x) = -x$ when x is small. Equation 6.32 is shown plotted in Fig. 6.17 for three values of starting orbit r_0 and three values of initial acceleration a_0.

It is seen that for any starting orbit, the time to a given distant orbit tends towards a limiting value for each value of initial acceleration and that this time is inversely proportional to the acceleration. This is also seen readily from Eq. 6.32, as very large values of r imply small values of the denominator, i.e.

$$\frac{a_0 t}{R_0}\left(\frac{r_0}{g_0}\right)^{1/2} \to 1$$

and $t \approx (R_0/a_0)[(g_0/r_0)^{1/2}]$, whence $t \propto 1/a_0$ for a given initial orbit.

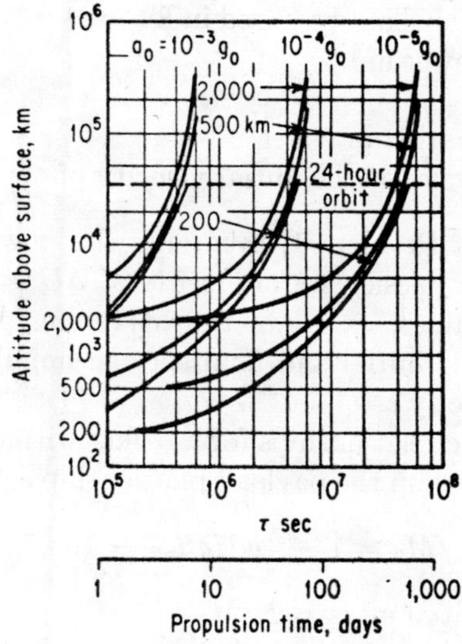

Fig. 6.17 Altitude vs time for a low thrust, spiral orbit (from Stuhlinger.)

For missions of this nature, it is important to be able to estimate the possible improvement of payload using a programmed variable thrust, rather than a constant thrust. This is indeed a complex problem, with numerical methods having to be used in order to get even a near-optimum solution. The general result is analogous to that for field-free space missions, namely that for short mission times, i.e. high velocity, there is some benefit from variable thrust but for slow journeys, the gain is very small.

The exact nature of the mode of operation of the initial phases of long-time missions is not clear and depends on many developments as yet in the future. It seems definite that the journey of the complete vehicle will have to start its powered, low-thrust phase from an earth orbit, but will it be boosted there as a whole or will it be built-up part by part from an orbiting workshop? Figure 6.17 shows the advantage of starting from a high orbit but this requires more initial energy to get it there. How important will the time of the journey be if it is a manned mission? Will it be better to reduce the time of spiraling by a high parking orbit and using a shuttle to build up the complete vehicle? It is more difficult in these circumstances to optimize mass fractions and initial acceleration. Analyses of the type resulting in Fig. 6.17 are essential in order to attempt an optimum mission.

Problems

6.1 A space vehicle is designed for an initial gross mass of 50 tons, with a power plant (including propulsion engine) having a specific mass of 50 lb/jet hp. The mission is to take 160 days and the velocity increment is to be 70,000 fps. For constant exhaust velocity, straight-line motion in field-free space, find:

(a) the specific impulse for maximum payload in sec.
(b) the amount of this payload in lb.
(c) the amount of propellant required in lb.
(d) the propulsive power in hp.
(e) the thrust in lb.
(f) the initial acceleration in ft/sec^2.
(g) the distance traveled if the initial velocity of the vehicle is 4×10^4 fps.

6.2 For the conditions of Problem 1, take only the power plant specific mass of 50 lb/jet hp and the mission requirements of $\Delta U = 70{,}000$ fps and $t_b = 160$ days. From these values, calculate the maximum payload fraction M_L/M_0, the acceleration a, the initial and final specific impulses, all for the optimum variable exhaust velocity c.

6.3 Using the simple basic relationships for a rocket with constant exhaust velocity in field-free space, develop the payload relationship

$$M_L/M_0 = 1 - (a_0 t/g_c I_s) - (\alpha a_0 I_s/2)$$

where a_0 = thrust/initial mass $= F/M_0$

Show that under certain conditions, the maximum payload is obtained when the specific impulse is $(2\eta_j t/g_c\alpha)^{1/2}$ and that its value is $1 - a_0 t/V_c - \alpha a_0 V_c/2g_c$ where $V_c = (2\eta_j t g_c/\alpha)^{1/2}$. State what the conditions are.

6.4 A space vehicle has been designed for a mission in field-free space with a velocity increment of 55,000 fps and a time of 200 days. The payload is to be 20,000 lb and the design specific power plant mass is 50 lb/jet hp. The specific impulse has been optimized for the given conditions. The engineers have done a pretty good job in attaining the necessary specific impulse and power level, but the power plant mass is 45% over design. Various compromises are possible but the velocity increment must be attained because this is essential for the mission. Take the exhaust velocity as constant in each case.

(a) Find the design values of specific impulse, power plant mass, propellant mass, initial mass and power output.
(b) Find the mission time for the increased plant mass, keeping the same payload and specific impulse. Compare the values of propellant and initial masses with the design values.
(c) Find the payload, propellant mass and initial mass if the specific impulse and mission are to be kept to the design values.
(d) If the engineers say they can alter the specific impulse without too much difficulty, find the payload, propellant and initial masses for the optimum specific impulse with the overweight power plant, keeping the time at its design value.
(e) What would the velocity increment be if the vehicle was flown as designed except for the overweight power plant?

Make a summary of the values of M_L, M_w, M_p, M_0, t_b, I_s and F for each case and discuss the consequences of each alternative. What would your decision be?

6.5 For a vehicle in field-free space with specified values of α, t and M_L/M_0, show that the maximum velocity increment ΔU for constant exhaust velocity c is given by the relation

$$\Delta U = \frac{2c}{1 + V_c^2/c^2}\left(\frac{1 - M_L/M_0}{M_L/M_0 + c^2/V_c^2}\right)$$

Show that this expression is equivalent to Eq. 6.8 and hence that the condition for maximizing payload is the same as for maximizing velocity increment, as stated in the text.

6.6 An electrically propelled rocket in field-free space is to have a mission time of 90 days with a velocity increment of 24,000 fps. The power plant has an α value of 16.5 lb/hp. There is a choice of propulsion engine between one of 2000 sec specific impulse with a conversion efficiency from input terminals to thrust of 30% and an engine of 4000 sec with a conversion efficiency of 50%.

(a) Which engine will allow the greater payload and what is the value of it?
(b) If the minimum value of α/η_j is 30 lb/hp, what is the optimum specific impulse for maximum payload in this case?

6.7 Using the basic relationships for vehicle performance with constant exhaust velocity in field-free space, show that $M_p/M_0 = a_0 t/g_c I_s$ and $M_w/M_0 = \alpha a_0 g_c I_s/2\eta_j$, where a_0 = thrust/initial mass, $g_c F/M_0$. For specified values of α/η_j, a_0 and t, show that the maximum payload fraction occurs when $I_s = (2\eta_j t/g_c\alpha)^{1/2}$, i.e. when $c = V_c$.

CHAPTER 7

Chemical Rockets

7.1 Performance Parameters

The chemical rocket process is basically extremely simple—combustion followed by simple nozzle expansion. Because the propellant may initially be a solid or a liquid or two liquids or a combination, a single T–s or p–v diagram to illustrate the process is not feasible. However, we might note that for the common liquid-propellant rocket, the process parallels the Rankine cycle in that a liquid is pumped to a high pressure, heated (by internal combustion) and evaporated, and then expanded down to a low pressure. There is no cycle in the thermodynamic sense, of course, even to the extent of the airbreathing engine process, as the working substance is completely different from the atmosphere.

For a fixed mass flow rate, the thrust depends on the magnitude of the jet velocity V_j and this is obtained from the steady-flow energy equation for simple nozzle expansion. Thus with Q and W_s zero, Eq. 3.3 yields

$$V_j = [2g_c(h_0 - h_j)]^{1/2} \tag{7.1}$$

where V_j = jet velocity
h_0 = combustion chamber stagnation enthalpy
h_j = nozzle discharge or jet enthalpy

The process then depends on the initial enthalpy h_0 being large, i.e., high temperature and high specific heat, and the discharge enthalpy h_j being low, i.e., low temperature consequent on a large expansion. Although not exactly true, the working substance behaves generally like a perfect gas after evaporation and the combustion and expansion processes are customarily analyzed in terms of a perfect gas. Thus we may write, using a constant specific heat,

$$V_j = [2g_c c_p(T_0 - T_j)]^{1/2}$$

then substituting for $c_p = R\,(c_p/R) = Rc_p/(c_p - c_v) = Rk/(k-1)$ and with $R = R_0/M_w$ where R_0 is the universal or molal gas constant and M_w is the molecular weight of the gases,

$$V_j = \left(2g_c \frac{k}{k-1}\frac{R_0}{M_w}(T_0 - T_j)\right)^{1/2} \tag{7.2}$$

For a reversible adiabatic expansion we may write

$$T_0 - T_j = T_0[1 - (T_j/T_0)] = T_0[1 - (p_j/p_0)^{\epsilon}]$$

where $\epsilon = (k-1)/k$, and hence

$$V_j = \{2g_c[k/(k-1)](R_0/M_w)T_0[1 - (p_j/p_0)^\epsilon]\}^{1/2} \tag{7.3}$$

Let us examine this expression to see how the different variables may influence V_j.

First of all, the factor $1 - [(p_j/p_0)^\epsilon]$ may be called the ideal cycle efficiency of the expansion process, as it goes to unity as $p_j \to 0$, and when p_j has a finite value, the final velocity V_j is less than the ideal. Second, the value of the specific heat ratio has a considerable effect, both as the coefficient $k/(k-1)$ and as the exponent $(k-1)/k$. The greater the number of degrees of freedom of the molecule, the lower is the value of k, hence k is higher for bimolecular substances, such as O_2, N_2, H_2, and lower for heavy molecules. For $k = 1.4$, as for air at normal temperatures, $k/(k-1) = 3.5$, whereas for $k = 1.2$ which is more representative of rocket exhaust products, $k/(k-1) = 6$. Similarly for the effect of the exponent $(k-1)/k$, if the temperature ratio T_j/T_0 is $\frac{1}{2}$, then the corresponding pressure ratio p_j/p_0 is 1/11.3 for $k = 1.4$, and 1/64 for $k = 1.2$. For overall effect on V_j, other factors being fixed, k should be low for the highest value of V_j.

The effect of M_w can be very considerable, because the range of molecular weight can be very large. It is seen that $V_j \propto 1/(M_w)^{1/2}$ and so light molecules are much more suitable than heavy ones, and where there can be a choice as in nonchemical rockets, propellants such as hydrogen or helium are indicated. For chemical rockets, the optimum mixture of fuel and oxidizer may not be the stoichiometric, because a mixture giving compounds of lower molecular weight may give a higher jet velocity than the mixture giving the highest temperature. Finally we see that other factors constant, $V_j \propto T_0^{1/2}$, favoring a high combustion temperature. Some properties of propellants are given later in Table 7.1 and it will be seen that those of highest flame temperature do not necessarily yield the highest specific impulse.

V_j has been taken as the criterion of performance because it represents specific thrust or specific impulse. Although the propellant is seldom fully expanded, the criteria for maximizing V_j are still valid for incomplete expansion. The degree of expansion as represented by the pressure ratio p_j/p_0 is a function only of the area ratio A_j/A_0, as discussed in Sec. 3.10 and shown graphically in Fig. 3.22. It is the area ratio which is the important design parameter, because although it should be as large as possible for maximum performance (Fig. 3.22), the length and weight are controlling considerations.

The two major processes of the rocket, combustion and expansion, can be analyzed separately for their effect on the overall performance. Direct measurement of the combustion process is extremely difficult owing to the very high temperature and the necessity for sampling and analyzing the gases for their degree of reaction. Hence it would be advantageous to see if the combustion performance could be gauged in other terms. This can be done with the aid of a parameter called the characteristic velocity and denoted by C^*.

Starting with the continuity equation, we may write

$$\dot{m} = A\rho V = A^*\rho^*V^*$$

where superscript * indicates the nozzle throat values where the Mach number is unity. Substituting $\rho^* = p^*/RT^*$ and $V^* = (g_c kRT^*)^{1/2}$ and writing A^* as A_t to emphasize that this is the throat area, then

$$\dot{m} = \frac{A_t p^* g_c k}{(g_c k R T^*)^{1/2}}$$

Substituting the critical temperature and pressure relationships for $M = M^* = 1$, i.e.

$$T^* = T_0[2/(k+1)] \quad \text{and} \quad p^* = p_0[2/(k+1)]^{k(k-1)}$$

then

$$\dot{m} = \frac{A_t p_0 [2/(k+1)]^{k/(k-1)} g_c k}{(g_c k R T_0)^{1/2} [2/(k+1)]^{1/2}}$$

$$= \frac{A_t p_0}{a_0} g_c k \left(\frac{2}{k+1}\right)^{(k+1)/2(k-1)} = \frac{A_t p_0}{a_0} g_c f(k) \tag{7.4}$$

where a_0 is the acoustic velocity at the stagnation or combustion temperature T_0.

The characteristic velocity C^* is defined as

$$C^* = \frac{a_0}{f(k)} = \frac{(g_c k R T_0)^{1/2}}{k[2/(k+1)]^{(k+1)/2(k-1)}} \tag{7.5}$$

and

$$\dot{m} = A_t p_0/(C^*/g_c) \quad \text{or} \quad C^* = g_c(p_0 A_t/\dot{m}) \tag{7.6}$$

From Eq. 7.5 it is seen that C^* is a function only of k, R and T_0, i.e.

$$C^* = f(k, R)(T_0)^{1/2}$$

and so it is independent of nozzle parameters and dependent only on the nature of the propellant and the combustion process itself. Thus it is characteristic of the merit of a propellant and the combustor performance. The theoretical value may be found from Eq. 7.5 and the experimental value evaluated by $A_t p_0/\dot{m}$, the latter three quantities being much more easily measured than temperature at these levels. The effectiveness of the combustion process can then be assessed by the ratio of the measured to the experimental value of C^*.

For the nozzle performance, we start with the thrust equation

$$F_n = (\dot{m}/g_c)V_j + A_j(p_j - p_a) \tag{7.7}$$

and substitute for $\dot{m}$ and V_j. Thus the throat or choked value of $\dot{m}$ is given by

$$\dot{m} = \frac{p_0 A_t}{(T_0)^{1/2}} \left(\frac{g_c k}{R}\right)^{1/2} \left(\frac{2}{k+1}\right)^{(k+1)/2(k-1)} \tag{7.8}$$

and from Eq. 7.3,

$$V_j = \{2g_c[k/(k-1)]RT_0[1 - (p_j/p_0)^\epsilon]\}^{1/2} \tag{7.3}$$

Substituting these in Eq. 7.7 there results

$$F_n = p_0 A_t k\{[2/(k-1)][2/(k+1)]^{(k+1)/k-1}\}^{1/2}[1-(p_j/p_0)^{\epsilon}]^{1/2} + A_j(p_j - p_a) \tag{7.9}$$

F_n is the total thrust and is independent of T_0, as $\dot{m} \propto 1/(T_0)^{1/2}$ and $V_j \propto (T_0)^{1/2}$, and is also independent of R. It is a function of the pressures p_0, p_j and p_a and sizes A_t and A_j.

We may form a thrust coefficient C_F as follows:

$$C_F = F_n/p_0 A_t$$

$$= k\left\{\left(\frac{2}{k-1}\right)\left(\frac{2}{k+1}\right)^{(k+1)/(k-1)}\right\}^{1/2}[1-(p_j/p_0)^{\epsilon}]^{1/2} + \frac{A_j}{A_t}\left(\frac{p_j}{p_0} - \frac{p_a}{p_a}\right) \tag{7.10}$$

p_j/p_0 is a function of A_j/A_t and thus C_F depends only on k, p_a/p_0 and A_j/A_t (or p_j/p_0). It is independent of T_0 and R and dependent on nozzle design theoretically (A_j/A_0) and on nozzle efficiency practically. F_n, p_0 and A_t are among the most readily measurable quantities and hence for an actual nozzle, the ratio of actual C_F to calculated or ideal C_F can be found. A plot of C_F is given in Fig. 7.1[1] for $k = 1.2$ as an example and it is seen that values for operation in space $(p_0/p_a = \infty)$ lie in the range 1.7–1.9.

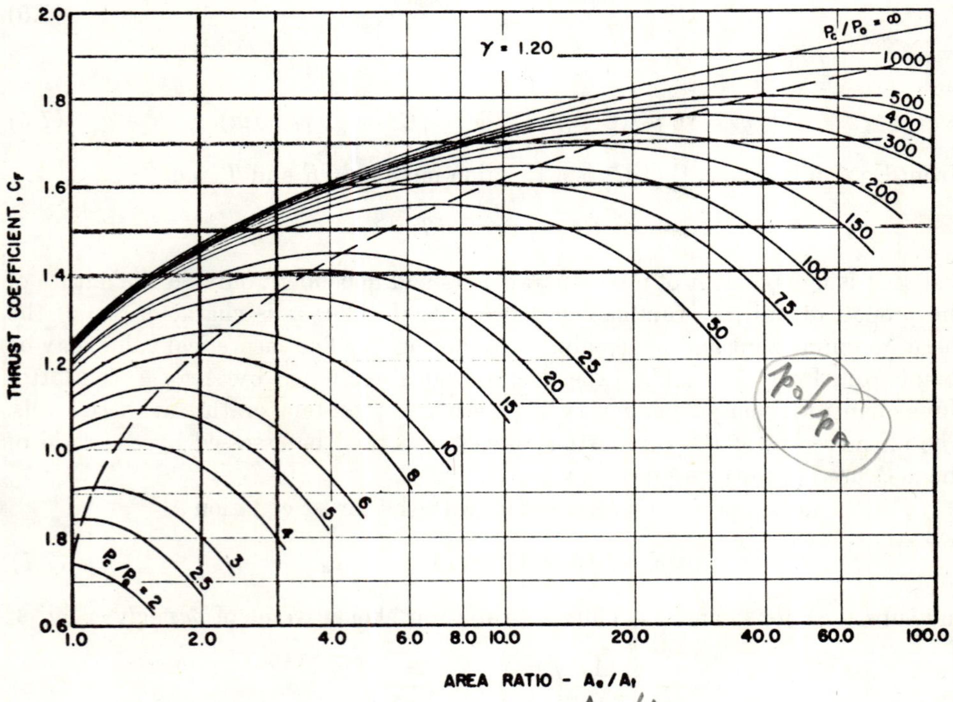

Fig. 7.1 Nozzle thrust coefficient vs area ratio (from Morrison).

[1] From R. B. Morrison, Ed., "Design Data for Aeronautics and Astronautics," Wiley, New York, 1962.

With C^* giving the combustion performance and C_F giving the expansion performance, the two together should give the overall performance. The product then gives

$$C^*C_F = g_c \frac{p_0 A_t}{\dot{m}} \cdot \frac{F_n}{p_0 A_t} = \frac{F_n}{\dot{m}/g_c} = I_s \tag{7.11}$$

that is, the specific impulse, which was established earlier as the criterion of rocket performance.

7.2 Propellant—General

The term propellants is all-inclusive of the gaseous discharge at nozzle exit. It includes what is usually thought of as a "fuel," that is the substance providing the thermal energy when chemically combined with what is generally called an "oxidizer," although the latter need not contain any oxygen. However, propellant action need not necessarily result from a chemical combination of substances but may take place as the result of a decomposition of a compound when in the presence of a catalyst or when subjected to a given temperature. Such a propellant consisting of only one substance is known as a *monopropellant.* One of the best-known examples is hydrogen peroxide which decomposes into water vapor and oxygen, releasing thermal energy in the process.

The great majority of propellants are *bipropellants,* i.e. fuel and oxidizer, which are stored separately in liquid form and mixed only when introduced into the combustion chamber. Solid propellants are a separate category and they usually consist of fuel and oxidizer which are stored completely mixed in solid form and requiring a certain minimum ignition temperature before a self-sustaining reaction occurs. Usually liquid propellants require a high temperature ignition source but a few react spontaneously on mixing of fuel and oxidizer. The latter are known as *hypergolic.* (Propellants requiring ignition are called diergolic, although usually it is assumed that they are not hypergolic unless so specifically stated).

7.3 Liquid Propellants

The ideal propellant has a large number of desirable characteristics. The first is a high specific impulse and this is almost preeminent. That is, very considerable efforts are made to overcome undesirable characteristics if there is promise of improved specific impulse. For example, liquid hydrogen–liquid oxygen (lox) is an often-used propellant nowadays as it has I_s values up to 450 sec, the best of any yet readily available, in spite of the very considerable difficulty of handling and storing cryogenic materials of this sort. However, there are some other properties which may be overriding for a particular application. Thus high values of density and boiling point, low values of freezing point, good stability, insensitiveness to variations of ambient pressure and temperature, easy ignitability, nontoxicity, and convenient long-term storability are all desirable. Density is very important because low values increase the size of the rocket and hence the structure weight and probably the vehicle drag. With a bipropellant such as liquid H_2–O_2 of very disparate densities, a considerable range of propellant density is possible and allows some flexibility in optimization for a particular mission.

Glassman[2] presents an interesting discussion of propellant chemistry showing the possibilities of various elements in combination. He plots the heat of combustion against the atomic number (the latter just a convenience in representation of the elements) and Fig. 7.2(*a*) is reproduced for the case of liquid oxygen as oxidizer. The heating value of gasoline is included as a comparison level and it is seen that it lies between the values for the composite elements of hydrogen and carbon, as must all hydrocarbons. Among the elements yielding higher heats of combustion are the groups of metals Al, Mg, and Si, but these would have to be prepared as slurries. The combustion products would consist of metallic oxides and would be in solid form and thus a loss would occur in the nozzle expansion process (Sec. 3.12). Elements having even higher heats of combustion are the very light elements, lithium, boron, and beryllium. Lithium combusts spontaneously and hence would be extremely difficult to handle other than as a compound, usually the hydride. Beryllium is highly toxic and scarce but it has the best heating value as seen from the figure. Boron would have to be used as a compound and the products of combustion are again solids which cause expansion difficulties in performance (particle drag) and also as deposits. Carbon and hydrogen compounds then appear to be the practical answer including the nitrogen-hydrogen compound hydrazine N_2H_4, as well as hydrogen itself. A factor that must be considered is dissociation, with CO_2 and H_2O having considerably more propensity towards dissociation at a given temperature than CO and N_2.

It would thus appear that using pure O_2 as oxidizer, then pure H_2 is a very suitable fuel, although highly cryogenic. Remembering that specific impulse depends not only on temperature but on molecular weight and that H_2 is much lighter than O_2, the mixture giving maximum I_s is not necessarily the stoichiometric. Table 7.1 shows the performance and properties of various mixes.

This gives a clear instance of the effect of molecular weight and temperature on specific impulse, as at I_s max = 345, the flame temperature is 1300°F below the stoichiometric. On the other hand, the specific gravity of the mixture is low, the SG at I_s max being only about 0.6 of that for the stoichiometric mixture.

TABLE 7.1

Mixture ratio (lb O_2/lb H_2)	9	8	5	3.2	2.5
Flame temperature (°F)	5750	5800	5300	4500	3400
Mean molecular weight (products)	17	16	11.5	8.6	7.0
Specific impulse (lbf sec/lbm)	280	295	338	345	340
Specific gravity	.455	.426	.325	.248	.215
		Stoich	Max. I_s		

Source: Most data from Glassman.

[2] I. Glassman, "The Chemistry of Propellants," *Amer. Scientist,* 53 (1965), 508–524. *See also,* I. Glassman and R. F. Sawyer, "The Performance of Chemical Propellants," AGARDograph 129, Technivision Services, Slough, England, 1970.

Oxygen is not necessarily the best oxidizer, as the criterion is valence. Glassman lists the following in order of decreasing performance:

$$O_3, F_2, F_2O, NF_3, O_2, ClO_3F, ClF_3, H_2O_2,$$
$$N_2O_4, H_2O_4, C(NO_3)_4, HNO_3, NH_4ClO_4, BrF_5$$

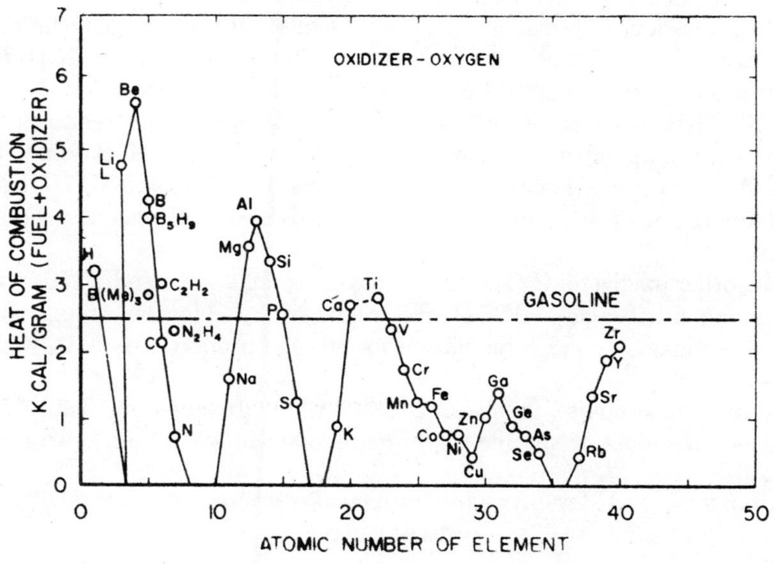

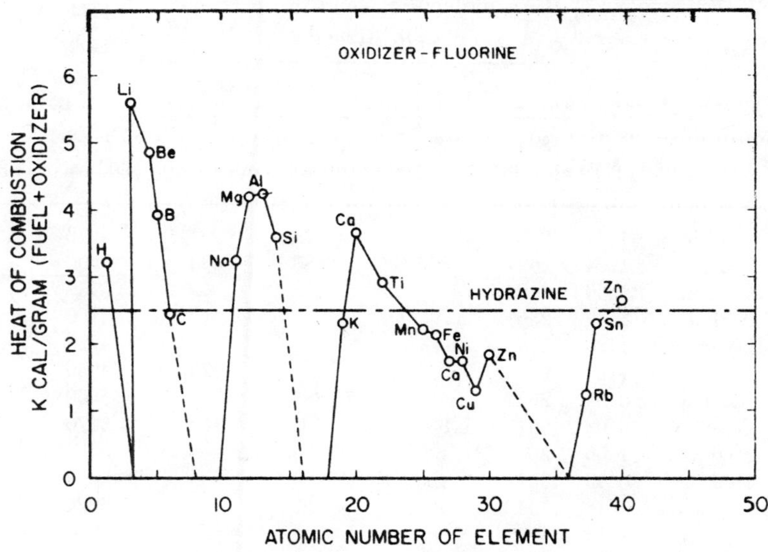

Fig. 7.2 Heats of combustion of the elements (from Glassman): (*a*) oxidizer—oxygen; (*b*) oxidizer—fluorine.

The prevalence of the halogens will be noticed, with fluorine the top performer. Figure 7.2(*b*) shows possible fuel combinations with fluorine as oxidizer, with hydrazine N_2H_4 given as a comparison, since this is an existing fuel in use. Advantages of fluorine are its low dissociation tendency compared to oxygen compounds and the fact that many of the metal fluorides, such as BF_3, are gaseous rather than solid. Disadvantages are its highly corrosive properties and, in conjunction with H_2, the low bulk density of the propellant mixture.

Hydrazine is a useful fuel, as it is denser than H_2, is not cryogenic and does not form CO_2 which is highly susceptible to dissociation. In Fig. 7.2(*a*), N_2H_4 is placed below the gasoline line, because the figure does not take into account dissociation, and actually N_2H_4 gives a better specific impulse. It has a high freezing point and if this is controlling, unsymmetrical methyl hydrazine (UMDH) may be used although the carbon content reduces its performance. N_2H_4 and UMDH are both hypergolic with nitrogen peroxide and nitric acid, and N_2H_4 can also be a monopropellant.

Of the other oxidizers, H_2O_2 is not stable in storage. N_2O_4 and HNO_3 (probably as red fuming nitric acid, RFNA) are used, with the former giving somewhat better performance. A good combination for storage properties is N_2O_4 with N_2H_4-UMDH.

Some specific impulse values and other data are given in Table 7.2 from Glassman. Four values of specific impulse are given in some cases, two combustor

TABLE 7.2

Optimum Specific Impulse Values for some Liquid Propellants

			I_s, lbf/(lbm/sec)					
			500 psia		1000 psia			
Oxidizer	Fuel	SG	Sea level	Vacuum	Sea level	Vacuum	T, °F 500 psia	Mol. wt. products
H_2O_2	Gasoline	1.28	248		273		4830	21
	N_2H_4	1.24	262	325	288	330	4690	19
HNO_3	Gasoline	1.30	240		255		5150	25
	Aniline	1.39	235		258		5100	21
	Hydrazine	1.26	255		261		4728	19
N_2O_4	N_2H_4	1.20	263	320		325	4950	19
O_2	Alcohol	0.99	259		285		5560	22
	Gasoline	0.98	264	335	290	340	5770	22
	Hydrazine	1.06	280		308		5370	18
	Hydrogen	0.43	364	450	400	455	4500	9
F1	Ammonia	1.16	306	400	337	405	7224	19
	Hydrazine	1.30	310		348		7940	19
	Hydrogen	0.32	373	465	410	470	5100	9

Source: Data from Glassman.

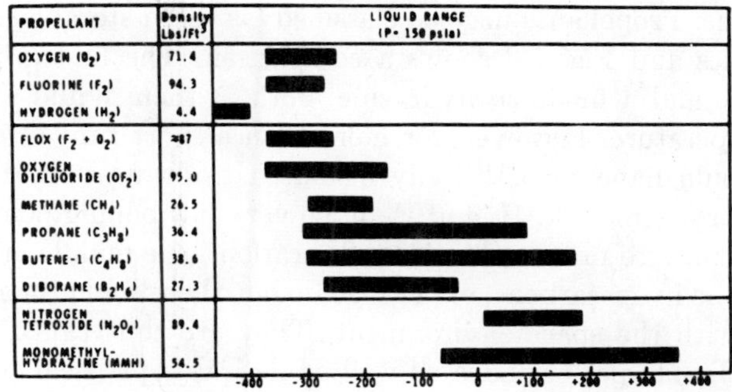

Fig. 7.3 Comparison of propellant liquid ranges (from Douglass.)

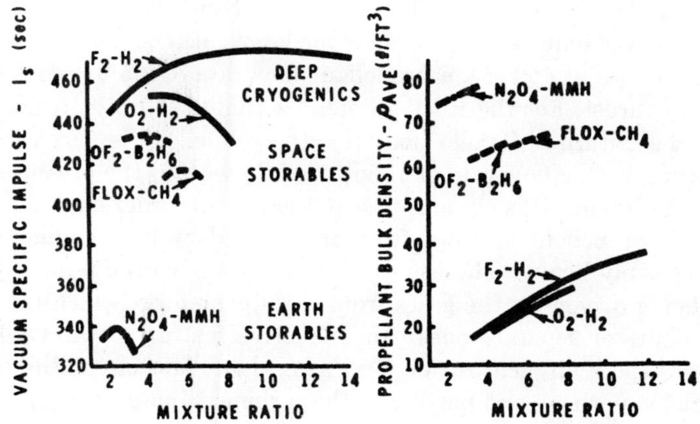

Fig. 7.4 Performance characteristics of some propellants (from Douglass.)

pressure levels being quoted, 500 psia and 1000 psia, and two back pressures, sea level and vacuum (space). Higher combustion pressure is desirable for I_s, but it may add weight for stress reasons.

The increase in I_s in space over sea-level operation is of the order of 15% at 1000 psia, nearer 20% for 500 psia. The values of specific impulse given are quoted as "optimum," presumably at the best mixture ratio and equilibrium expansion. Differences in specific impulse for nominally the same propellants occur in the literature and it must be remembered that pressure level, back pressure, mixture ratio and purity all have effects, and quoted values may not agree exactly.

Monopropellants all have lower values of I_s although they allow very simple operation of the rocket as a whole. H_2O_2 (90%) has a specific impulse of only about 135 sec (300 psia—sea level) but has the advantage of producing about 40% oxygen in the decomposition products which can then be used for further reaction with a fuel. N_2H_4 has an I_s value of about 175 sec for the same conditions, and nitromethane CH_3NO_2 a value of 218 sec.

Most of the best propellants are cryogenic, that is, they have very low temperatures in the liquid phase. Thus there is a considerable problem in storage for any length of time and specific impulse may have to be sacrificed for storage capability in certain missions. Propellants may be classified[3] as earth storable, space storable, or deep cryogenics and Fig. 7.3 shows a comparison. The first group consists of nitrogen peroxide and a modified hydrazine, both of them liquid at 150 psia and atmospheric temperature. However, for more than a short time in space, the propellant tanks would have to be heavily insulated to prevent freezing. The third group, the deep cryogenics O_2, H_2 and F_2, have very low boiling points and require insulation at all times to prevent loss by evaporation. The middle group, the F_2–O_2 compounds and the hydrocarbons, are cryogenic but their liquid ranges overlap and are compatible with the space environment. They are thus called space storable and have a reasonable performance as well as desirable physical properties (Fig. 7.4).

Liquid propellants have to be pumped up to a high pressure, injected into the combustion chamber in a finely divided state, mixed and evaporated before burning. The pumping power required is enormous for large rockets. Thus even assuming a high specific impulse of 400 sec, a propellant flow rate of 2.5 lb/sec is required for each 1000 lb of thrust. For the Rocketdyne F–1 engine of 1.5×10^6 lb thrust (five required for the Saturn V Apollo booster), the propellant is an oxygen–kerosene (RP–1) mixture, with a flow rate of about 6000 lb/sec and the pressure about 1800 psi, requiring 55,000 hp. It is obvious that powers of this order have to be generated directly from a propellant and not from any secondary heat engine cycle, which would be impossibly heavy. The usual method is to drive the pump by a turbine, the turbine being driven by the gases from a gas generator, which may utilize the rocket propellants or separate ones. An attractive feature of the O_2–H_2 rocket is that sometimes the hydrogen can be used as coolant, first circulating through the exhaust nozzle, vaporizing and passing to the turbine driving the propellant pumps. This regenerative action is desirable but not always possible.

For small rockets, the turbine may be driven by a monopropellant such as H_2O_2 or a solid propellant. Where turbines and pumps are not warranted, the propellants can be pressurized by a gas, either air or preferably an inert gas. The weight of pressurizing gas plus the increased structural weight of the propellant tanks due to pressurization has to be weighed against the turbine-pump system.

A pressure considerably higher than the combustion chamber pressure is required in order to inject the liquids at high velocity through a very large number of small orifices. Methods of mixing are many, impinging streams, spray injection, shower head, parallel stream pattern, coaxial fuel-oxidizer streams, and so forth. For H_2–O_2 propellant, oxygen in a central stream surrounded by an annular stream of hydrogen has been found very suitable, with the relative velocity of the two being an important factor in combustion performance and stability.

7.4 Solid Propellants

Solid propellants are intimately mixed combinations of fuel and oxidizer and there are two main types. The colloidal or *homogeneous* propellant has the fuel and oxidant in the same molecule or phase and is typically a *double-base* propellant, such as

[3] H. W. Douglass, "Cryogenic Rocket Propellants," ASME Preprint 67-WA/AV-9, 1967.

combinations of nitroglycerin and nitrocellulose, each of which contains a carbohydrate or hydrocarbon in the nitrate form. Ideally the constituents consist of very small particles that tend to approach molecules and form a chemical solution. Such a propellant usually contains a small percentage of additives to promote smoother burning and to improve physical properties. The other main type is the *composite* or *heterogeneous* propellant which is a mechanical mixture of relatively large particle size compared to molecules. Here *oxidizing* elements or compounds are cemented together by a *fuel binder*. Nitrates and perchlorates are common oxidizers, with the latter being preferred as usually giving better performance and handling quality.

Binders, which are the fuel element, may be asphalts, resins, or polymers. Asphalt was used in some of the earliest rockets but has poor physical properties due to lack of elasticity with temperature variation. Many different polymers are now used, mostly of the elastomeric kind, that is, having elastic, rubberlike characteristics. In fact, synthetic rubbers replaced asphalt but have now largely given way to other polymers such as polyurethane, and polybutadiene of various sorts. Powdered metals are sometimes added, aluminum being the most widely used, another being beryllium. The latter is toxic but can be used for upper-stage units. While performance as specific impulse is very important, the physical and handling qualities of a solid propellant can impose considerable restraint. A solid rocket may have to undergo temperature cycling and be required to operate satisfactorily under extremes of atmospheric temperature. Being a solid, it is subject to impact effects that can cause cracking or even breakage and thus yield a very irregularly burning propellant.

Specific impulses of solid propellants are typically lower than for liquid propellants but are being improved. Propellants in wide use have a range of I_s from about 180 to 250 sec, the compositions being generally as discussed above. Advanced solid fuels using lithium or lithium hydride in a polybutadiene binder and used with liquid oxidizers are possible, yielding I_s values up to 400 sec.

The physical mass of a solid propellant has come to be referred to as the *grain*. This does not refer to the particle geometry such as in "grain size" but the whole body of the propellant, as in grain length or diameter. The geometrical disposition of the material is known as the *grain configuration*. Once a grain is ignited, there is little possibility of control of the combustion process in a straight, solid propellant-type engine and burning characteristics must therefore be built into the grain originally. It is thus imperative to know the main parameters that control burning. First, of course, is the chemical composition and particle size of the constituents and then for a given propellant, the three factors most important are the chamber pressure, the initial grain temperature and the velocity of the burning gases over the surface of the grain.

The rate of propellant burning is proportional to the surface area of burning propellant A_b and to the specific rate of burning of the propellant normal to the surface, U_b, length per unit time (*cf.*, flame speed, Chapter 3). The mass rate of propellant consumption is then

$$\dot{m}_p = \rho_p A_b U_b \tag{7.12}$$

where ρ_p is the propellant density.

The effect of pressure level on U_b is found empirically and can be expressed simply as

$$U_b = \alpha p_0^n \tag{7.13}$$

where α is dependent on the particular propellant and initial temperature, p_0 is the chamber pressure and n is an exponent between zero and unity, its value being a function of the particular propellant but almost independent of temperature. The temperature of the reaction, that is the combustion temperature, is mainly dependent only on the chemistry of the propellant, thus we may say T_0 is a constant for a particular grain. The coefficient α may be expressed by

$$\alpha = C_1/(C_2 - T_i) \tag{7.14}$$

where C_1 and C_2 are constants and T_i is the initial propellant temperature. This is an empirical expression and carries the idea that if T_i is close to C_2, then the burning rate may increase sharply.

Now for a steady state in the combustion chamber, the mass rate of propellant consumption $\dot{m}_p$ must equal the mass rate of flow of gas $\dot{m}_g$ from the nozzle. The nozzle will be choked during the major part of the burning as pressures are of the order of several hundreds of psi, and hence, with a constant T_0, $\dot{m}_g$ is dependent only on p_0. Thus if the rate of burning increases, pressure increases and this in turn increases U_b and so on. Gas outflow rate $\dot{m}_g \propto p_0$ and propellant consumption or gas production rate $\dot{m}_p \propto U_b \propto p_0^n$. Hence $d\dot{m}_g/dp_0 =$ constant and $d\dot{m}_p/dp_0 \propto p_0^{n-1}$. If $n > 1$, then the process is unsteady because increase of pressure will increase $\dot{m}_p$ more than $\dot{m}_g$, theoretically to explosion, while decrease of pressure will cause cessation of reaction. If $n < 1$, then increase of pressure will cause $\dot{m}_g$ to increase more than $\dot{m}_p$, hence reducing pressure and stabilizing the reaction; similarly for decrease of pressure. With n close to 1, the reaction is very sensitive to burning rate and overpressures could readily result. Thus low values of n are desirable, but again not too low as then combustion may too easily be quenched. Values of n range generally from 0.2 to 0.7, with those for homogeneous propellants being higher than those for the composites as a rule. Burning rates U_b can vary widely, with the majority falling in the range 0.1 to 0.8 in./sec. Although pressure controls the rate of burning, or vice versa, and hence the burning time of a given mass of propellant, the total impulse as thrust $\times$ time F_t is approximately constant. Total impulse is a significant performance parameter for a rocket.

The equivalency of $\dot{m}_p$ and $\dot{m}_g$ can be carried further to give a relationship for the chamber pressure. Equation 7.12 gives $\dot{m}_p$, and $\dot{m}_g$ for choked flow is given by Eq. 7.8, so

$$\dot{m}_p = \dot{m}_g$$

$$\rho_p A_b U_b = A_t p_0 [(g_c k/RT_0)[2/(k+1)]^{(k+1)/(k-1)}]^{1/2}$$

from which

$$\frac{A_b}{A_t} = \frac{p_0 f(k)}{\rho_p U_b (RT_0)^{1/2}}$$

Replacing U_b by αp_0^n from Eq. 7.13,

$$\frac{A_b}{A_t} = \frac{p_0^{1-n} f(k)}{\rho_p (RT_0)^{1/2}}$$

which for a given propellant gives

$$p_0 \propto (A_b/A_t)^{1/(1-n)} = K_n^{1/(1-n)} \tag{7.15}$$

where K_n is the ratio of propellant burning surface to nozzle throat area. Equation 7.15 shows that the burning surface must remain constant in area for a constant chamber pressure and that again small values of n are required for stability.

The influence of initial temperature was noted in the empirical equation 7.14 and a *temperature sensitivity* factor π_p is defined as

$$\left(\frac{\partial \ln p_0}{\partial T_i}\right)_{K_n} = \left(\frac{1}{p_0}\frac{\partial p_0}{\partial T_i}\right)_{K_n} \tag{7.16}$$

Equation 7.16 gives the rate of change of pressure with initial temperature at constant A_b/A_t. The effect can also be expressed as rate of change of burning rate with temperature, thus

$$\pi_b = \left(\frac{\partial \ln U_b}{\partial T}\right)_{K_n} = \frac{1}{U_b}\left(\frac{\partial U_b}{\partial T}\right)_{K_n} \tag{7.17}$$

This temperature sensitivity can be large, say of the order of 0.5%. Thus an atmospheric temperature change of 100°F can mean 50% change of burning rate. Furthermore, the physical characteristics can change markedly with temperature, a propellant becoming brittle and crack-prone at low temperatures, and becoming nonrigid at higher temperatures. There is then a considerable premium on binders in which these effects are minimized, particularly for military applications where launching times and places cannot be controlled.

The requirement for control of burning area if control of pressure and thrust are required implies that the grain configuration is important. Constant area, pressure and thrust requires a grain with *neutral burning characteristic*, while increasing and decreasing values of these parameters require *progressive* and *regressive* burning characteristics, respectively. The only simple type of grain with a single burning surface yielding a neutral characteristic is that with end burning or "cigarette" burning. This does not yield a high thrust, as cross-sectional area is limited and hence it is used only for small rockets. A simple hollow cylinder to give longitudinal burning results in progressive burning, as the area continually increases, but the addition of a central cylindrical grain or provision of a key slot throughout the length can be made to give a neutral characteristic. A star or multifin type of grain can also give neutral burning and this general type of configuration is popular, as the grain has structural strength and requires no additional supports except in the very largest sizes. Small variations of the fin shape can yield progressive or regressive burning to a program such as two-step burning. Some grain geometries are shown in Fig. 7.5. Portions of grain surfaces which are not required to burn but which may

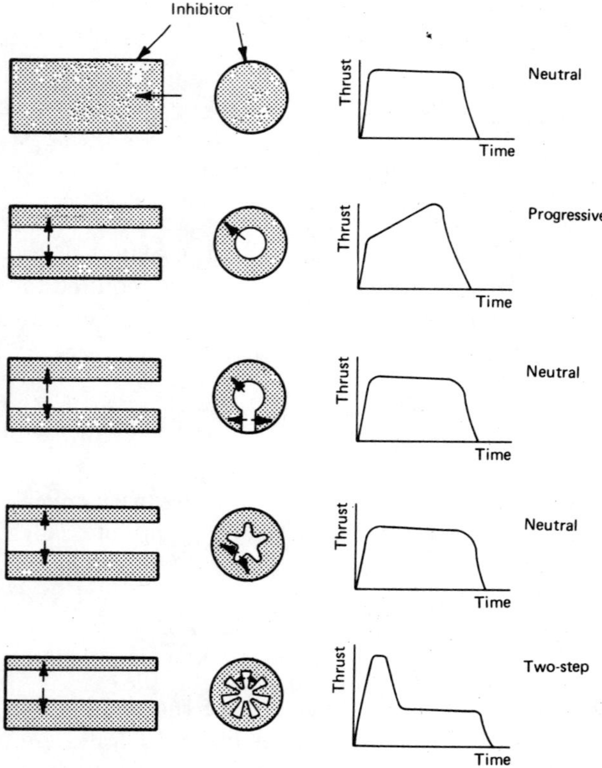

Fig. 7.5 Some solid-propellant grain configurations.

ignite adventitiously can be inhibited by coating the surface with an inert or very slow-burning substance. The outside of the grain at the wall must have an inhibitor and a layer of noncombustible insulating material to protect the casing.

Grains for the smaller rockets may be made separately and then inserted into and fastened to the casing but large rockets have to be filled with liquid mix which then has to be dried and hardened, i.e. such grains are cast and cured in place.

A third factor in solid propellant burning is *erosive* burning, that is, accelerated combustion due to the flow of burning gases along the surface of the propellant. At ignition, the cross-sectional area for the flow of the gas within the grain is a minimum. It is larger than the nozzle area of course and increases with time as the propellant burns. There is thus a significant gas velocity during the early part of the combustion period, with the gases sweeping over the burning surface. The effect is one of a boundary layer, with high velocities producing a thinner layer, with increased heat to transfer to the propellant. It is largely this heat transfer and consequent preheating of the propellant which increases the burning rate and leads to "erosive" burning. Thus a fast-burning propellant will be less subject to erosive burning than a slower one and hence the phenomenon is associated with the propellant composition, at least up to high pressures (≈ 1000 psi). Theoretical analyses may be made using a boundary layer approach and are helpful in delineating the major parameters but cannot give a complete answer. Experimental values of erosive burning rate have been made and these are used in design to obtain as uniform a rate of combustion as possible over the whole propellant surface.

Ignition of solid propellants is not an easy problem and represents a considerable subdivision of the technology. Ignition should take place as rapidly as possible over a large area of the propellant surface so that the pressure builds up rapidly but at the same time it must not cause shock. Ignitors must be stable, insensitive to ambient conditions including radiation, strong but of low weight, easy to install and operate, and above all, must be completely reliable. Liquid igniters have been used but are not common as they tend to be complicated in containment. The materials are invariably hypergolic and chosen from the chemicals previously discussed. Types of solid ignitors are legion, but nearly all are based on a pyrotechnic substance electrically ignited. Major types can be classified as *confined*, in which the initial reaction is contained until rupture occurs, and *unconfined* in which the rocket chamber is the only confinement. An intermediate class seems to offer the best promise, a type in which the ignition reaction is controlled gasdynamically by ports. Extinction of solid-propellant burning may be carried out by injections of inert gases or by opening ports to reduce pressure, but neither method is simple. Sudden extinction rather than burning fadeout may be necessary for certain mission functions, as in staging operations. For this, burstable diaphragms can be located at one or more points and operated by a command signal. In addition to causing a rapid lowering of pressure, the direction of the exhausting gases can give opposing thrust for a brief time.

7.5 Hybrid Rockets

Hybrid rockets, that is, a combination of a solid fuel and a liquid oxidizer, can offer some attractive features. This is particularly true where a rocket for a variable mission is required, as for controllable range in military application. The control of one liquid only is attractive in conjunction with the simplicity of a solid fuel. Successful hybrids have been used using a polymer substance and powdered metal as fuel with a nitrogen-bearing oxidizer, e.g., nitric oxide, nitrogen tetroxide, or nitric acid. Most such hybrids have been small rockets but there appears to be no particular barrier to large sizes and possibly even some cost advantages by suitable optimization. Even a so-called "tribrid" has been suggested and given some initial testing, tribrid implying three propellant constituents. In this case, the basic combination proposed is solid lithium as fuel and fluorine as liquid oxidizer, with hydrogen injected at nozzle entrance to reduce the very high Li–F reaction temperature and also to add energy.

7.6 Combustion Instability

The combustion chambers of all types of chemical rocket are subject to instabilities of burning resulting in oscillations of pressure and general vibrational phenomena which can lead to disastrous failures. The problem is perhaps more serious in solid-propellant motors, because of the interaction of pressure and burning rate and the greater complexity of the phenomena. Although an important part of the technology of chemical rockets, the topic can only be alluded to briefly here.

In liquid-propellant motors, a low-frequency oscillation known as "chugging" caused by interaction of the chamber pressure and the propellant pump pressure can be analyzed[4] and is often cured by increase of pump pressure so that normal variations of combustion pressure are relatively small and do not lead to coupling. High-frequency oscillations of an acoustic nature that result in screaming,[4] howling or screeching are much more difficult to analyze. They occur in various modes, (radial, transverse, etc.) and are akin to similar phenomena in the corresponding high energy input combustion process of afterburners in jet engines.

In solid-propellant motors, two of the major types of instability may be recognized as (1) acoustic resonances and (2) nonacoustic instabilities. The former is relatively well understood and a theoretical framework exists for the analysis of linear instabilities. A more-or-less empirical cure has been found in the addition of aluminum particles which provide damping when they are of appropriate size. In general, the instability is reduced or eliminated by introducing elements that tend to break up a regular pattern of oscillation, such as the insertion of an axial non-burning rod into the chamber or provision of radial perforations in the grain to give transverse gas jets to destroy wave patterns. Nonlinear acoustical instabilities and nonacoustical instabilities are much more difficult to account for and treat on a rational basis leading to design parameters. A comprehensive treatment with bibliography is given in the reference cited below.[5]

7.7 Status and Development of Chemical Rockets

Chemical rockets of all sizes are now extremely reliable although there may be occasional failures of upper-stage elements in the more complex missions.

Liquid-propellant engines are now operational in very large sizes, notably the Rocketdyne F–1 booster for the Saturn V of the Apollo spacecraft. The Apollo first stage has five F–1 engines each of 1.5×10^6 pounds thrust, with kerosene-lox as propellant. The figures of propellant flow rate and energy level are staggering—1250 gal/sec of kerosene (RP1) and 2083 gal/sec of liquid oxygen, or a total of 900 tons/min for 160 sec, requiring a turbine of 55,000 hp for driving the pumps. The five F–1 engines are equivalent to 85 Hoover Dams in power generation, each of 1345 MW. The Aerojet M–1 (now shelved) likewise had 1.5×10^6 lb thrust using H_2–O_2 propellant, thus requiring considerably less propellant owing to its higher specific impulse. Many hundreds of launches have been made by the Thor and Atlas boosters. As an instance of the rapid development, one of the early H_2–O_2 engines was the Pratt and Whitney 15,000 lb thrust RL–10 used on the Centaur. Within a few years, the RL–20 H_2–O_2 engine has been built, which is about the same size as the RL–10 but has 250,000 lb thrust, sixteen times as great.

[4] L. Crocco and S. Cheng, "Theory of Combustion Instability in Liquid Propellant Rocket Motors," AGARDograph No. 9, Butterworths Scientific Publications, 1956.

[5] F. A. Williams, M. Barrere and N. C. Huang, "Fundamental Aspects of Solid Propellant Rockets," AGARDograph No. 116, Technivision Services, Slough, England, 1969.

Solid-propellant rockets come in all sizes, from very small units of a few hundred lb thrust to extremely large ones of several million lb thrust. They are very convenient, as they do not need "fueling" and can be transported complete. They can cover a wide range of duties as boosters, upper stages, sounding rockets, guidance and control, and in particular perhaps, as ballistic missiles (Minuteman, Polaris). As example one might take the Scout motor, which now has a considerable history and many launchings. It is interesting because it is capable of satellite launching, having four stages, all solid propellant. The thrusts of the four stages are 203,000 lb, 62,000 lb, 13,000 lb and finally 2800 lb, the last putting a 300-lb payload into a 300-n mi orbit. It is used to put scientific payloads into orbit and for international launchings.

The largest solid-propellant engine is an experimental Aerojet SL model, 260 in. in diameter, of which the third model, the SL–3, has 5.4×10^6 lb thrust with a burning rate of 22,000 lb/sec. The first unit of this type, the SL–1, had over 3.5×10^6 lb thrust, using 1.673×10^6 lb of propellant, burning for about 130 sec with an average chamber pressure of 489 psia. The propellant was a composite of polybutadiene binder, aluminum powder and ammonium perchlorate oxidizer.

An interesting development for solid rockets is to fire them from guns, which then effectively become a reusable first stage. This does limit the size of rocket which can be used, as 16 in. is one of the largest guns, but for small sounding rockets, the method has considerable possibilities in economic operation and in extending range.

Comparison of solid-vs-liquid-propellant motors shows that the former has advantages of simplicity, ease of construction, lower development cost, storability (up to a point) and availability at short notice. The major disadvantage is lower specific impulse, as one cannot use the high-energy liquid propellants, and demands of mechanical properties of the grain prevent use of the optimum mixture compositions for maximum performance. Firing times are apt to be short and cooling is difficult as there is no liquid which can be circulated.

Problems

7.1 An oxygen-JP4 rocket has a combustion chamber pressure of 300 psia, temperature of 5470°F, and a nozzle throat diameter of 10 in. The propellant gases are taken as an ideal gas of molecular weight 23, with $k = 1.22$ and $R = 67.2$ ft lb/lb °R. If the nozzle is designed for complete expansion to an ambient pressure of 5 psia, assuming ideal flow throughout, find:

(a) nozzle discharge area
(b) mass flow rate
(c) throat velocity
(d) jet velocity
(e) thrust
(f) thrust coefficient
(g) characteristic velocity C^*
(h) specific impulse

7.2 Show that for a rocket with a convergent nozzle operating at an altitude where the atmospheric pressure is negligible, the specific thrust is given by the expression

$$\{2[(k + 1)/g_c k]RT_0\}^{1/2}$$

7.3 A rocket has a combustion stagnation temperature of 5260°R and is designed to have a nozzle exit Mach number of 4. Find the specific impulse of the rocket when operating in space. Assume reversible adiabatic flow with $k = 1.2$ and $R = 85$ ft lb/lb °R.

7.4 For a single-stage rocket having a fixed initial mass M_0 and structure fraction s (assume as 0.05), and a required velocity increment ΔU to place the payload into a satellite orbit, show that in the range of $I_s \approx 400$ sec, a change of I_s of 1% causes a change of payload fraction l of about 6% (hence the emphasis on maximum values of I_s).

7.5 Discuss all the possible effects of increasing the operating combustion chamber pressure of a liquid-fueled rocket and then estimate the degree of importance you might attach to instigating a development program for higher pressure.

7.6 A boost rocket which places a payload into a circular orbit at a height of 200 miles has an initial mass of 100,000 lb, a structure mass of 4000 lb, a specific impulse of 350 sec, and a burning time of 200 sec. Assuming the same values of overall mass, structure mass and burning time, find the increase of payload possible for a 10-sec increase in the specific impulse. Assume no drag and a vertical trajectory.

CHAPTER 8

Electrothermal Engines

8.1 Performance Factors

It has been seen that chemical rockets are likely to be limited to a specific impulse of about 500 sec at the very best. The limit is placed by (1) the temperature attainable in a chemical reaction of an oxidizing nature and (2) the molecular weight (or specific heat) of the propellants which are combustion products. Accepting the idea of a separate power plant whose energy output can be used in many different ways, one of the simplest propulsion methods could very well be the heating of a propellant to a higher temperature than is possible by combustion, with the propellant being chosen to give the maximum temperature drop in expansion for a given energy supplied. This is the basis of *electrothermal propulsion,* which is the use of electrical energy to heat a suitable propellant at a pressure high enough to give an expansion ratio sufficient to extract most of the thermal energy available.

From Eq. 7.1 we have for a perfect gas,

$$V_j = [2g_c(h_0 - h_j)]^{1/2} \equiv [2g_c c_p(T_0 - T_j)]^{1/2}$$

and this can also be written as

$$V_j = \{2g_c[k/(k-1)](R_0/M_w)T_0[1 - (p_j/p_0)^{(k-1)/k}]\}^{1/2} \tag{7.3}$$

From these relationships we see that the specific heat c_p of the propellant should be high and the molecular weight low. The latter is modified by the factor $k/(k-1)$ but as this has a minimum of 2.5 for a monatomic perfect gas and a maximum of about 7 for a gas mixture, while the molecular weight can vary from a value of two upwards to high values, it is M_w which is the major influence. However, c_p gives the simplest and most direct answer. We might note the order of specific impulse we could expect by using hydrogen, the element of lowest molecular weight, at a temperature of 6000°R as the absolute maximum for any material to withstand. We assume that the expansion is sufficient so that $T_j \to 0$ or is at least negligible compared to T_0. Then $I_s \approx 1400$ sec. Actually developed electrothermal units using hydrogen vary from about 800 sec to 2000 sec with the higher figure being attained by so shaping the gas stream that its temperature near the containing walls is much less than its temperature in the main body of the flow, which can then be much greater than the material-limiting temperature of 6000°R. Thus, we could expect electrothermal units to cater for the low specific impulse range above the limit of chemical rockets. The factors in the performance then are those of (1) selection of a propellant of optimum properties, (2) production of high temperatures, and (3) the containment of the hot gas to provide long life.

8.2 Propellants

With respect to the propellant, it is not a straightforward matter of selecting hydrogen as having the lowest molecular weight because of the effect of dissociation and ionization at the very high temperatures indicated and the subsequent possibility of "frozen flow." Frozen flow implies that the thermal energy is not regained as directed kinetic energy in the expansion time available. Other propellants that might be considered, because of their relatively low molecular weight, are helium (≈ 4) and lithium (≈ 7), or compounds of hydrogen such as ammonia NH_3 (≈ 17) or hydrazine N_2H_4 (≈ 32). The last two have much higher apparent molecular weights but as some dissociation is bound to occur, the heated composition yields a lower value. The balance lies in the degree of dissociation and the speed or otherwise of reassociation.

Pressure has a considerable effect on dissociation and one mode of attack is to use higher pressure levels. However, this may introduce construction difficulties of containment, higher rate of wear due to erosion and increased heat transfer losses or cooling problems because of the higher density. This is largely a matter of experience and at the present time a pressure of a few atmospheres seems to be the limit, although this can change as time goes on.

Helium as a monatomic gas does not dissociate and, therefore, is much better than hydrogen with respect to frozen-flow loss, although it does ionize at sufficiently high temperature. Lithium has a fair specific heat but has low recombination rates. Ammonia gives good results, almost as good as hydrogen, in spite of its higher equilibrium molecular weight.

It is possible to form a "frozen-flow efficiency" which is defined as the ratio of the kinetic energy of the jet with completely frozen flow to the energy supplied. Jack[1] has done a comprehensive study of propellants for electrothermal propulsion and Fig. 8.1 is reproduced from his results. It shows clearly the advantage of

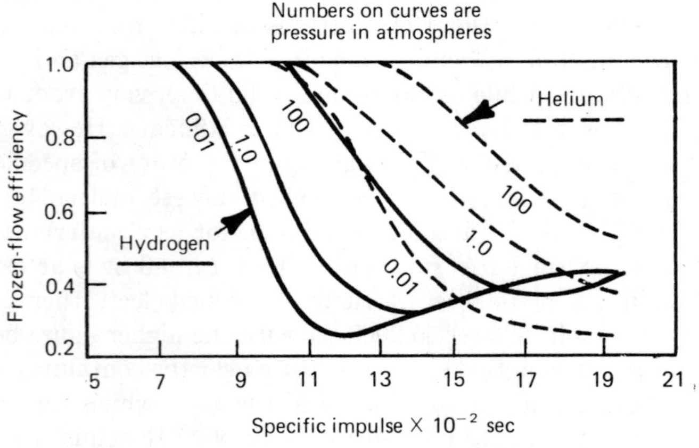

Fig. 8.1 Comparison of frozen-flow efficiency for H_2 and He (from Jack).

[1] J. R. Jack, "Theoretical Performance of Propellants Suitable for Electrothermal Jet Engines," *ARS Jour.*, 31, (1961), 1685–89.

helium over hydrogen and the effect of pressure. At the higher specific impulses it is most probable that the flow will be completely frozen and therefore the calculated figures are not pessimistic.

As usual, the choice of propellant is guided not only by its thermodynamic performance but by its suitability for storage and supply and by its corrosion properties, both in storage and in reaction. Unfortunately, helium is a truly cryogenic substance with a liquid temperature of about 7°R at one atmosphere pressure and the difficulties of storage for large lengths of time are considerable. Ammonia is corrosive although readily storable, and the same applies to hydrazine. Lithium is a very active element and does not appear attractive. All in all, it would seem that hydrogen is the most suitable substance, particularly as its handling is now routine. Most of the practical development work has been done with hydrogen, although ammonia has also been used in a few units. However, in very small thrusters, such as are used for station-keeping, efficiency is low and times are long and the amount of propellant required is relatively large. The tank for keeping hydrogen for this length of time is very heavy and so ammonia is preferred.

The heating of the propellant has been accomplished by two different basic methods, resistance heating and arc heating, while a third, electrodeless discharge heating, has been experimented with.

8.3 Resistance Heating

The simplest method is to pass the propellant over an element heated to a high temperature by passing electric current through it. This type of resistance-heating unit is called a *resistojet* and a schematic arrangement is shown in Fig. 8.2. The heating element may be coils of wire or the duct walls themselves may be heated. Because of difficult geometries, the approach to design of heaters and minimization of radiation losses is largely empirical. To a considerable extent, the problem is one of materials, to withstand very high temperatures and to resist erosion. The latter point is important, because nozzle throat diameters are small and velocities very high (10–15,000 fps for hydrogen).

Resistojets so far have been used in very small sizes with thrust starting at micropounds and going up to a few pounds. The corresponding powers are small,

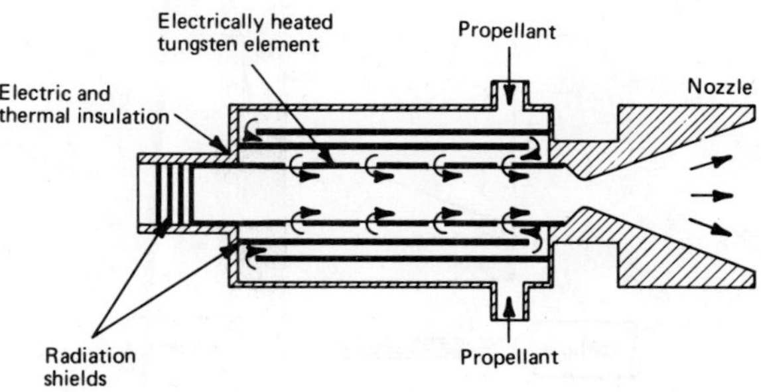

Fig. 8.2 Resistojet—diagrammatic.

up to 30 kW with efficiencies from 50% to 85%. The physical dimensions are small, a few inches in diameter, usually less, and up to a foot in length, weighing only a few pounds for the thrust unit itself.

The resistojet fulfills a useful function, namely a compact, reliable, low-thrust device for the various auxiliary but very important jobs in connection with satellite operation, station-keeping, drag neutralization, attitude control, and so forth. Remembering that the power input is proportional to $\dot{m}_p I_s^2$, the low specific impulse can be advantageous because the power-thrust ratio is low, i.e. the power required is relatively low. More propellant is used and has to be taken along but there are many situations where the power requirement is a governing factor.

8.4 Arc heating

The *arc jet* or *plasma jet* utilizes the very high temperatures in arcs to heat the propellant. The latter becomes partly ionized, but remains neutral overall, hence the term plasma, and the operation is not "electric" in the sense that the propellant is accelerated by an electric field, but by expansion.

Gaseous discharges between electrodes take many forms, but here we are concerned with what might be called the "fully developed" discharge when the cathode becomes sufficiently hot to emit electrons by thermionic means, photovoltaic means (from the intense light of the gas) and by field emission (electrons emitted by virtue of the ambient electric field). Typically, an arc has a relatively low voltage and a high current, values of each in volts and amps being of the order of the low hundreds or less.

The potential across the electrodes is divided up into a cathode fall region of the order of 10–20 V, an anode fall region of variable amount but generally of the order of the cathode fall, with the remainder of the potential being across the *positive* or *arc column*. The linear dimensions of the cathode and anode fall regions are small, with the positive column being the predominating feature and the region where the propellant gas is heated (Fig. 8.3).

An arc has a valuable feature called the *pinch effect*, a self-constriction which tends to confine it to a cylindrical shape instead of having the arc spread out irregularly. The constriction is brought about by the creation of a magnetic field produced by the arc current itself reacting with the current that produces it, to yield a force which counteracts a radial gradient in the gas pressure in the arc column tending to

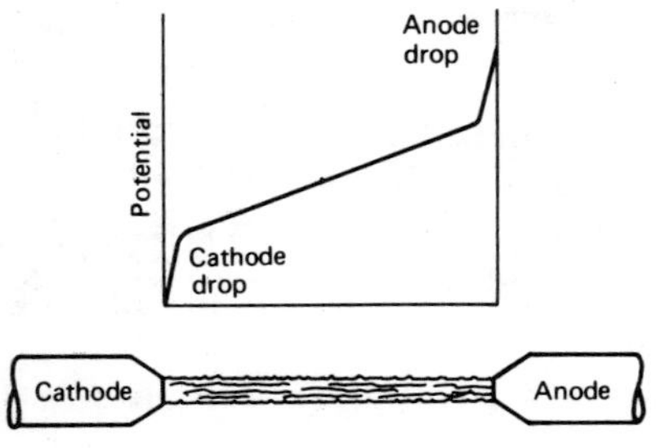

Fig. 8.3 Arc potential.

disperse it. In equilibrium, the arc column is ideally cylindrical. This confinement of the arc allows higher current density in the core and hence higher temperatures. However, at the high current densities which are useful for arc thrusters, an inherent instability phenomenon is present which it is necessary to counteract. If some external agency produces a bending of the discharge, the pinch effect is greater due to the reduced radius of curvature. Thus a momentary "waisting" of the column produced by some disturbing effect is intensified and may lead to breaking of the discharge. Similarly, deflection of the gaseous column in the form of solid-column buckling may be amplified and lead to an irregular discharge. Thus some means must be used to neutralize this instability effect.

Perhaps the obvious one is to furnish an external magnetic field which contains the discharge, but it is equally obvious that this is an undesirable complexity. Two simple techniques are possible, which may be used together although this does not appear necessary in many instances. The first method is to swirl the propellant around the arc by introducing it tangentially. The vortex motion of the fluid produces a radial pressure gradient as follows. Assuming a free (potential) vortex which obeys the relation $V_u r$ = constant where V_u is the tangential velocity, then inserting this in the Euler equation gives the radial pressure gradient. Thus in general,

$$dp + \rho V \, dV = 0$$

and

$$(dp/dr) + \rho V(dV/dr) = 0$$

From $V_u r = k$, $V_u = k/r$ and $dV_u = -k \, dr/r^2$ with $V = V_u$,

$$(dp/dr) - (k^2 \rho/r^3) = 0$$

and

$$dp/dr \propto 1/r^3 \tag{8.1}$$

The swirling motion also confers the advantages of improving the heat transfer from the walls and the electrodes and so keeping them cooler, and also of improving the mixing and heat transfer of propellant and arc. Thus the propellant is heated more uniformly to a higher temperature and the arc is consequently cooled more efficiently. A lower arc temperature increases its resistance (reduced ionization) and allows higher voltages and power inputs. Another possible positive feature is that the swirl action may help to spread the discharge over a greater area of the cathode and thus reduce locally severe erosion. However, swirl can lead to inefficient units if the propellant and arc do not mix sufficiently, that is, the fluid cools the outer part of the arc but leaves a hot core which simply wastes power. Swirling the propellant is simple to provide in practice, although it complicates analysis.

Another method is to simply flow the propellant axially through a tube of small diameter and long length, with the arc being maintained along the axis. The confinement of the propellant likewise confines the arc and stabilizes it. It seems that adequate heat transfer is obtained to cool the walls and heat the gas so that vortex flow does not appear necessary. Nevertheless, it is sometimes used in conjunction with the constriction pattern as it does not necessarily complicate the design unduly.

It is possible to set down the relevant physical relationships and bounding conditions of arc behavior, but it is necessary to choose a very much simplified model to which to apply them. Such analytical procedures have some value in elucidating certain features of operation, but the arc process itself is extremely complicated. As they are mostly concerned with aspects of the detailed phenomena rather than giving general performance relationships, they will not be pursued here.

A typical arcjet then looks schematically as shown in Fig. 8.4. The main body consists of an anode block forming a constrictor and a diverging nozzle. The cathode is pointed and the arc discharges from its tip, forming a column in the constrictor and then spreading out radially to attach itself to the anode. The propellant is injected close to the cathode, which it cools, with or without swirl as development dictates. The anode walls are kept cool by a radial temperature distribution in the arc, with a very hot central core to give the velocities leading to specific impulses of 1500 sec or more. Tungsten is commonly used for both electrodes, with nitrides for insulation between anode and cathode.

8.5 Electrodeless Discharge

The wear of electrodes associated with "conventional" arc jets for the long periods of time likely with propulsion units has led to efforts to develop heating methods without electrodes or with cold electrodes. Basically the idea is to heat the propellant gas inductively by ac power at radio frequency or even in the microwave range. The energy is absorbed by free electrons and transferred to the neutral atoms of the propellant by collision. Elastic collision of this sort is poor for transferring energy and inelastic collisions may result in dissipation by radiation. To be classified as electrothermal acceleration, the process must be one of simple heating and some of the proposed methods also indicate a magnetic field which renders the operation partly electromagnetic. At the present time, methods of coupling the electrical energy and the propellant are poor and the efficiencies are low. The type of power unit required is likely to be heavy, thus only experimental devices have been tried. The attraction lies in providing an accelerator in which wear and erosion of electrodes is absent or negligible.

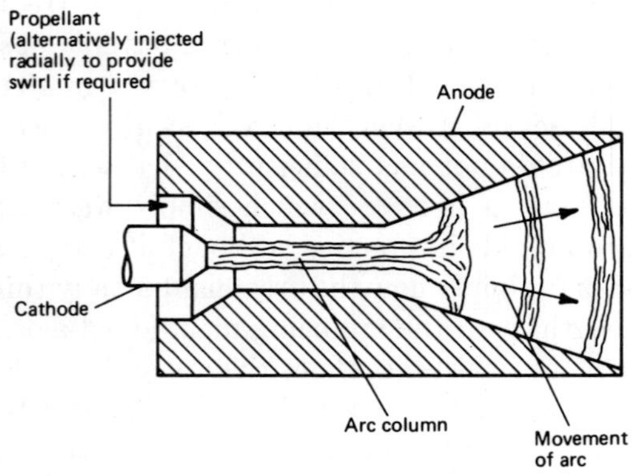

Fig. 8.4 Arc jet—diagrammatic.

CHAPTER 9

Ion Propulsion

The basis of electrostatic or ion propulsion is simply the acceleration of charged particles by an electric field. Figure 9.1 shows schematically a typical ion thrust system. The propellant source feeds neutral atoms to an ion source and the positive ions generated are accelerated by one or more sets of electrodes which are maintained at zero potential, the ion source being at a high potential. The negative ions or electrons have to be returned eventually to the positively charged exhaust stream to maintain a neutral beam. The propellant can in principle be any substance capable of ionization, solid, liquid or gas, and may be in atomic, molecular or particle form. Our object here is to look at the problems involved and to analyze the operation from the point of view of understanding the performance and major design factors.

9.1 Performance Parameters

Consider a single particle of mass μ and unit charge e moving in a homogeneous electric field of strength E provided by potential V across the distance s (Fig. 9.2). The force on the particle F_e is eE and work done on it is $eEs = eV$, and this must

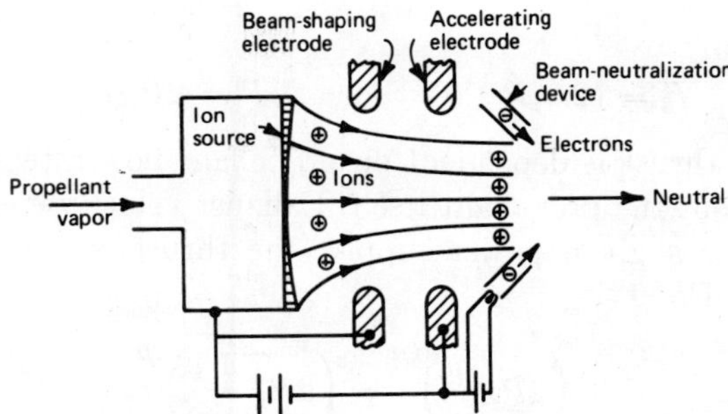

Fig. 9.1 Ion thrust system—schematic.

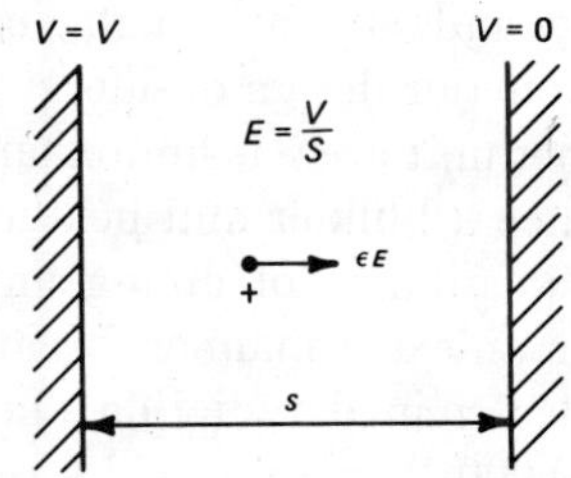

Fig. 9.2 Charged particle in a plane-parallel field.

equal its change in kinetic energy from anode to cathode. Assuming it leaves the anode with negligible velocity, then the kinetic gain is $\mu c^2/2$. Hence

$$\frac{\mu c^2}{2} = eV$$

and

$$c = [2(e/\mu)V]^{1/2} \tag{9.1}$$

and

$$I_s = c/g_c = (1/g_c)[2(e/\mu)V]^{1/2} \tag{9.2}$$

The velocity of all particles under these conditions will ideally be the same and the thrust F will be $\dot{m}_p c$ where $\dot{m}_p$ is the total mass flow of ions (propellant flow rate). The total current I is the total charge per unit time, i.e.

$$\text{Current} = \frac{\text{charge}}{\text{time}} = \frac{\text{charge}}{\text{particle}} \times \frac{\text{particle}}{\text{mass}} \times \frac{\text{mass}}{\text{time}}$$

or

$$I = e\dot{m}_p/\mu \qquad \text{and} \qquad \dot{m}_p = \mu I/e \tag{9.3}$$

Substituting for $\dot{m}_p$ and c in the thrust relation,

$$F = \dot{m}_p c = (\mu I/e)[2(e/\mu)V]^{1/2} = I[2(\mu/e)V]^{1/2} \tag{9.4}$$

The kinetic energy rate is $\dot{m}_p c^2/2$ and this is the *beam power* P_j, which is also equal to the electrical power VI, thus

$$P_j = \dot{m}_p c^2/2 = VI$$

and with $F = \dot{m}_p c$,

$$F = (2\dot{m}_p P_j)^{1/2} \qquad \text{or} \qquad F = 2P_j/c$$

This states that the thrust is dependent on propellant flow rate and beam power, and says nothing about the propellant itself. Exhaust velocity c is specified by the mission as I_s and for a given power supply, the thrust is fixed. However, substituting $P_j = VI$ in Eq. 9.4,

$$F = \left(2P_j I \frac{\mu}{e}\right)^{1/2} = \left(2P_j^2 \frac{\mu/e}{V}\right)^{1/2} \tag{9.5}$$

Thus for a given power and thrust, a propellant of high μ/e (heavy ions) requires a low current and a high voltage and, similarly, light ions require a high current and low voltage. This is one of the major design desiderata for ion engines. We shall see that current *density*, current per unit area, is limited and thus a high current implies a larger cross section and hence a bulkier and heavier motor. On the other hand, voltages are limited by the possibility of arcing and electrical breakdown. The latter cannot be determined in an exact manner, because the vehicle will operate in space and be subject to undetermined radiation and particle reaction. It would appear that perhaps about 50,000 V, with a field strength of 10^7 V/m, might be a practical limit in the foreseeable future.

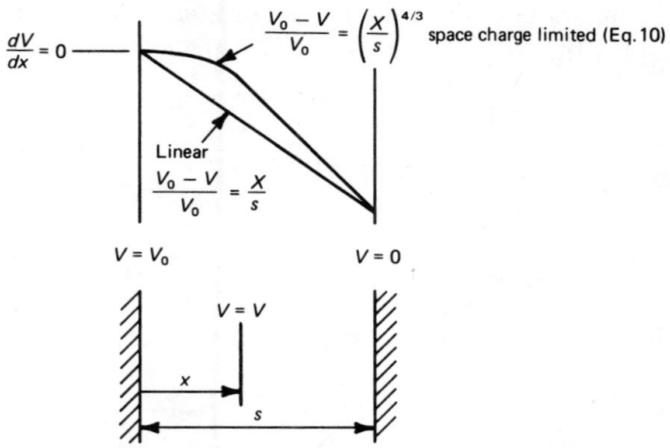

Fig. 9.3 Potential distribution in a plane-parallel field—effect of space charge.

It was suggested that there was a limitation to the current density which could be utilized and as this is fundamental to the ion engine, the relationship will be developed in some detail. As the positive ion density increases, an electrostatic field is created which tends to counteract the applied accelerating field. The extracting agent at the anode is the potential gradient dV/dx, where x is the distance outward from the anode in the s-direction. If the field built up by the ion flow reaches the point where its effect neutralizes the applied potential, there is no longer an electric field to move the ions from the anode, i.e. $dV/dx = 0$ at $x = 0$. This represents the condition of maximum current. It is called the *space charge* effect and is a major criterion in design of ion motors. A quantitative relationship was developed by Child in 1911 in early studies of ion emission behavior and was given a little later independently by I. Langmuir. It is known as Child's law, or the Child-Langmuir law and can be developed as follows for a geometry of two infinite parallel planes as electrodes (Fig. 9.3).

The basic relationship is Poisson's equation, $\nabla^2 V = \rho_e/\epsilon_0$, which in turn stems from Gauss's law relating field strength E and charge density, ρ_e, which for our one-dimensional case is $dE/dx = \rho_e/\epsilon_0$, where ϵ_0 is the permittivity or dielectric constant of free space in mks units. Combining this with the definition $E = -dV/dx$, then Poisson's equation becomes

$$d^2V/dx^2 = -\rho_e/\epsilon_0 \tag{9.6}$$

To introduce the current density, j, we have $j = \rho_e u$ (charge per unit volume times velocity = charge per unit area per unit time = current density), with u = charge velocity at any point. To eliminate charge velocity u in terms of the ion characteristics μ and e and potential difference V, we can use the energy relation $\mu u^2/2 = e(V_0 - V)$. Combining these three relationships we get

$$dV^2/dx^2 = -(j/\epsilon_0)[\mu/2e(V_0 - V)]^{1/2} \tag{9.7}$$

If this equation is integrated, using the limiting boundary condition of $dV/dx = 0$ at $V = V_0$ (supposing no initial KE of the ions), then we will have the limiting current density in terms of potential, electrode separation and ion parameters.

The current density j is independent of x for steady state and it is convenient to replace the constant factor $j(\mu/2e)^{1/2}/\epsilon_0$ by K, thus

$$d^2V/dx^2 = -K(V_0 - V)^{-1/2}$$

To integrate, we multiply each side by $2(dV/dx)$,

$$2\frac{dV}{dx}\frac{d^2V}{dx^2} = -2K(V_0 - V)^{-1/2}\frac{dV}{dx}$$

and then the left-hand side can be rearranged to give

$$\frac{d}{dx}\left(\frac{dV}{dx}\right)^2 = -2K(V_0 - V)^{-1/2}\frac{dV}{dx}$$

This can be integrated immediately to

$$(dV/dx)^2 = 4K(V_0 - V)^{1/2} + C_1 \tag{9.8}$$

Applying the limiting condition $dV/dx = 0$ at $x = 0$ and $V = V_0$, then $C_1 = 0$, and thus

$$dV/dx = 2K^{1/2}(V_0 - V)^{1/4} \qquad \text{or} \qquad dV/(V_0 - V)^{1/4} = 2K^{1/2}\,dx$$

A second integration gives

$$\tfrac{4}{3}(V_0 - V)^{3/4} = 2K^{1/2}x + C_2 \tag{9.9}$$

For C_2, we have $V = V_0$ at $x = 0$, hence $C_2 = 0$. Resubstituting for K, and rearranging we get

$$V = V_0 - (\tfrac{3}{2}x)^{4/3}(\mu/2e)^{1/3}(j/\epsilon_0)^{2/3} \tag{9.10}$$

For electrode spacing $x = s$ and potential difference V_0, Eq. 9.10 yields for the current density,

$$j = \tfrac{4}{9}(2)^{1/2}\epsilon_0(e/\mu)^{1/2}(V_0^{3/2}/s^2) \tag{9.11}$$

This is the maximum current density we can accomplish due to space charge effect and we can use it as a design parameter.

For example, we can now investigate the maximum thrust per unit area with this in mind. We have

$$f = F/A = \dot{m}c/A$$

Now

$$\frac{\dot{m}}{A} = \rho c \qquad \text{and} \qquad \text{mass density} = \frac{\text{charge}}{\text{volume}} \times \frac{\text{mass}}{\text{charge}}$$

or

$$\rho = \rho_e(\mu/e)$$

Hence

$$f = \rho_e c(\mu/e)c = j(\mu/e)c \tag{9.12}$$

and with $c = [2(e/\mu)V_0]^{1/2}$ and j from Eq. 9.11,

$$f_{\max} = \tfrac{8}{9}\epsilon_0(V_0/s)^2 = \tfrac{8}{9}\epsilon_0 E^2 \tag{9.13}$$

Thus the maximum thrust per unit area depends only on the square of the field strength. Hence from this, we would like to use high voltages and close electrode spacing. We can obtain a measure of the motor size from Eq. 9.13. Taking a potential of 5000 V and a spacing of 0.5 cm, $f_{\max} \approx 7.85$ N/m² = 0.165 lb/ft², which means a large cross-sectional area for large thrust. From the exhaust velocity expression, Eq. 9.1, the value of c for 5000 V and a cesium ion of charge-mass ratio of 7.24×10^5 C/kgm is about 8.5×10^4 m/s or 280,000 fps. The corresponding specific impulse is 8700 sec. The current density from Eq. 9.11 is 67 amp/m² = .043 amp/in.².

From the energy relationship, $\mu c^2/2 = eV_0$ and $c = g_c I_s$,

$$f_{\max} = \tfrac{2}{9}\epsilon_0(\mu/e)^2(g_c I_s/s^2)^4 \tag{9.14}$$

I_s will be fixed by the mission and s will be as small as possible. Hence for high thrust per unit area, heavy particles are required, with a large potential gradient E.

It will be appreciated that the optimum design is not an easy matter. In general, heavy ions are desirable and it might be noted how this is just the opposite conclusion reached for thermal rockets, in which hydrogen is a desirable propellant.

The example previously given was fairly conservative, with $V = 5000$ V and $E = 10^6$ V/m. If we accept what appear to be probable outside limits of 50,000 V and 10^7 V/m, then for cesium ions the specific impulse is about 27,400 sec.

9.2 Efficiency of Ion Engines

The power efficiency of an ion engine is defined as the ratio of beam power P_j to beam power plus losses P_L, i.e.

$$\eta_j = P_j/(P_j + P_L)$$

Very little power is required directly for the ionization process itself but the ionization mechanism requires high temperatures and the heat loss from radiation can be large. There are certain other losses in the propellant conditioning process but these vary and in general the thermal loss is controlling.

The thermal loss is very largely radiation and we can express this by

$$P_L = \epsilon\sigma A T^4$$

where ϵ is the emissivity, σ is the Stefan-Boltzmann constant, A is the effective area of the ionizing source, and T is its absolute temperature. This expression supposes that the apparatus is completely surrounded by free space at a temperature T_0 of zero or so low as to make T_0^4 negligible compared with T^4.

We can express P_j as $\dot{m}_p c^2/2g_c$ and using Eq. 9.3,

$$\frac{1}{2g_c}\dot{m}_p c^2 = \frac{1}{2g_c} I \frac{\mu}{e} c^2 = \tfrac{1}{2} jA\left(\frac{\mu}{e}\right) g_c I_s^2$$

Thus

$$\eta_j = \frac{\dot{m}_p c^2/2g_c}{\dot{m}_p c^2/2g_c + \epsilon\sigma A T^4} = \left[1 + \frac{\epsilon\sigma A T^4}{\frac{1}{2} j A (\mu/e) g_c I_s^2}\right]^{-1} = \left[1 + \frac{2\epsilon(e/\mu)\sigma T^4}{j g_c I_s^2}\right]^{-1} \tag{9.15}$$

Hence for a given propellant at a particular source temperature, the power efficiency increases with increasing current density and specific impulse. Also, other factors equal, heavy ions are advantageous both directly and for the fact that they are associated with high current density.

It thus appears that low specific impulse may lead to high losses, as it is the square of I_s which appears in Eq. 9.15. This is indeed the case and Fig. 9.4 is a representative illustration. It would seem that for values of I_s under about 6000 sec, the power efficiency can be very poor. The power efficiency η_j is the conversion efficiency used in the performance analysis of Chapter 6 and the mission parameters will be considerably affected by low values. A value of about 85% may perhaps be considered as a representative target.

9.3 Design Factors

The major performance criteria have been established and now we will discuss briefly the design problems encountered in fulfilling them. The main areas may be given as follows.

1. Production of ions, including choice of propellant.
2. Production of a uniform, parallel ion beam accelerated to the necessary level.
3. Neutralization of the positively charged beam.

9.4 Production of Ions

There are at present three main methods of generating ions: (1) surface contact, (2) electron bombardment and (3) electric arc. The first-named method is historically the first, as it was developed and used to produce ions for scientific investigation of beams of particles, mostly of electrons.

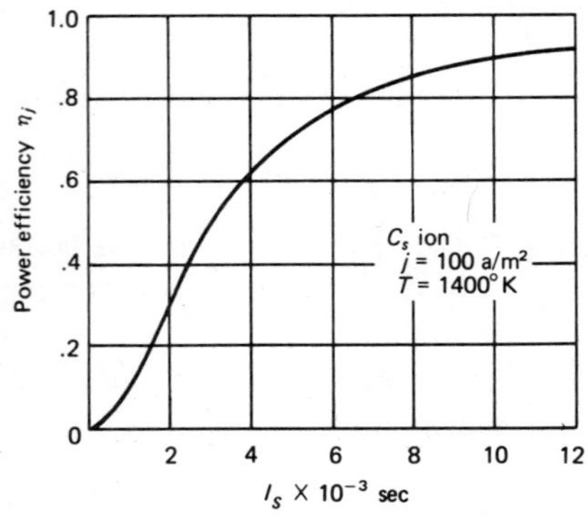

Fig. 9.4 Representative relationship of power efficiency and specific impulse.

The basic model for contact ionization is that of a metal surface removing an electron from a vapor in which it is immersed. The picture of a metal is that there are free electrons moving about, the conduction electrons. The energy to move such an electron to infinity is called the *work function*, ϕ. The conduction electrons near the surface create a field which is similar to that of half the field between unlike charges, i.e., the electron creates a field as if there were an image charge of equal and opposite nature outside the surface. An atom from the vapor of another element striking the surface of the metal adheres to it by polarization due to the field of the conduction electron acting on the nucleus. An electron in the outermost ring of the vapor element (the valence electron) may be detached and join the metal. Now if the vapor atom is caused to leave the surface, i.e., by evaporation if the surface temperature T_s is sufficiently high, it may leave the electron behind. The essential condition for this is that the *ionization potential* v_i of the vapor atom, that is the energy to remove the electron from the atom, is less than the work function of the metal. Thus the necessary conditions are that $\phi > v_i$ and that T_s be high enough to evaporate the vapor atoms rapidly.

The work functions of many possible metals are quite similar, platinum being about 5.4 eV, uranium 5.3 eV, nickel 5.0 eV, carbon 4.6 eV, tungsten 4.5 eV and chromium 4.4 eV, for example. The ionization potentials of elements which can be vaporized fairly readily vary quite considerably. Among the elements having the lowest values of v_i are the alkaline metals, lithium at 5.36 eV, sodium at 5.1 eV, potassium at 4.3 eV, rubidium at 4.16 eV and cesium at 3.87 eV. Mercury has a value of 10.39 eV. The practical criterion then becomes the ability of the ionizing metal to withstand continuous high temperature and to resist corrosion for long periods (months and even years). Platinum would seem desirable, but its ability to withstand continued high temperature is relatively poor, and while carbon also looks suitable, its mechanical properties are poor. The most suitable material to date seems to be tungsten. The required properties of the propellants are ability to be handled and to be readily prepared for ionization, i.e., storage, vaporizing, corrosion properties, and so on.

From the point of view of behavior in the ion engine, the alkali metal with the highest atomic weight is desired and, on this count, cesium is best at 133. Although it is not abundant and is costly to produce, it is the choice to date. Hence the tungsten-cesium combination is the one which has undergone most development in the contact-ionization process.

The ionization efficiency is important, that is, the ratio of ion production to reflection of neutral atoms from the metal surface. The neutral atoms can interfere seriously with the engine operation by upsetting the beam collimation process due to collision with ions and to impingement on the accelerating grid and elsewhere, leading to corrosion and erosion. The ionization efficiency is a function of temperature of the ionizing metal. The surface of this metal gets covered with vapor atoms and they must be ionized and removed quickly. A tungsten temperature of 1300–1400°K (1900–2050°F) is necessary and it will be appreciated that such incandescent temperature can lead to high radiation losses.

The design of ionizing geometries has been given considerable attention and much of it is empirical, as the atomic nature of the process at pressures where the mean free path of the cesium is only a few millimeters in length precludes exact modeling. The most successful method would appear to be diffusion of the cesium vapor through a very thin porous plate or wafer of tungsten formed by sintering tungsten powder. The object is to get the cesium atoms in contact with the tungsten surface and to avoid un-ionized atoms from passing through the pores and interfering with the accelerating process. In practice this seems to have been accomplished with success.

The electron bombardment technique depends on collision of electrons with propellant atoms in a chamber followed by preferential extraction of the atoms ionized by this process by means of electrodes at one end of the chamber. The successful Kaufman type of ionizer is shown schematically in Fig. 9.5. Electrons are emitted by the heated cathode filament and are attracted by a surrounding cylindrical anode. However, an axial magnetic field is applied so that they follow a spiral path in the space between cathode and anode and before they collide with a propellant atom. A fraction of such collisions produces ions and those ions near the end electrode are extracted. This electrode may also be the accelerating grid.

Ionization potential of the propellant should still be low for good efficiency but it need not be matched with a work function of some other material and thus elements of higher mass/charge may be used. Mercury with atomic weight 200 is a common choice.

The drawbacks of this method are that a magnetic field is required and that the cathode is subject to erosion and contamination. The former is not serious and the field may be supplied by a permanent magnet if the presence of a solenoid seems to require too much power. The cathode problem may be alleviated by using a pool of mercury as in conventional equipment, if the zero-gravity problem can be over-

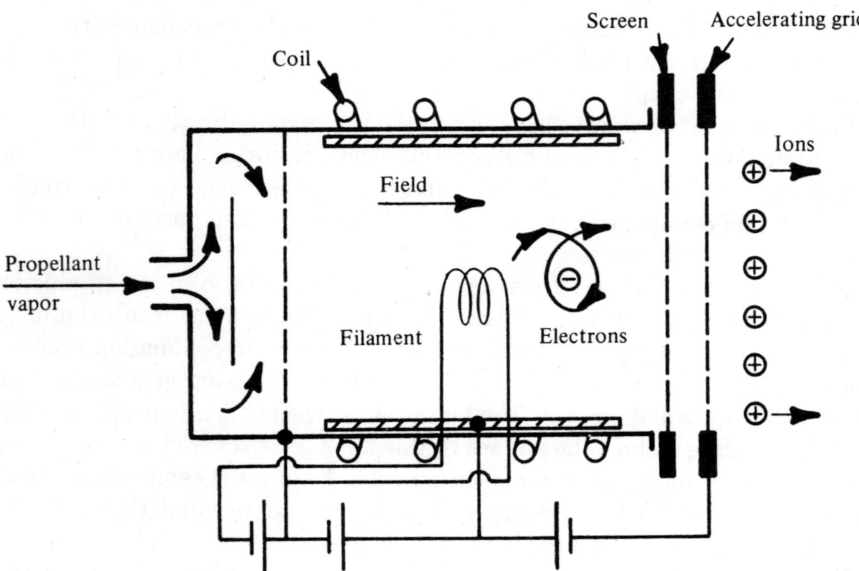

Fig. 9.5 Kaufman-type electron bombardment ionizer—schematic.

come, or by using a cesium compound for the filament together with cesium as propellant. The latter reduces contamination and sputtering but, of course, it limits the propellant choice.

Both contact and bombardment ionization methods seem to have about the same efficiency at present. The former can ionize a somewhat higher proportion of the propellant flow but suffers a higher thermal loss by radiation. The bombardment method is less efficient in the ionizing process and may produce more doubly ionized atoms which upset the beam process, but requires less power.

The arc-type ionizer known as a *duoplasmatron* is not so advanced in development for flight as yet, although it is not a new scheme. Basically it consists of a low-voltage arc through which the propellant flows, the arc being constructed by mechanical arrangement and by a magnetic field which allows a dense ion flux to be attained. The ionizing efficiency is very high but the power efficiency is low and the size and weight are large.

9.5 Acceleration of the Beam

The object is to assemble the ions from the ionizing source into a coherent parallel beam accelerating to discharge. The process is known as *beam optics*.

An ionized beam by itself tends to diverge because of the repulsion effect of like charges. There must be a minimum of impingement of the ions with any surface because of the severe erosion at the velocities in question and, of course, loss of useful effect. Impingement of the highly energetic ions on a surface causes sputtering—the ejection of several atoms for one incident ion. Over a long time this erosion can be quite destructive related to the small physical size of ion thruster components. Beam optics have been studied for many years in connection with other electronic devices, largely electron guns, and one of the major works in the field is that of Pierce,[1] which provides a guide to the problem for ion engines.

The solution of the theoretical problem is given by a sort of inverse reasoning, namely to find the field in charge-free space produced by a parallel, rectilinear beam of given properties and boundary conditions. The equipotential lines thus found may be represented by real electrodes and then it is supposed that a real beam passing through such would behave in the desired manner.

Within the beam and as a boundary distribution, we have the condition represented by the Poisson equation, Eq. 9.6, solved in terms of potential in Eq. 9.10. Outside the beam in charge-free space, we have the potential distribution in terms of the Laplace equation, i.e., $\nabla^2 V = 0$. A solution for this plane geometry is possible by utilizing the properties of complex analytic functions, but there is no closed solution for the cylindrical case. However, the latter can be handled by means of physically simulating the situation by the electrolytic-tank technique. The general result for either geometry shows that the equipotential surfaces are planes forming an angle with the axis as shown in Fig. 9.6.

For application to ion engines, the Pierce-type beam-shaping electrodes would have to be large, as they are many times the cross-sectional area of the beam. However, the ideal solution can be approximated by shaping the ion source as a

[1] J. R. Pierce, "Theory and Design of Electron Beams," Van Nostrand, Princeton, N. J., 1954.

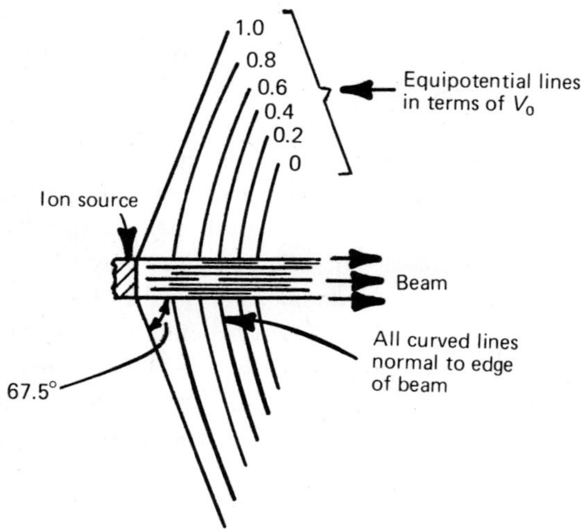

Fig. 9.6 Equipotential lines for ion beam shaping (from Pierce).

"dish" so as to provide the initial condition and to shape the accelerator grid in similar fashion. The simplest and most suitable geometries can be determined by analogue computers and direct test. It may be noted that the beam-shaping technique can be applied to thrust vectoring for maneuvering purposes by additional electrodes or by partial application of the full potential.

Two parameters have been found useful in characterizing beam geometry. One is the *aspect ratio* $\mathcal{R}$ and the other, allied to it, is the *perveance* $\mathcal{P}$. From Eq. 9.13 for the maximum specific thrust, we can obtain the total thrust by multiplying by the beam area taken in this instance as a circle of diameter d. Thus

$$F = f_{\max}A = \tfrac{8}{9}\epsilon_0(V_0/s)^2(\pi/4)\,d^2 = (2\pi/9E_0)\,V_0^2\mathcal{R}^2 \tag{9.16}$$

where $\mathcal{R} = d/s$, the aspect ratio. An aspect ratio of about unity is representative, as larger values would present difficulties in beam optics with simple electrodes.

The perveance is a parameter expressed in electrical quantities as the ratio $I/V_0^{3/2}$ which, in terms of aspect ratio, is given by the current-limiting equation (9.11) and the area again as $\pi d^2/4$, thus

$$\mathcal{P} = \frac{I}{V_0^{3/2}} = \tfrac{4}{9}\epsilon_0\left(\frac{2e}{\mu}\right)^{1/2}\frac{1}{s^2}\cdot\frac{\pi}{4}\,d^2 = \frac{\pi}{9}\,\epsilon_0\left(\frac{2e}{\mu}\right)^{1/2}\mathcal{R}^2 \tag{9.17}$$

Thus aspect ratio may be used in terms of given engine requirements of thrust, type of propellant, and specific impulse.

9.6 Beam Neutralization

The positive ions after leaving the plane of the accelerating electrode would, if left to themselves, slow down and at some point reverse their direction and return to the engine, due to the accumulation of the opposite negative charge on it. A plane-parallel beam of infinite extent would theoretically reverse itself in a distance from the accelerator less than that from source to accelerator. The reversal effect is mitigated by very small aspect ratio, $\mathcal{R} < 1$, but this is of little practical help because an actual engine of significant thrust will either have $\mathcal{R} > 1$ or be composed of many small thrusters of $\mathcal{R} < 1$ in parallel, effectively giving a large aspect ratio. It can be shown that the rate of charge build-up without neutralization is effectively instantaneous, as the vehicle will have a very low capacitance.

This problem of beam neutralization was a major one for a long time because of the complexities of theoretical analysis and the ambiguity of practical test, as it is very difficult to simulate the necessary free-space conditions for a length of time adequate for such a test in a laboratory on earth. The presence of surrounding walls, stray fields and particles, etc., no matter how good the supplied vacuum, renders such tests inconclusive. It now appears, however, that neutralization can be accomplished successfully.

The ideal solution is to inject a stream of electrons of the same uniform velocity as that of the ions, and of the same charge density, immediately behind the accelerator electrode. Unfortunately, this is not easy in practice, as electrons emitted from a heated filament have random velocities most of which are very much higher than ion velocities. Furthermore, direct placement of an emitter in the ion stream to obtain rapid mixing leads to problems of erosion due to ion impingement and again lifetimes of months and years will be required.

Jahn[2] distinguishes four methods that have been used for neutralization (Fig. 9.7). One is the direct filament insertion as stated above, (*a*) of the figure. A second method is to place the emitter to one side and have the electrons drawn into the beam by a potential difference, Fig. 9.7(*b*). This can be satisfactory if sufficient electric current is drawn into the beam at a potential sufficiently low so that any ions attracted to the emitter have low enough velocity to cause negligible damage. This is difficult for beams of low peripheral density. A third method is to use an electron gun which shoots a stream of the required density and velocity into the

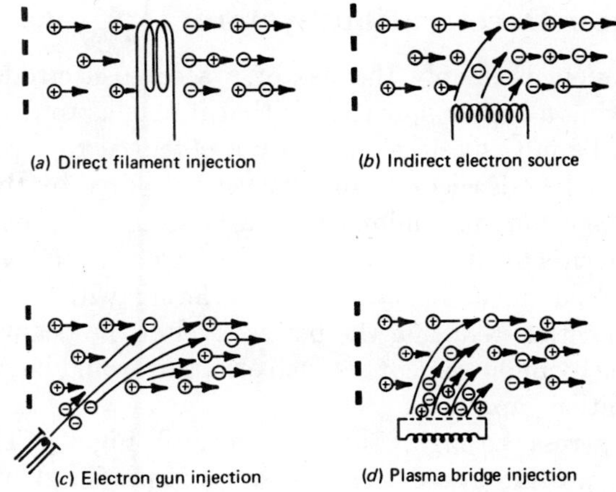

Fig. 9.7 Methods of beam neutralization (from Jahn).

[2] R. G. Jahn, "Physics of Electric Propulsion," McGraw-Hill, New York, 1968.

beam, Fig. 9.7(*c*). This has the drawback of additional complexity and power requirements. Finally, the electron source may simultaneously vaporize atoms of the propellant used, that is, an "autocathode" containing, for example, atoms of cesium or mercury in some form is used. The electrons ionize these atoms forming what is called a "plasma bridge" between emitter and beam. The particular design and operating conditions are factors in making one of these methods more suitable than others.

There was much speculation before 1964 as to the difficulty of beam neutralization or even the possibility of it under actual space conditions, but in that year the first ion engine was launched and operated. The beam thrust was indicated unequivocally by using it to alter the spin rate of the spacecraft by being tangentially mounted. The tests showed conclusively that beam neutralization was possible and, most importantly, the agreement of flight data with bench data showed that at least enough significant information on neutralization can be obtained in ground tests.

9.7 Optimum Specific Impulse

It has been implied that although in theory the propellant can be chosen to suit the mission requirements, in practice there is a severe limitation, with cesium or mercury being the alternatives. High voltages are desirable to obtain high current densities and heavy particles are desired to obtain the most efficient ion engines together with high values of specific impulse. If the particles are not sufficiently heavy, an optimum accelerating voltage will produce beam velocities higher than are required for the mission. Reducing the voltage reduces the efficiency and effectiveness of the motor. It is apparent that development of heavy-ion engines is required but this will take time. In the meantime, and perhaps always useful, the acceleration-deceleration principle can be used.

9.8 Acceleration-Deceleration System

The *accel–decel* system is simply the use of a second electrode beyond the accelerating grid having a higher potential so that it acts as to decelerate the beam. Thus the ions can be initially accelerated by a potential to give the desired high current density and then decelerated to a velocity required by the mission.* There is also the secondary but not unimportant advantage of providing a potential gradient which prevents upstream migration of electrons from beam neutralization. Such migration would have the effect of interfering with the potential in the acceleration gap and thus reducing the power level of the beam and also of damaging the ion source by impingement, as such electrons would be greatly accelerated once in the acceleration gap.

The potential across the engine then appears as in Fig. 9.8. The thrust per unit area is given by Eq. 9.11. The current density j is produced by the full applied

* One can make the analogy with the combustion process, in which a rich primary zone is created to provide the high temperature necessary for the maximum rate of combustion, with diluting air added thereafter to give the desired turbine inlet temperature.

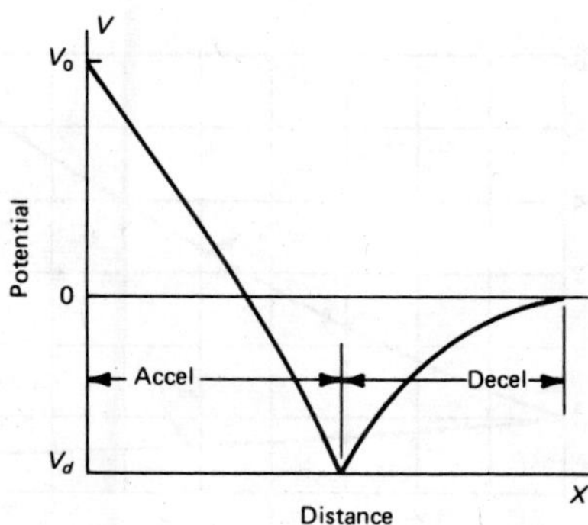

Fig. 9.8 Potential vs electrode spacing, accel-decel system.

potential from $x = 0$ to $x = s$, which is $V_0 + V_d$, and thus from Eq. 9.11,

$$j = \tfrac{4}{9}(2)^{1/2}\epsilon_0(2e/\mu)^{1/2}[(V_0 + V_d)/s^2]^{3/2}$$

The velocity c from Eq. 9.1 is given by the net potential, V_0, i.e.,

$$c = [2(e/\mu)V_0]^{1/2}$$

Hence

$$f_{ad} = j(\mu/e)c = \tfrac{8}{9}\epsilon_0[V_0^{1/2}(V_0 + V_d)^{3/2}/s^2] \tag{9.18}$$

Comparing this with the thrust produced for continuous acceleration through the full potential $V_0 + V_d$, which is

$$f_a = \tfrac{8}{9}\epsilon_0[(V_0 + V_d)/s^2]^2$$

and the ratio is

$$f_{ad}/f_a = [V_0/(V_0 + V_d)]^{1/2} = [(1 + V_d/V_0)^{-1}]^{1/2} \tag{9.19}$$

However, if the acceleration had been through the potential V_0 only, to give the desired velocity directly, then the ratio of thrusts would be

$$\frac{f_{ad}}{f_a'} = \left(\frac{V_0 + V_d}{V_0}\right)^{3/2} = \left(1 + \frac{V_d}{V_0}\right)^{3/2} \tag{9.20}$$

These values, which are ideal and will not be fully realized in practice, are shown in Fig. 9.9. The lower exhaust velocity induces more beam spreading and increases the problem of beam neutralization. However, it is definitely worthwhile as the figure shows.

9.9 Heavy Ions

The provision of suitable ions of elements is severely limited by charge-mass ratios and ionization potential. Hence there is interest in "synthetic" ions, i.e. particles of vapor, liquid or solid, ideally of suitable mass for the required duty given unit charge by special means. They are often referred to generally as *colloids*, although this

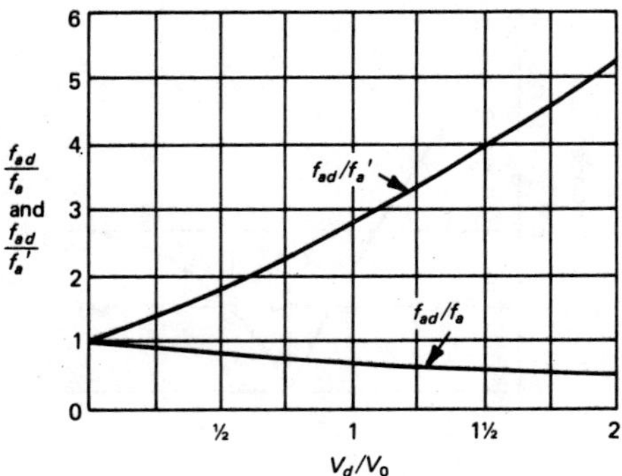

Fig. 9.9 Comparison of thrust per unit area for accel-decel and single potential systems.

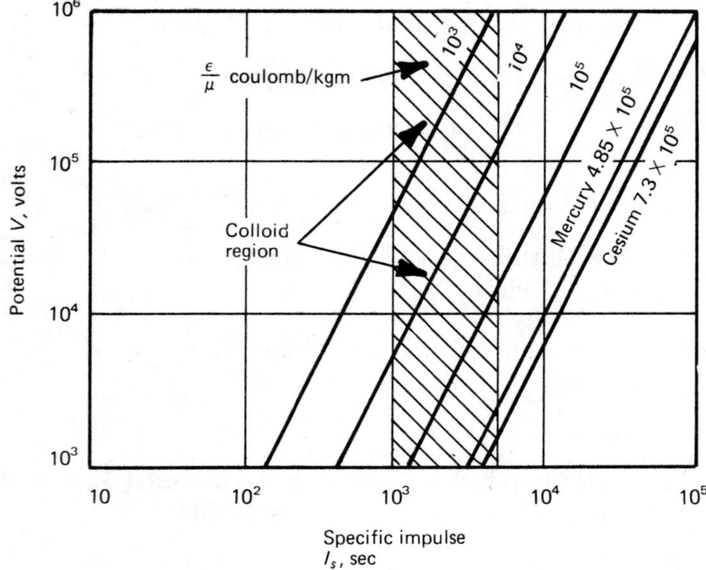

Fig. 9.10 Charge-mass ratio, potential, and specific impulse (from W. F. Braerman, "Electrical Propulsion Handbook," Vol. III, "Characteristics of Ion Propulsion," The Boeing Company, Seattle, Wash., 1963.)

covers only a part of the field, as aerosols are equally possible. Figure 9.10 shows Eq. 9.2 relating specific impulse and acceleration potential for values of charge-mass ratio. The field for heavy particles is shown shaded and it is apparent that there is need of propellants other than the presently used cesium and mercury.

From Eq. 9.14,

$$f_{\max} = \tfrac{2}{9}\epsilon_0(\mu/e)^2(g_cI_s/s^2)^4$$

With I_s fixed by the mission and s by the engine design, then the thrust per unit area is proportional to the square of the particle mass. A useful relationship can be obtained[3] by using the approximate optimum value of exhaust velocity from Sec. 6.2 as $c = (2)^{1/2}V_c/2 = (t/\alpha)^{1/2}$, together with an assumed limiting value of potential, then substituting the value for c into Eq. 9.2,

$$c = [2(e/\mu)V]^{1/2} = (t/\alpha)^{1/2}$$

and

$$(\mu/e)\big|_{\text{opt}} = 2\alpha V/t \tag{9.21}$$

Heavy particles are required when the specific mass of the power plant is large, voltage high and mission time short. For $\alpha = 10$ lb/hp, a time of one year and a limiting voltage of 50,000, then $\mu/e \approx .54 \times 10^5$ C/kg or an ion of atomic weight about 1800 (Cs = 133, Hg = 200). Note that for the future, α may be reduced, V may possibly be increased, and t may well be larger as more experience is gained and more adventurous trips are undertaken. This will reverse present requirements and tend to make lighter atoms more suitable.

Many methods of providing larger particles are available for trial, such as an oil mist, condensation of supersaturated vapor, grinding of solid material, and so forth—there are countless possibilities. Apart from the provision of relatively simple means of preparing, delivering and charging particles, there are three major restrictions.

The first such restriction is concerned with preparation and is that the particles must have a high degree of uniformity of charge-to-mass ratio. Nonuniformity is less serious for vehicles of low velocity increment and large payload ratio, but has an increasingly severe effect as velocity increases and payload decreases. The other two restrictions are concerned with charging the particles. A solid or liquid particle is subject to a mechanical stress when charged, the stress acting radially outward and tending to break up the particle. In solids, this stress is resisted by the yield strength of the material and in liquids by the surface tension. The third restriction is that of emission of electrons or positive ions from the particle if the adjacent field strength induced by the charge exceeds certain values. The limiting field strength is greater for positive ions than for electrons, for which it has a value of the order of 10^7 V/cm. The charge, particle size and limiting field strength for a particle of a given material are related, and the minimum mass-charge ratio is proportional to the particle diameter.

Colloid engines tend to be more efficient than either the contact-ion or bombardment type of engine. Referring back to Sec. 9.2, contact-ion thrusters have high radiation loss at low specific impulse and with light ions. Electron bombardment thrusters can lose power through recombination of ions. The colloid type of engine suffers neither of these losses and its efficiency can be higher. Also because the specific impulse required is relatively low in proposed uses of small thrusters in the near future, the ratio of power to thrust is low (Sec. 6.1). Both these factors tend to reduce the power supply requirements for a given mission and thus in the present state of development of such power plants, the colloid engine appears attractive on a system basis.

[3] E. Stuhlinger, "Ion Propulsion for Space Flight," McGraw-Hill, New York, 1964.

One example of a colloid thruster[4] has glycerol pumped through capillary tubing, the glycerol containing sodium iodide to increase its electrical conductivity. A positive potential of several thousand volts applied to the needle creates an intense field at the tip and a combination of this and the pressure and surface tension of the liquid produces a continuous stream of very fine charged particles. Considerable flexibility of the charge-mass ratio is achieved by variation of the potential and the conductivity of the glycerol.

[4] E. Cohen and M. N. Huberman, "Research on Charged Particle Electrostatic Thrusters," Air Force Aero. Propulsion Lab., AFAPL-TR-66-94, 1966.

CHAPTER 10

Electromagnetic Propulsion

The third group of electrically powered rockets consists of that wherein the acceleration is produced directly by electromagnetic fields, the $\mathbf{j} \times \mathbf{B}$ effect. The number of different types is legion and analysis most complex. In fact sometimes the effect defies analysis except in the very broadest terms because of the geometry of the device. The group is less well developed than the other two and apart from the very basics of electromagnetic thrust, there are no fully established methods of analysis one can say are representative. Because of this lack of a settled technology and supporting theory, the presentation here will not attempt to pursue the particular theoretical analyses of performance which have appeared for certain classes of unit, because it is not clear at this time which elements are valuable and relevant to the field as a whole. Where detail is given, it is mainly to show the type of problem encountered and the kind of analysis needed.

Two initial conditions must be met. The first is that the propellant must be electrically conducting and the second is that the propellant density be sufficient to allow a continuum type of analysis to be applied. Both these conditions are met if the propellant is a *plasma,* a mixture of molecules, atoms, positive ions and electrons, neutral overall. The main complexity of analysis lies in the fact that the electrical conductivity of a plasma is a tensor property and not a simple scalar. Thus in addition to the straightforward one-dimensional conduction in the direction of the electric field, there is conduction mutually orthogonal to the mutually normal components of the electric and magnetic fields, giving rise to *Hall currents.*

10.1 Fundamentals of E-M Propulsion

A very brief review of the fundamentals of the behavior of charged particles in electric and magnetic fields will be given, sufficient for an understanding of the main propulsion phenomena. First we will look at the behavior of an individual charged particle, the microscopic viewpoint, and then at the behavior of the plasma as a whole, the macroscopic viewpoint.

Consider a single electron moving in a plane perpendicular to a steady uniform magnetic field. As shown in Fig. 10.1, the magnetic field of intensity $\mathbf{B}$ (measured in W/m^2 or G) is in the z-direction and the electron has a velocity U_e in the xy-plane. The interaction of moving charge and magnetic field sets up a force on the electron

$$\mathbf{F} = \epsilon(\mathbf{U}_e \times \mathbf{B}) \tag{10.1}$$

that is, normal to both $\mathbf{U}_e$ and $\mathbf{B}$. As a result, the field changes the direction but not the magnitude of the velocity (no work is done) and the electron follows a circular

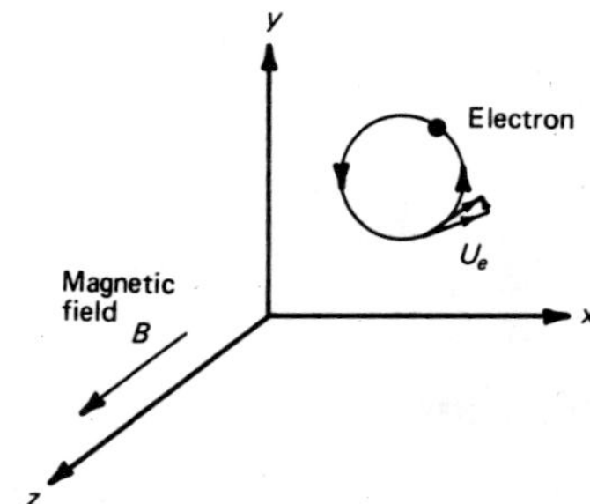

Fig. 10.1 Motion of a single moving electron in a magnetic field.

path at constant speed. We can find the path characteristics by equating the magnetic force and centrifugal force due to acceleration toward the center of curvature. Thus for the simple case shown, in scalar terms,

$$m_e U_e^2/r = \epsilon U_e B$$

and

$$r = m_e U_e/\epsilon B \tag{10.2}$$

With the angular velocity $\omega_e = U_e/r$, then

$$\omega_e = \epsilon B/m_e \tag{10.3}$$

which is called the *cyclotron* frequency (sometimes the gyro frequency). ω_e is important because it determines, among other factors, the effective conductivity of the plasma.

If now an electrostatic field of strength E is applied in the y-direction, an electron at rest experiences a force in the negative E or $-y$-direction. A moving electron, however, already has the circular velocity discussed above and the addition of the electrostatic field E results in a spiral motion in the x-direction, normal to both E and B. The force on the electron is given by

$$\mathbf{F} = \epsilon[\mathbf{E} + (\mathbf{U}_e \times \mathbf{B})] \tag{10.4}$$

and the resulting motion is usually validated by substituting for $\mathbf{U}_e$ the vector sum of a circular motion U' and a linear velocity $(\mathbf{E} \times \mathbf{B})/B^2$, thus

$$\mathbf{F} = \epsilon \left[\mathbf{E} + (\mathbf{U}' \times \mathbf{B}) + \frac{1}{B^2} (\mathbf{E} \times \mathbf{B}) \times \mathbf{B} \right] \tag{10.5}$$

Now $\mathbf{E} \times \mathbf{B}$ is in the x-direction (Fig. 10.2), and $(\mathbf{E} \times \mathbf{B}) \times \mathbf{B}$ is in the negative y-direction. Thus $(\mathbf{E} \times \mathbf{B}) \times \mathbf{B} = -B^2\mathbf{E}$ and

$$\frac{1}{B^2} (\mathbf{E} \times \mathbf{B}) \times \mathbf{B} = -\mathbf{E}.$$

Hence in Eq. 10.5, the $\mathbf{E}$-term is cancelled out and we are left with

$$\mathbf{F} = \epsilon(\mathbf{U}' \times \mathbf{B}) \tag{10.6}$$

which is similar in form to Eq. 10.1 representing the cyclotron motion. However, we have superposed the velocity **E** $\times$ **B** in the x-direction, so that the total velocity is the cyclotron motion plus the *drift* velocity in the x-direction. The result is a spiral motion with axis parallel to x.

The above is for a single undisturbed electron. In the plasma, the electron is continually colliding with atoms (collisions with positive ions are generally rare in the weakly ionized plasma) and this produces the general effect shown in Fig. 10.3. A collision instantaneously destroys the electron momentum and the electron being stationary, it is acted upon by the E-field in the $-y$-direction. However, immediately it acquires velocity it assumes the circular motion and acquires a component of velocity in the x-direction. Another collision repeats the process and *on the average,* the electron is moving so that it has an x-component as well as a y-component. This means that there is an x-component of current as well, and this is the so-called *Hall* current which is an outstanding feature of magnetohydrodynamic motion. Of course, it occurs in solid materials as in electric motors but then the plasma is replaced by individual wires, each insulated from its neighbor, so that net transfer in the x-direction is negligible. The significance of the Hall effect is that it complicates the behavior of a plasma, because here we have taken the simplest possible viewpoint and in reality, velocities and electrostatic fields may have components in more than one coordinate direction, not having the simple mutual perpendicularity depicted here. The result is that the current is three-dimensional or expressed differently, the electrical conductivity of a plasma is a

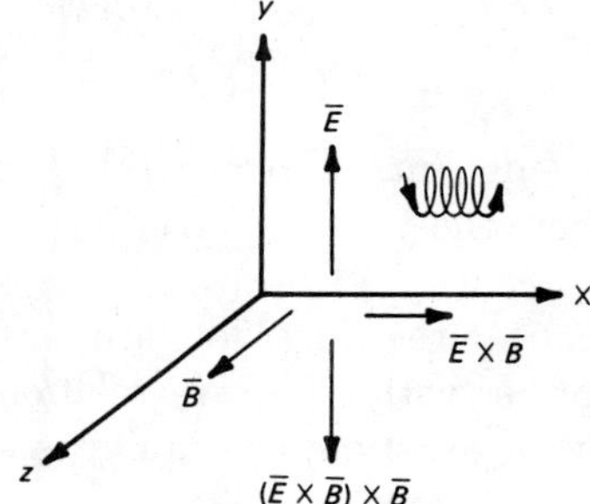

Fig. 10.2 Motion of a single moving electron in a combined magnetic and electric field.

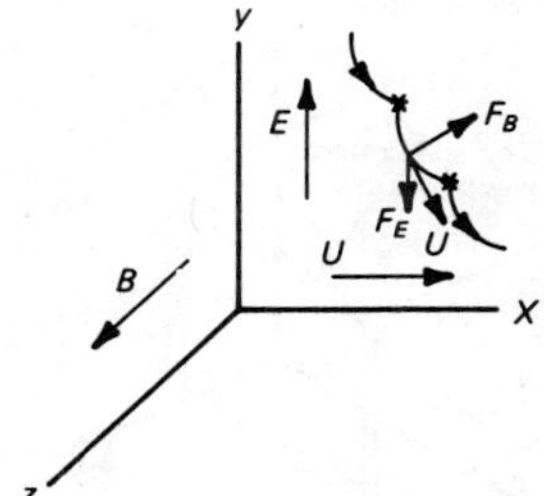

Fig. 10.3 Effect of collisions on a moving electron.

tensor, rather than a simple scalar. The Hall effect can be troublesome, but under certain circumstances it may be allowed for and a device may operate mainly with a Hall current rather than the normal current.

From the macroscopic view, the plasma experiences a force in the positive x-direction under the circumstances of the $\mathbf{E} \times \mathbf{B}$ field of Fig. 10.2. The *plasma* is accelerated because the electrons on the average are losing their momentum as shown in Fig. 10.3 and giving it to the atoms which are moved in the x-direction. No word has been said about the positive ions in the electromagnetic field, although they likewise experience similar effects. However, their mass is so much greater than that of the electron that their cyclotron radius is very large and their frequency very low, thus they play only a negligible part in the process as described above.

10.2 Steady-Flow, Channel Accelerator

A simple steady-flow channel accelerator based on the crossed-field principle, is shown schematically in Fig. 10.4.

The rectangular channel is formed of a top and a bottom electrode with side walls of electrically insulating material, permeable to the transverse magnetic field. $\mathbf{E}$, $\mathbf{U}$ and $\mathbf{B}$ are mutually perpendicular and the vector products can be replaced by simple scalar terms when necessary. The net electric field is $\mathbf{E} + (\mathbf{U} \times \mathbf{B})$, or $E - UB$ and this gives rises to a current of density $\mathbf{j}$ (amp/m^2) in the y-direction if the assumption is made that the electrical conductivity of the plasma is a scalar, i.e., parallel to $\mathbf{E}$. With this assumption, which we have seen from the analysis of electron behavior above is not strictly true, then we can write Ohm's law as

$$j = \sigma(E - UB) \tag{10.7}$$

where σ is the conductivity (mho/m) or reciprocal of resistivity. The body force per unit volume on the plasma, the Lorentz force, is given by $\mathbf{j} \times \mathbf{B}$ or jB for this case.

The governing equations for the duct flow are as usual those of continuity, momentum and energy, together with necessary relations of state and properties. For the one-dimensional case in question, continuity is expressed by

$$\rho U = G = \text{constant} \tag{10.8}$$

Again idealizing the flow as frictionless, momentum is expressed by the Euler

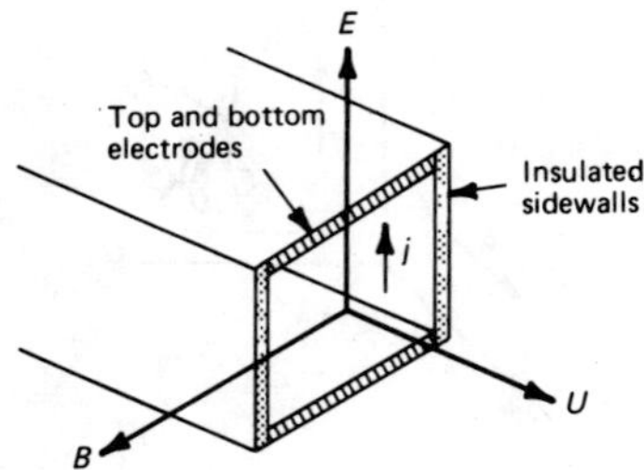

Fig. 10.4 One-dimensional, steady-flow channel accelerator.

equation plus a term for the Lorentz force thus,

$$\rho U(dU/dx) + (dp/dx) = jB \tag{10.9}$$

Since we are dealing with a gas, no gravity term is required.

For energy, we have the steady-flow energy equation. Assuming no heat loss, the flow is adiabatic in the sense that there is no Q-term. However, we shall see shortly that part of the electrical input is equivalent to a transfer of heat. There is no energy transfer as shaft work, so we have enthalpy, kinetic energy and electrical energy constituting the equation. Enthalpy is represented by c_pT for a perfect gas and electrical energy per unit volume by jE, hence

$$\rho U(d/dx)[c_pT + (U^2/2)] = jE \tag{10.10}$$

Note that although a magnetic field is present, it contributes no energy as the force on a charged particle is always normal to its velocity and hence can do no work.

Assuming the plasma behaves as a perfect gas as it is only very lightly ionized, we have $p = \rho RT$. Because of the temperatures involved, the specific heat is not constant but may be approximated as a mean value over restricted ranges. Likewise the thermal conductivity, while considered scalar, is dependent on pressure and temperature and either needs a state relationship or the use of mean values. Equations 10.8, 10.9 and 10.10 for continuity, momentum and energy, together with Ohm's law and relationships for c_p and ρ if required, represent the statement of conditions for this simplified crossed-field accelerator.

Because of the resistance of the plasma, as represented by Ohm's law, the energy input jE is only partly accounted for by useful work in accelerating the plasma directly, with the remainder being dissipated as "ohmic" or joule heating. The useful part is the Lorentz force jB times the velocity U and the dissipated part corresponding to the "I^2R" loss is j^2/σ, hence

$$jE = jUB + (j^2/\sigma) \tag{10.11}$$

A further simplification that is implicit in the preceding development is that the magnetic field B is a purely external field of arbitrary amount. Because of current flow in the plasma, internal magnetic fields are generated and Maxwell's field relations are needed. In this case, the dynamical equations of momentum and energy are coupled with the Maxwell equations and make solutions more complex. In order to gauge the relative importance of the induced field, a *magnetic Reynolds* number R_m is used which physically represents the rate of induced to applied magnetic field strength, B_i/B_a, and which may be shown to be proportional to the plasma conductance σ, the plasma velocity U and a characteristic dimension L, the channel length. Thus, with μ_0 the magnetic permeability of free space,

$$B_i/B_a \propto R_m = \mu_0\sigma UL \tag{10.12}$$

It has been found that for values of R_m less than unity approximately, B_i can be neglected relative to B_a.

The governing equations for steady, linear duct flow (10.8, 10.9 and 10.10), although representing very much simplified conditions, can still be solved analytically only with further restrictions such as constant Mach number or velocity, isothermal or adiabatic flow, constant electric or magnetic field, etc. Inasmuch as simple duct flow accelerators do not appear very promising for propulsion, the development will be pursued here only in outline as representative of the crossed field principle.

The theoretical results show that isothermal flow does not produce substantial acceleration and of course adiabatic flow in the sense usually used in these circumstances requires that the Joule heating due to plasma resistance be negligible. Variable field strengths along the channel produce a much better result but are difficult to carry out in practice. A diverging channel is beneficial, as constant area causes rapid heating of the plasma from energy dissipation and hence a choking effect. Increase of area is then helpful as in simple heating in a duct (Rayleigh process).

Hall effects are of course present and may detract from the performance. It is possible to attempt to neutralize the Hall current by dividing the electrodes axially with insulation between the segments and applying a small axial potential difference in addition to that in the y-direction. Another possible arrangement is to use the Hall current for acceleration by again segmenting the electrodes and applying only a streamwise potential gradient. The current is still in the y-direction, due to $\mathbf{U} \times \mathbf{B}$, and the force is still in the x-direction, due to $\mathbf{j} \times \mathbf{B}$, Fig. 10.5(a). In this case a coaxial rather than a rectangular geometry is helpful [Fig. 10.5(b)] as the current is re-entrant and electrode wear is therefore reduced.

Whatever the geometry, real fluid effects cause considerable departure from the one-dimensional conditions postulated. Both viscous and thermal boundary layers are developed which produce marked effects in the generally small sizes of channel in question. The viscous effects produce a nonlinear velocity profile and the thermal effects produce a temperature gradient and hence variation of electrical conductivity. These variables are interdependent and thus the property distortions can be large. Furthermore there can be strong end effects, as at inlet and outlet the abrupt discontinuities of electric and magnetic fields distort the flow pattern and as L/d ratios for accelerators should be small, these effects are proportionately severe.

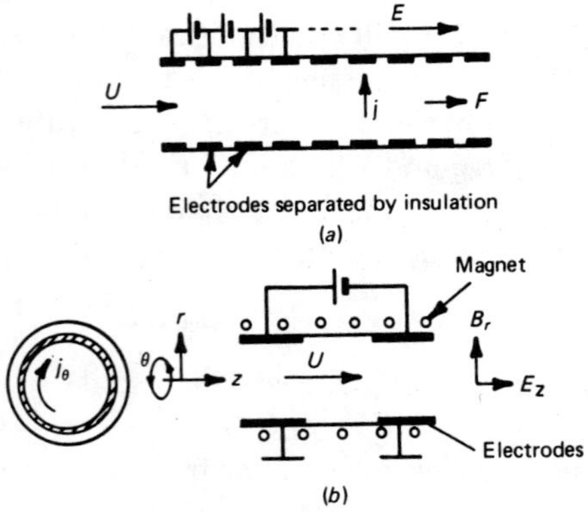

Fig. 10.5 Hall current accelerators: (a) linear model; (b) coaxial model.

Wall and end effects, together with the necessity of providing an external magnet and auxiliary power supply thus seem to discourage simple linear crossed-field accelerators.

The idea of operation by self-induced magnetic fields is attractive because it eliminates the necessity of providing an external magnetic field. Also, without an external magnet, shorter channels are possible and hence a reduction of wall losses. Some interest is centered in what is called the *magnetoplasmadynamic arc*, since its development arose from the electrothermal arc discussed in Chapter 8.[1] A unit of the latter type, if operated at a very low pressure and high current density can give exhaust velocities of the order of 3×10^5 fps, much higher than can reasonably be accounted for by simple plasma heating because the temperatures would have to be impossibly high. Although there may be some element of this, the main contribution must come from a $\mathbf{j} \times \mathbf{B}$ effect by high-intensity arc currents self-generating a magnetic field. There are indeed many unexplained phenomena and it is difficult to give a straightforward analytical explanation of the effect in quantitative terms. Figure 10.6 shows a model of this type of acceleration and it will be seen that the geometry makes analysis difficult. Specific impulses of several thousand seconds have been attained with efficiency up to 50% and without the electrode erosion which might be expected at such levels of power density.

10.3 Pulsed Plasma Accelerators[1]

The idea of unsteady electromagnetic acceleration stems from a number of considerations. Among them are (1) the possibility of very high current density and thus good efficiency, with intermittent operation which limits the power supply and minimizes electrode erosion and (2) to take advantage of self-induced magnetic induction. Also perhaps there was originally a background of experience in intermittent plasma devices used in different circumstances but which were able to provide a starting point for experiment with some analytical background.

The basic concept, which results in many different geometries, is (1) that of building up electrical energy which discharges at a given level as from a capacitor, this energy being concentrated in a small region to provide high intensities; (2) acceleration of this discharge region under the action of the $\mathbf{j} \times \mathbf{B}$ force auto-

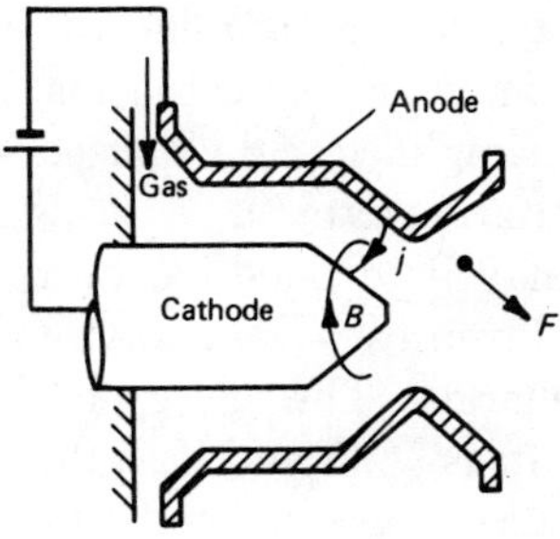

Fig. 10.6 Magnetoplasmadynamic arc accelerator.

[1] See R. G. Jahn, "Physics of Electric Propulsion," McGraw-Hill, New York, 1968.

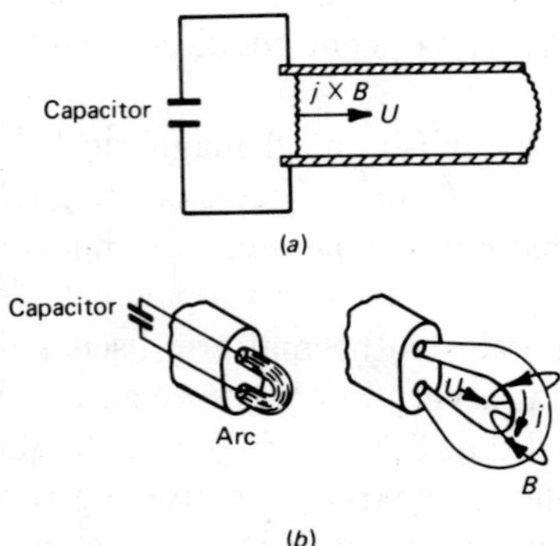

Fig. 10.7 Pulsed plasma accelerators: (a) arc-rail type; (b) button-gun type.

generated by the discharge; and (3) breaking of the circuit by free escape of the individual discharge. Perhaps the simplest manifestation to understand is the "arc-rail" accelerator shown schematically in Fig. 10.7(a). Discharge of the condenser initiates an arc as shown and the high-density current in the arc induces the associated magnetic field. The resulting $\mathbf{j} \times \mathbf{B}$ or Lorentz force causes the discharge region or arc to travel to the right, accelerating as it goes. When it leaves the electrodes, another arc is struck and the phenomenon is repeated at a high frequency to give a finite thrust. Another geometry representative of the same general behavior is the "button gun" type shown in Fig. 10.7(b).

If the propellant can be localized at the arc discharge region, the process can be modeled as that of a *slug* of constant mass during the discharge process. Such an idealization is closely approximated by the exploding wire device, in which the initial, very large surge of current vaporizes a wire running between the rails or plates. On the other hand, a discharge across the space between the rails initially embodies a small amount of propellant gas but which accumulates further mass as it travels through the propellant to escape. This is called a *snowplow* model. This can be extended to a gas dynamic model, treating the discharge as a piston which creates a shock wave traveling ahead of it, and thus existing shock technology can be used as a diagnostic tool. Such models as these can be used to elucidate certain overall phenomena and can be used as a guide to development. Sometimes they show where limitations may arise or where the maximum gain is to be attained. However, inevitably they are simplified and idealized representations and cannot be expected to yield exact quantitative data.

An overall efficiency η may be defined as the rate of thrust power P_j to the electric power supplied. P_j is given by $\dot{m}_p c^2/2$, where $\dot{m}_p$ is the mean mass flow rate over a finite number of discharges. The energy of a condenser of capacitance C charged to a potential of V volts is $\frac{1}{2}CV^2$ and for a rate of energy input, this must be multiplied by the number of discharges per unit time, ν. Thus

$$\eta = \frac{m_p c^2/2}{\nu C V^2/2}$$

This can be put in more readily observable quantities by substituting thrust

$F = \dot{m}_p c$, hence

$$\eta = F^2/\nu \dot{m}_p C V^2 \tag{10.13}$$

This overall efficiency is made up of several component efficiencies relating to the various processes in any one particular type of accelerator and it is difficult to change one without affecting another. That is, they are to a large extent interdependent. Up to now, then, improvement of efficiency is very largely an empirical matter. One may separate out three major processes in the operation of a pulsed accelerator, and discuss them in turn although recognizing the previous remarks about mutual dependence. These are initiation or establishment of the discharge, its acceleration and its discharge or exhausting.

Regarding initiation, using the linear rail model as example, closing the circuit causes at first a glow discharge over the whole rail. The current density increases together with an electric field component opposing the applied potential, produced by the rate of charge of the accompanying induced magnetic field. These effects are more pronounced at the channel entry, with the result that the arc proper occurs first there and localizes the effect. Again it is difficult to predict quantitatively the process because of the variation of electrical conductivity of the plasma which is so dependent on its thermodynamic state, which in turn is a function of the electrical charges occurring in it. The current density established is a function of the circuit constants, that is, the applied field and circuit inductance. Desirable rates and intensities of current are a balance between effective propagation and loss due to electrode damage. Some of the energy supplied is used for ionization, but this is not great. One trouble is that in pulsed operation the cathode does not get very hot, which is good from some points of view (radiation loss for example) but does inhibit the emission of ions by thermionic effect. A result may be that electrode damage occurs because very high local fields are necessary to produce the required electrons. This feature is instanced as an illustration of the type of difficulty which may occur in a particular instance and which it is difficult to predict, at least as to degree of severity. Since there are so many of these phenomena, development is very much of a test-bed nature.

The acceleration of the plasma is subject to certain inherent dynamical limitations and to account for these requires the use of a definite model as discussed earlier. Using the snowplow model, we have initially a quantity of plasma contained in the discharge but when the column moves under the action of the Lorentz force, it experiences a drag due to inelastic collision of the particles in the rest gas with those of the moving discharge column. One can define a dynamical efficiency η_d which relates the rate of increase of the kinetic energy of the plasma used for propulsion to the dissipation effect of the inelastic collisions. Then one can examine velocity programs to establish trends in this efficiency and hopefully look for an optimum. Higher values of efficiency are found for accelerating rather than constant velocity programs, for channels of decreasing area, and for decreasing density in the flow direction. The energy dissipated to thermal energy is not necessarily all lost, as part of the unit may operate as a pure expansion device, wherein thermal energy is useful. There can be an effect analogous to "reheat" in turbines.

The ejection process is of interest for its possible effect on upstream processes and for the extent to which it directly contributes to the total thrust. It has been found that relatively large exhaust openings can be used without deleterious effects on the development upstream of initiating the discharge and of accelerating the main body of flow. Currents and magnetic fields do extend out into the exhaust "plume" and do contribute thrust, and some of the energy turned to enthalpy can be recovered.

It has been found that practical pulsed units are more effective if designed so that the plasma sheet and induced magnetic field are closed within the channel. With a linear channel the magnetic field has to close externally to the working area and distortions of the current sheet lead to nonuniform forces on the plasma. Radial symmetry can improve these aspects and there is opportunity to arrange the current, the induced field and the consequent acceleration in a number of ways. One of those currently in development is the *coaxial gun*, Fig. 10.8(a), with radial electric field and current, circumferential magnetic field and axial acceleration. This is a convenient geometry, but j and B both vary as $1/r$ and thus the thrust density varies as $1/r^2$. Another type is the *pinch accelerator*, Fig. 10.8(b), with axial electric field and current circumferential magnetic field and radial thrust. The last-named is inconvenient as the flow has to be turned from the radial to the axial direction but provides a uniform thrust over the area and is proved to be more stable. The turning of the flow can be accomplished quite readily and in fact it appears that it will occur naturally by the induced field forming a magnetic nozzle for it.

10.4 Traveling-wave Accelerator

It remains to note a third differentiated class of accelerator known as the traveling-wave type. It is based on the idea of inductively coupling the electrical energy with the gas, thus eliminating, or at least minimizing, electrode wear which is likely to be one of the major problems for long-time operation of any of the electric or electromagnetic engines.

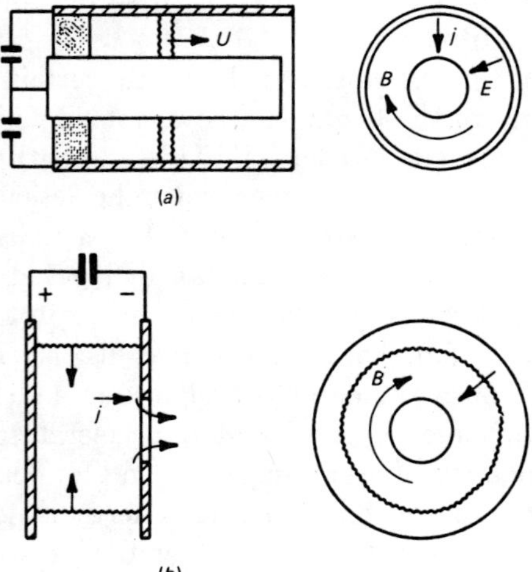

Fig. 10.8 Pulsed accelerators with radial symmetry: (a) coaxial gun; (b) pinch type.

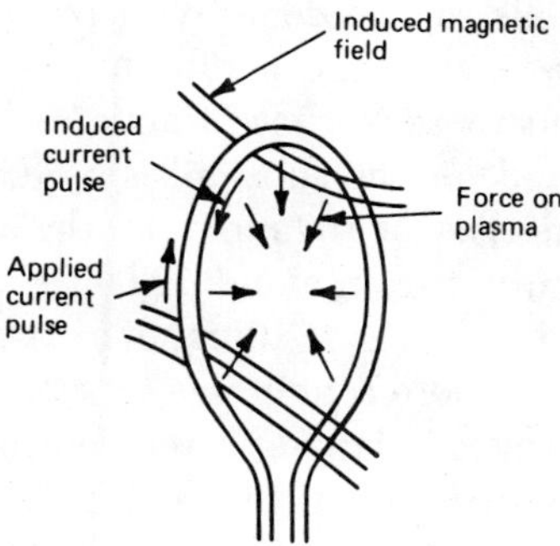

Fig. 10.9 Principle of the traveling-wave accelerator.

The traveling wave accelerator may best be understood by first considering a single pulsed inductive unit as shown in Fig. 10.9. A current pulse in the loop initiated by closing the circuit sets up an increasing induced magnetic field with lines of force axial (in the plane of the loop). By Faraday's law, the induced electric field allows a current to be established circumferentially in the plasma but oppositely directed to the applied current in the coil. Now, this induced current and the induced magnetic field give a $\mathbf{j} \times \mathbf{B}$ force to the plasma which is directed radially inward. This compression must result in an axial flow near the axis and hence thrust in a suitable geometry.

As the plasma moves away from the plane of the loop, the induced current and applied current move apart and the coupling necessary to produce the effect is inherently weakened. The efficiency thus tends to be low. To overcome this deficiency, a number of such coils may be placed in series with a suitable external switching or pulsing device to allow the magnetic field to "push" the active sheet of plasma successively along the axis. This then becomes the basic idea of a traveling wave accelerator. A considerable diversity of geometries has appeared, some with fixed magnets, some with ferrite yokes to help to contain the magnetic flux. Conversion efficiencies are low, of the order of 20–40%, and troubles can arise from wall effects and very high local plasma temperatures from Joule heating.

10.5 Propellants

Little has been said of the propellant in the foregoing discussion beyond the fact that it was to be a plasma, that is, a lightly ionized but overall neutral gas. Various gases have been used, often hydrogen, helium or argon, but all have to be made conducting in order to allow current flow. The ionization has been assumed in the above but in practice it is one of the most difficult aspects of electromagnetic propulsion. This is because there is extremely little ionization of such gases below temperatures of several thousands of degrees. If we take 5000°F as an absolute top temperature for container materials, the conductivity of argon, for example, is almost negligible and it would have to be of the order of 10,000°F to be suitable for most accelerators. Since this is clearly impossible, it is necessary to induce conductivity in other ways. This can be done by introducing a small quantity of an easily ionizable substance, one of the alkali metals, for example, discussed in Chapter 9. It turns out that there is a seed fraction which gives maximum ionization, hence conductivity, and for the substances used, this fraction is very small. Thus for argon for a conductivity of 100 mhos/m, the addition of 1% of potassium halves the required temperature from about 10,000°F to 5000°F. For comparison, the conductivity of copper is about 6×10^7 mho/m at room temperature.

The plasma can be made more conductive by causing it to be in a nonequilibrium state with the free electrons having a temperature above that of the other constituents. This can be brought about by passing a discharge through it prior to entering the accelerator but this requires low mass densities for the electron temperature to remain at a higher level for any useful period of time.

10.6 Performance of Electromagnetic Accelerators

To date, electromagnetic accelerators have been developed only to a laboratory stage and there are many problems to be overcome. It might then be asked if development is worthwhile in view of the more advanced and apparently successful state of the ion engine and of the arc accelerator for the lower velocities. One of the main arguments in favor is the possibility of much higher thrusts per unit area. Although the propulsion is essentially by electrical energy, there is no space charge effect with its consequent limiting of current density. We can make a comparison of the order of thrust density levels quite simply. The Lorentz force is $\mathbf{j} \times \mathbf{B}$ and from a form of Ampere's law we have

$$\mathbf{j} = (1/\mu_0)(\nabla \times \mathbf{B})$$

Thus the force becomes

$$(1/\mu_0)(\nabla \times \mathbf{B}) \times \mathbf{B}$$

which by the rules of vector calculus for triple products becomes

$$(1/\mu_0)[(\mathbf{B} \cdot \nabla)\mathbf{B} - \tfrac{1}{2}\nabla\mathbf{B}^2]$$

For the one-dimensional case of simple channel flow, $\mathbf{j} \times \mathbf{B} = jB$ and the triple product becomes simply $(1/2\mu_0)(d/dx\ B^2)$, i.e.

$$jB = -(1/2\mu_0)(d/dx\ B^2)$$

Recalling the Euler equation (10.9), we then have

$$\rho U \frac{dU}{dx} + \frac{dp}{dx} = jB = -\frac{1}{2\mu_0}\frac{d(B^2)}{dx}$$

Integrating,

$$(\rho U^2/2) + p + (B^2/2\mu_0) = \text{constant} \tag{10.14}$$

The term $B^2/2\mu_0$ has the dimensions of a pressure and is called the *magnetic pressure*. It is representative of the thrust per unit area.

It will be recalled that for the electrostatic accelerator, it was found that the maximum thrust per unit area was $\frac{8}{9}\epsilon_0 E^2$. Using the reasonable value of E of 10^6 V/m, this gives a thrust of 7.85 N/m^2. The magnetic pressure, however, even with a conservative flux density of 0.1 Wb/m^2 (1000 G), has a value of about 4000 N/m^2, which is the order of 500 times greater. Even if E is increased to the suggested limit of 10^7, the magnetic accelerator provides a greater thrust even with a conservative value of B. We might note that the electrostatic type is space-charge limited whereas there is no similar limitation on the electromagnetic type. For very small units, the *microthrust* type used for station-keeping, etc., the size of the thrust unit itself is inconsiderable. However, as the propulsion unit for flights to distant planets, where possibly several hundred newtons of thrust may be needed, then the ion engine becomes of very large cross section.

Because electromagnetic accelerator units reported have nearly all been bench models to test out methods rather than usable designs, it is difficult to give representative performance data. Power efficiencies are generally low for many types, 20–50%, although up to 70% has been reported. Small units of the steady magnetoplasmadynamic type have operated with thrusts of the order of one newton at specific impulses from about 800 to 6500 sec, with efficiencies from about 12% to 36%.

It must be remembered that it is the *system* performance that is critical for the vehicle and so the power plant mass and the propellant mass are the necessary parameters. The accelerator itself may be only a small part of the whole but, on the other hand, its efficiency, type of propellant and required tankage may be controlling. Another consideration is the form of the input power required. That is, power at say 100–200 V might be more acceptable than at 2000 V, as it could be used directly from the generator or alternatively a different type of generator might be possible. Thus any *power-conditioning* equipment must be taken into account in the overall balance.

At this time it is not possible to make any firm assessment of the possibilities of electromagnetic plasma accelerators. They appear to be more suited in larger sizes where frontal area is important and where self-induced magnetic fields can be utilized. The manifold ways of utilizing the $\mathbf{j} \times \mathbf{B}$ effect have still not been exhausted in spite of the many designs initiated and discarded. It seems highly likely that suitable models will be developed even if empirical development outstrips full theoretical understanding.

CHAPTER 11

Space Power Generation

11.1 Requirements

The mission analyses of Chapters 5 and 6 showed that a separately powered engine of very low thrust and propellant consumption is a necessity for journeys of long duration such as those of interplanetary travel. Chapters 8, 9 and 10 discussed types of engine—electrothermal, electrostatic and electromagnetic—which appear to have potential for these missions. All of them require a supply of electric power but the characteristics of this power may vary over a wide range. Direct current appears most probable but by no means exclusive. Voltage and current requirements may vary from perhaps 100 V at several hundred amperes to 50,000 V and milliamps of current. Thus there may well be a range of power sources to provide these extremes or, alternatively, power-conditioning equipment may be an important element of the system. Nuclear fission power is possible either as a source of heat for indirect transfer to a propellant, as in a conventional land-based solid-core reactor plant, or as a gas-core reactor in which the propellant is heated directly by the fission reactions of the gaseous nuclear fuel.

As pointed out earlier, the higher the required specific impulse (exhaust velocity), the greater is the power required for a given thrust. Thus, in review,

$$\text{Power } P_j = \frac{\dot{m}_p c^2}{2g_c} = \frac{\dot{m}_p c}{g_c} \cdot \frac{c}{2} = \frac{F g_c I_s}{2}$$

and for a given thrust, power is directly proportional to specific impulse. Using a thrust of 1 lb with a specific impulse of 5000 sec as standard values for a basis of comparison,

$$P_j = \frac{1 \times 32.2 \times 5000}{550 \times 1.341} = 218 \text{ kW/lb thrust at 5000 sec}$$

With a thruster efficiency η_j of 75%, this is close to 300 kW_e for each pound of thrust at 5000 sec and, as power conversion equipment may have overall efficiencies as low as 6–7%, the provision of power for even a modest main propulsion unit can be a formidable undertaking. In addition the power plant must be capable of continuous operation without maintenance for long periods, perhaps months on end, although there are many applications in which power may be required on demand over a long time but only intermittently, as in orbit correction, for example. The basic specific energy of the energy source is an important consideration but only as a guide to possibilities, because it is the final packaged system that counts, that is, the efficiencies of conversion, the shielding, the rate of energy release (power), and so forth, which determine the mission suitability.

11.2 Energy Sources

We may divide energy sources into onboard (or in-vehicle) types and outboard (or ex-vehicle) types.

For the former, there are chemical sources and nuclear sources. Chemical processes in which thermal energy is released to heat the propellant directly are not considered here as the so-called chemical rocket has been dealt with fully in Chapter 7. However, electrochemical processes can be a major source of power in this connection either in the form of batteries or fuel cells. Nuclear sources can be either in the form of a fission reactor, a fusion reactor, or a radioisotope generator. The fission type can have a solid core and deliver thermal energy directly to a propellant or to a substance which acts as heat transfer agent to the working fluid of a dynamic heat engine or directly to such a working fluid. Alternatively, it can have a gaseous core with the fission products mixing with the propellant. The fusion type of reactor has the greatest potential for energy release but the problems in successful control of the reaction appear so great as to make its feasibility remote for the foreseeable future, particularly for space applications where size and weight are at a premium and reliability is paramount. Radioisotopes, on the other hand, are in use for space power, mainly in conjunction with thermoelectric equipment to generate electric power directly, although the thermal energy consequent on the conversion of the kinetic energy of the emitted particles can be used for a dynamic heat engine.

The only energy source outside the vehicle contemplated at the present time is the sun and this again can be used directly to generate electricity or indirectly as thermal energy for a conversion system. Some of the possibilities for onboard energy are given in Table 11.1. These are taken from Szego[1] and represent only a selection from his comprehensive data to show the possible range.

The table shows an astonishing range of specific energy of sources, many of which are in use. Thus both batteries at about 80 W-hr/lb and radioisotopes at

Table 11.1

Energy Source	W-hr/lb
Batteries	
Ni–Cd, slow discharge	17
Ag–Zn, slow discharge	80
H_2O_2 decomposition	510.3
Gasoline-air	1260
H_2–O_2 combustion	1692
Li–F_2 combustion	2961
Hydrogen—free radical recombination	27,000
Radioisotope-polonium	3×10^8
Uranium fission	10^{10}
Nuclear fusion	5×10^{10}
Matter annihilation	10^{13}

[1] G. C. Szego, "Space Power Systems", Chap. II in "Space Power Systems", Part I, AGARDOgraph 123, Propulsion & Energetics Panel, AGARD, 1969, p. 52.

3×10^8 W-hr/lb are in service. That such extremes could both be utilized is a consequence mainly of the specific *power*, that is, the rate at which energy can be released, together with the ancillary equipment necessary to supply the energy in the form required.

The second group of sources shown Table 11.1 are all chemical and are included for comparison. It should be noted here that the figures are the basic energy available and additional equipment may be necessary to obtain it as electrical energy in the form required. Radical recombination energy stems from its converse, where as dissociation it is generally troublesome (*vide* Chapter 3). If the radicals could be stored as such, then the reassociation energy release could be used for an energy source, as the values per unit mass are high compared with conventional reduction–oxidation reactions. Thus from the table, the reaction $H + H \rightarrow H_2$ gives some 16 times the specific energy that the $H_2 + O_2$ combination reaction yields. However, the difficulties of storage and extensive equipment required for stabilization of radicals are such that the idea is not thought to be promising at this time. The highest concentrations of free radicals yet successfully stabilized are much below 1%.

The final group, the nuclear sources, have the highest energy potential by orders of magnitude but are limited in uses by heat transfer, power output or the ancillary equipment needed for practical utilization.

Some of the existing energy sources and the more promising schemes under development will now be discussed by type and then followed by some analysis of the general characteristics of conversion devices to produce electricity for the various types of thruster.

11.3 Batteries

Although Table 11.1 shows that batteries have the lowest specific energy of any energy source used, they have provided much of the electrical energy in space missions to date (nearly all for auxiliary power, not for propulsion). They will continue to be necessary pieces of equipment even when more sophisticated sources are readily available, partly because of their versatility in rate of supply, instant availability and ruggedness, but mostly because there will probably always be a need for storing energy for later use.

It is customary to distinguish primary, or nonrechargeable, cells and secondary cells that are rechargeable. The distinction is basically electrochemical reversibility but the distinction is not always clear as some "primary" cells are rechargeable to a limited extent, while some secondary cells are in effect primary because of inefficient rechargeability.

Some cells deteriorate in storage or have limited "shelf life" as there are self-discharge reactions between the electrode materials and the electrolyte. The term "reserve" battery is used when the electrolyte is added immediately prior to use. A reserve battery will have higher specific energy but it is seldom possible to arrange this in space applications without further complications of automatic activation.

All cells are temperature sensitive, a fact which is known to most of us who have had automobile starting troubles in cold weather. There is a minimum and maximum temperature at which the cell will operate and the specific energy and power are functions of temperature. Thus a typical zinc-silver oxide battery has a useable top temperature of about 165°F, while the minimum useable temperature depends on the discharge rate required. Between 80°F and 32°F, the rate diminishes sixfold and from 32°F to 0°F, it is reduced a further thirtyfold. Thus there are problems of environment for use in space vehicles.

Of the so-called primary batteries, the major type in use is the Zn–AgO, which is used as a "reserve" battery. Its normal operating voltage is 1.4 V and its specific energy varies from about 40 W-hr/lb for very fast discharge ($\approx$30 min) to about 100 W-hr/lb for slow (50 hr) discharge. If the silver oxide is replaced by pure oxygen, forming a kind of battery-fuel cell hybrid, the specific energy can be considerably increased, up to the order of 150 W-hr/lb but this requires a supply of oxygen to be carried with it. Because alkali metals have greater electrochemical potential than the metals usually used as anodes, such as zinc and lead, attention is being given to their use. As they react with aqueous electrolytes, it is necessary to use organic liquids. Higher specific energy cells have been developed, but shelf life is very short and discharge rates are low.

The two main types of secondary battery are known as nickel–cadmium and silver–cadmium, both nickel and silver being in the form of oxides. The former has been used most extensively, in spite of its low specific energy, because of its stable voltage, storeability, high rate chargeability and low temperature capability. It has extremely good cycle life, up to 3 years with 60-min charge periods every 90-minute orbit. Ag–Cd batteries have higher specific energy, about twice that for Ni–Cd, but require longer charging time and this may be out of place on low orbit vehicles. One overriding capability is that of being nonmagnetic, an attribute that is imperative for some other equipment. There is a place for both types of cadmium battery dependent on mission characteristics, chiefly charging time allowable.

11.4 Fuel Cells

Fuel cells are electrochemical devices in which fuel and oxidant are fed continuously to produce electricity. A fuel such as hydrogen is fed to an anode where it ionizes with the H^+ ions going into solution in an electrolyte containing the anode and a cathode. Oxygen is fed to the cathode, where it combines with water in the electrolyte and electrons from the electrode to form hydroxyl ions, OH^-. These are neutralized by H^+ ions to form water. If the electrodes are connected externally, a current flows and the reaction proceeds continuously as long as the circuit is closed and the reactants supplied and consumed. Figure 11.1 shows this schematically. Output is about 1 V, with current densities of about 100 mA/cm^2 for sustained use at the present time, with hope for improvement of the latter.

Fuel cells are operated isothermally and are not heat engines, hence their efficiency is not limited by the second law of thermodynamics in the way that heat engines are, although they have their own irreversibilities. Efficiencies are high compared with other power sources, 60% overall based on ideal potential performance being a good figure. Although schematically simple, there are many problems in providing stable operation continuously for long periods. The system

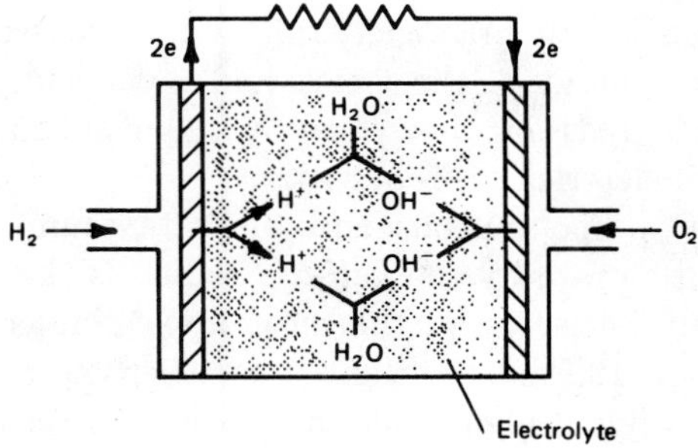

Fig. 11.1 Fuel cell—schematic.

has to provide for regulation of the flow of reactants, removal of the products, removal of heat to maintain constant temperature, reactant purification, storeability, quick response and so forth. Some fuel cells are operated under pressure and at temperatures several hundred degrees above atmospheric. The problems with fuel cells per se are largely those of materials, e.g., degradation of the membranes or matrix which holds the electrolyte or of the cathode which may suffer from overpotential and hence cause inefficient oxygen reaction.

H_2–O_2 fuel cells have been used successfully in the Gemini and Apollo vehicles, although certain difficulties such as regulation were reported. However, there is no doubt of their usefulness for longer-term supplies of low voltage power of small-to-medium amount. One useful characteristic is that the product of the reaction can be highly useful instead of being a nuisance, as is so often the case. This product, water, is completely pure and potable, and therefore potentially valuable on a manned mission. As an indication of the state of the art (1970) the Apollo fuel-cell system had a cell output of about 36 W at 0.97 V, with a current density of 92 mA/cm^2. The module size had 1 kW output and weighed 268 lb.

11.5 Energy Limitations

Batteries and fuel cells may be called *energy* limited sources, because there is only a finite amount of energy available as stored chemical energy. Nuclear energy is *power* limited, that is, it is the rate of energy conversion which is limiting more than the total amount. Likewise solar power is unlimited but the rate is very definitely controlled by the area needed to capture it. Nuclear power may be via radioisotopes or reactors, which will now be considered in turn.

11.6 Radioisotopes

The energy of a radioisotope may come from decay or by spontaneous emission and the energy is in the form of γ-rays (high frequency, low wavelength), β-rays or particles (electrons or positrons), and α-particles (H_e^{++} or mass of helium atom minus two electrons). For space use, particularly for manned missions, isotopes with any degree of γ-ray emission are practically ruled out because of the great amount of shielding required. The energy of the β-rays and α-particles is experienced in the form of a temperature increase on absorption by a solid substance, and a conversion device is then needed if electricity is required.

The radioactivity is spontaneous and the rate cannot be controlled. It is customary to express this rate by the *half-life*, which is the time for the emission to be halved. The half-life is one of the most important criteria of performance because a low value (< 100 days) means a considerable decline of output for a mission of any length while a high value implies low power density. The absolute power level is of course also important. Presuming satisfactory availability and cost, then initial output, half-life and shielding problems become the main criteria and act to limit the possible choices out of many possibilities to only a few. Polonium 210 is the present favorite for short-to-medium length missions, with a half-life of 138 days, high power density (815 W/cm^3) and low shielding requirement. For long-term use, plutonium 238 with a half-life of about 88 years, a low power density (3.8 W/cm^3) and little shielding requirement seems the most likely.

The instantaneous specific power is given by

$$P_t = P_0 e^{-kt} \tag{11.1}$$

where

P_0 = initial specific power per unit mass
k = decay constant = ln 2/half-life
t = time

The total specific energy available in time t is

$$E_t = \int_0^t P_t\,dt = \int_0^t P_0 e^{-kt}\,dt = (P_0/k)(1 - e^{-kt}) \tag{11.2}$$

Then the specific energy available for the half-life period is

$$E_{1/2} = P_0 t_{1/2}/(2 \ln 2) = 0.725 p_0 t_{1/2} \tag{11.3}$$

Some shielding is always required and precautions are necessary to contain the material under impact. The container is hermetically sealed and there will be a pressure buildup as the radioisotope decays and helium is produced.

For low power levels, radioisotopes have the desirable performance characteristics of relatively low weight and volume, and they are self-contained, readily calculable, and dependable. Usually they are used with thermoelectric devices for electrical generation and these will be discussed shortly.

11.7 Nuclear Reactors

Nuclear technology is a complex subject which will not be enlarged upon here. It will have to suffice to assume that the general principle of fission reaction is known that is, the disintegration of certain heavy atoms by neutron bombardment with the resulting kinetic energy of the products on collision with other atoms and molecules producing heat in the reactor core. A certain critical mass of fissionable material is necessary before the reaction is self-sustaining and, for a reactor (as opposed to an explosive device), means are necessary for controlling the rate at the desired level. Although the energy per atomic reaction of fissionable material is tremendous compared with the molecular energy of chemical processes, only a fraction of the material can be utilized. The problem in nuclear reactors is largely one of materials, which limit the maximum temperature usable. There is the combined problem of nuclear compatibility with temperature, with new corrosion problems arising. The nuclear core has to be cooled and it is this heat carried away which is the useful energy output for propulsive use. Because it is purely thermal in character, conversion equipment is needed to furnish electrical energy. As yet there is no developed means of obtaining this directly although one of the most promising means for the future could be the use of the thermionic conversion principle (discussed later). For the foreseeable future, the reactor will have a solid core and the cooling medium will be an alkali-metal combination.

Although this chapter is concerned with generation of electric power for use in propulsive devices, a brief account of the use of the nuclear principle for thermal rockets is in order as it has not been broached before. For the reactor used as a direct propulsive device, the solid core can be cooled with hydrogen, which is ideal for propulsion because of its low molecular weight. In this case, the specific impulse is still limited by metal temperature, with a probable value of about 1000 sec. If a gaseous reaction core could be used, a much higher temperature would be possible, as propellant and nuclear particles would be in contact. Many problems arise with this, notably that of separation of nuclear reactants and propellant before discharge, as otherwise only very short life would be possible. Theoretically this separation could be made by a vortex flow pattern, as hydrogen and nuclear particles are at the opposite ends of the spectrum from the viewpoint of centrifugal effect. Practically it will be some years before such a reactor becomes a real possibility.

However, the nuclear rocket is an attractive possibility and work has been carried on for many years at a nonpriority level. Two previous projects were named Kiwi and Rover, undertaken to obtain some preliminary information. Development is now continuing on the NERVA (Nuclear Engine for Rocket Vehicle Application). A NERVA XE ground experimental engine has been successfully operated at its design output of 50,000 lb thrust under conditions simulating those of altitude. The XE engine was designed and developed by Aerojet, with the nuclear reactor by Westinghouse. Although approximating a flight configuration, it is still essentially a testbed setup. A nuclear stage as an upper stage in a large rocket such as Saturn V would greatly extend the possibilities of using existing rockets to carry statisfactory scientific payloads on extended interplanetary missions. It is possible this may come in time for a "Grand Tour" mission in the late 1970's. (The Grand Tour is a journey when the outer planets Jupiter, Saturn, Uranus and Neptune are so situated that a vehicle could fly-by all or a combination of them in a single mission, aided by the gravitational pull of each to accomplish the journey in a reasonable time.)

Returning to the nuclear reactor as power source for other propulsive methods, it appears to be the only possibility for powers at the hundreds of kilowatts level, that is for large earth-orbiting vehicles, such as laboratories, and for manned interplanetary missions. It can be made ruggedly and need not become operational, and hence a radiation hazard, until it has been boosted into an initial orbit well away from earth. Shielding is a drawback, as it is necessary in some degree for any vehicle but of course most particularly for a manned mission. Shielding entails a large mass increment, as the necessary material has to be thick, but this can be alleviated by "shadow shielding," that is, locating the life-support region well back along an elongated structure with the reactor in front. A plane shield, or approximately sc, immediately behind the reactor can then protect the manned unit, as out in space there is no atmosphere to scatter the radioactivity in all directions.

Nuclear reactors have been developed under the SNAP program (Systems for Nuclear Auxiliary Power), but only one is currently underway. This is SNAP 8, intended to give 35 kWe for a minimum of 10,000 hr. The complete system consists of the reactor and a Rankine-cycle mercury turbine which will be described later. The reactor is cooled by a flow of 49,000 lb/hr of Na–K alloy and the highest coolant temperature is 1330°F.

11.8 Solar Energy

Solar energy is attractive because it is "free" and does not have to be lifted from earth to a mission starting point. The power of solar radiation at any point in space (i.e., outside any planetary atmosphere in which absorption occurs) is a fixed quantity given by the expression

$$E_s = (3 \times 10^{25})/r^2 \quad \text{joule/(sec)(m}^2) \tag{11.4}$$

where r is the distance from the sun in meters. At the earth distance (1 a.u.), $E_s \approx 1342$ joule/(sec)(m^2) = 0.125 kW/ft^2 or 8 ft^2/kW. Figure 11.2 shows Eq. 11.4 as the collecting surface in square feet per kilowatt solar power for the relative distances of the nine planets. For Venus and Mars, the radiation intensity is very similar to that on earth but out beyond this distance, it falls off very quickly. The

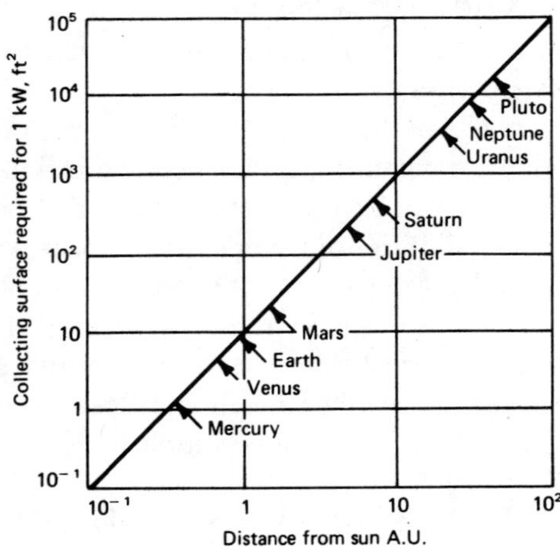

Fig. 11.2 Area-distance relationship for solar radiation.

rather low intensity at even the earth distance means that the collecting surface is large for any appreciable amount of power, with a corresponding mass penalty. If the collecting surface is flat, then such a plate-type collector would have a mass of the order of 2 kg/kW and yield a temperature of only about 900°F for the working substance receiving the energy. This specific mass and temperature are unacceptably low and it is necessary to focus the radiation. While parabolic reflectors are best for concentration on a small area and the highest temperatures, they are difficult to organize in space without large mass, at least in other than small power levels. A spherical collector, which can be obtained by an inflated material, is relatively easy to provide and the somewhat spread-out image would not be a great handicap at present-day working temperatures. One problem with a solar collector is to keep it always directed toward the sun. This is not difficult when once well away from the earth, but as a power unit for satellites there will always be a period when the earth cuts off the solar radiation. For such periods, energy storage means are needed, such as batteries that can be charged during the collecting periods. In general the use of solar energy appears attractive only for low power levels, although it has been used for several years and will probably continue to be used for particular duties, usually of an auxiliary nature in propulsion.

11.9 Energy Conversion Devices

Of the energy sources discussed, only batteries and fuel cells deliver electric current directly. Nuclear sources deliver thermal energy and the sun supplies radiation, which can be used thermally or electronically. Hence conversion devices are needed and at present four major groups can be distinguished. These are *thermoelectric, thermionic, photovoltaic,* and *thermodynamic.* Other possibilities exist, such as electrostatic and electrogasdynamic generators, and no doubt others will continue to be devised, but the above four groups cover the existing devices which appear to have real possibility in the near future.

The action of the first three types of conversion listed above are dependent on the electronic structure of atoms, and a very brief review of certain concepts and terms will be given as applicable to all three. First there is the concept of the *Fermi* level, which at any given temperature represents the average level at which electrons are situated or, more exactly, the Fermi level is that level which has a probability of occupation of one-half. Current is the movement of electrons and hence anything which causes such movement produces a potential and possibility of current flow. A temperature gradient in a material has the effect of exciting more electrons above the Fermi level in the hotter region than in the cooler one. Different materials have different Fermi levels and hence current flow is a function of both temperature and material. The *work function,* introduced in Chapter 9 in connection with contact ionization, may now be more exactly defined as the work needed to move an electron from the Fermi level to infinity. Figure 11.3(a) shows the difference of Fermi levels of two materials and the work function in relation to an energy level external to each. Figure 11.3(b) shows the *contact* potential due to a work function between two materials in contact, in which case the Fermi levels must be similar.

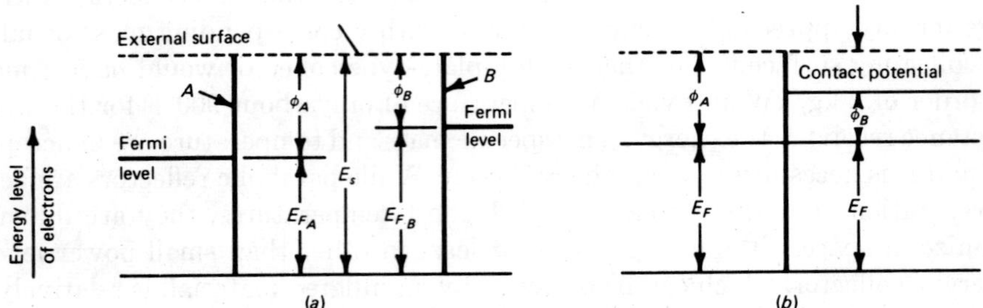

Fig. 11.3 Electronic energies: (*a*) Fermi level; (*b*) contact potential

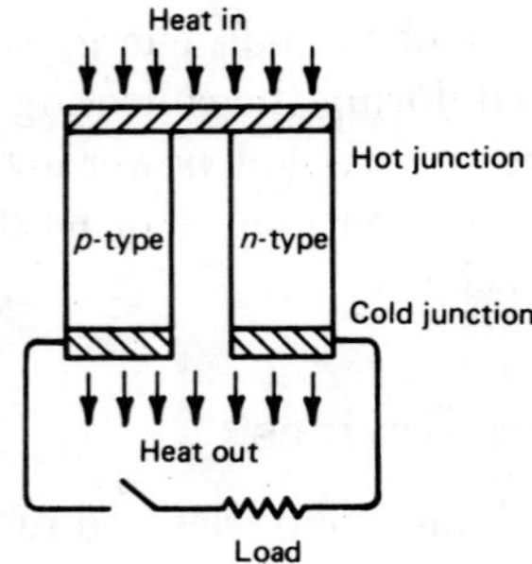

Fig. 11.4 Thermocouple action—schematic.

Another concept is that of p-type and n-type materials. Electrical conduction occurs by electron movement from atoms and this can be controlled to a certain extent in particular materials called *semiconductors* by the addition of small amounts of impurities, a process called *doping*. In brief, when impurity atoms with a higher valence than the lattice (host) atoms are added, their electrons are readily detached and may be easily moved into the conduction band of the substance. Such impurities are called *donors* because they donate electrons and result in conductivity being controlled by electrons, hence the resulting material is called n (negative)-type. Atoms introduced having a lower valence capture valence electrons and hence are called *acceptors*. They create positive "holes" in the atoms and the materials become p (positive)-type.

11.10 Thermoelectric Energy

The basic phenomenon of thermoelectricity, discovered 150 years ago by Seebeck, is that current will flow in a closed circuit consisting of two different conducting materials if one junction is kept at a higher temperature than the other. Until fairly recently, the effect was used only to measure temperature, for which thermocouples have proved a most valuable tool.

The main reason is that the EMF/degree is very low—the highest value for the metals used for thermocouples being of the order of 3 mV per 100°F temperature difference. It was the application of solid-state physics to the problem which allowed the development of more usable materials for power production, the semiconductors of p- and n-type.

Schematically, a thermoelectric unit is shown in Fig. 11.4. It will be realized from this that in addition to high sensitivity, the materials should have a low resistivity, to avoid electrical loss, and low thermal conductivity, to avoid thermal loss. Unfortunately, electrical and thermal conductivity are proportionally related in simple metals, which was the main stumbling block in development from Seebeck to the present day. The development of doped semiconductors, which within limits allow electrical conductivity to be separated from thermal, together with increased sensitivity, was necessary for the effect to be reconsidered.

These factors are combined in a parameter called the *figure of merit,* defined by

$$Z = S^2/\rho\kappa$$

where

S = Seebeck voltage coefficient, V/°C
ρ = electrical resistivity, ohm/cm
κ = thermal conductivity, W/(cm) (°C)

The efficiency of a unit may be shown to be about $\frac{1}{4}Z\Delta T$ where ΔT is the temperature difference of hot and cold junctions. The use of this parameter also brings in the actual voltage developed, because of the factor ΔT, which is higher for materials that can be operated at high temperature, other things being equal. On the basis of $Z\Delta T$, a silicon-germanium material is one of the most suitable for space use. Its figure of merit at about 0.0006/°C is not particularly high but it can operate up to 1000°C and has good mechanical properties at such temperatures. Efficiencies are low, even with the advent of semiconductors, values of possibly up to 10% being achieved for long-term use.

It will be realized that with the very low Seebeck coefficient, arrays of thermoelectric units have to be used in series and in parallel in order to obtain useful power. The design must be optimized to avoid unwanted thermal conduction and maximize output. Thus ideally the composition of each leg should vary to obtain maximum Z for the particular temperature range. This has been done to some extent by segmenting modules using different materials and by cascading, in which each material is packaged in a separate converter to operate in the required temperature interval.

Thermocouple elements have no moving parts, can be thoroughly ground-tested, and are very reliable if properly designed. However, they do have to reject heat, which in space means by radiation, hence a low cold junction temperature implies a large radiator area and hence a large mass. On balance they have proved quite popular for some space use. The radioisotope thermoelectric generator or RTG is useful for small powers, particularly for satellite orbit as it is operable continuously in comparison to solar-powered devices which are necessarily in shadow a fair time. Also they will give their steady power at any distance from the sun, so have a place in interplanetary probes.

The SNAP 10A project was quite ambitious, using a small nuclear reactor and thermoelectric unit, yielding about 500 W at 30 V. The nuclear reactor had NaK cooling giving an average hot junction temperature of 940°F, with an average radiator temperature of 640°F. Thus the Si–Ge thermoelectric module did not operate at near its maximum capacity but was limited by considerations of the power source and radiator size. The conversion efficiency was only about 1.4% which is not untypical of such systems. However it operated over 10,000 hr on a ground test and was thought to be good for 5 years or more. The flight test in 1965 operated for 43 days without trouble, then being terminated by a failure in the spacecraft which caused the reactor to be shut down.

11.11 Photovoltaic Power

The use of solar radiation in the photovoltaic effect is again an old device, as the phenomenon in selenium was observed about 100 years ago. Until the early 1950's, the main uses made of it were those requiring miniscule output, such as for the "photoelectric eye" used for operating a relay when a light beam is interrupted, and for the exposure meter. Efficiencies were very low, always less than 1%. The breakthrough came in the development of semiconductors, notably the silicon cell, doped to produce p and n regions.

The solar cell consists of a sandwich of two thin silicon plates, the top one being about 10^{-4} in. thick, the bottom more massive one being about 10^{-2} in. thick. Silicon has a valence of four, so doping the top wafer with boron of valence three and the other with arsenic of valence five produces p- and n-type layers. Photons striking the cell are absorbed near the layer junction and those above a certain energy level generate the electron-hole pairs and the charges move in the field this created, causing a current to flow in an external circuit. The activating influence is light quanta, not heat per se, and the solar cell is not a heat engine like a thermocouple. Although the underlying phenomenon of both devices is the p- and n-type regions, the mechanism is not exactly the same. Solar cells do not utilize radiation less than 1 wavelength and such radiation merely heats the cell.

The advent of the silicon material made the solar cell into a realistic conversion device, efficiencies of around 12% being realized at voltages of about 0.5 and slightly above. A complete cell requires antireflection and antiabrasion coatings and metallic contacts, the overall thickness being about 0.008 to 0.012 in. An individual cell is usually about one inch square. A solar battery requires many thousands of cells which must be directly exposed to radiation. They are mounted on a backing which can then be applied to the vehicle surface or on panels which can be extended when the vehicle has been launched into the space environment.

The solar cell has been used successfully on many vehicles and is indeed the workhorse for energy supply at the present time. Although the area required is large and hence provision of unfolding paddles creates a design problem, the power output is steadily being increased. Outputs of 1 kW were once regarded as a limit but values greater than this are feasible and up to 50 kW have been considered for interplanetary missions. Improvements are being made in thickness and specific masses of 50 lb/kW may be in prospect.

11.12 Thermionic Power

The basis of thermionic energy conversion is the use of heat to raise an electrode to a temperature sufficiently high to drive off many of the higher-energy electrons, followed by the collection of these electrons on a cool electrode. Connecting *emitter* and *collector* externally produces a circuit in which the only moving parts are the electrons in the gap between electrodes. It is necessary that there should be a difference of *work function* ϕ between emitter and collector, this difference ideally being the useful EMF in the circuit. A schematic thermionic *diode* is shown in Fig. 11.5(a), with the idealized potential energy diagram in (b) of the figure.

The space between electrodes can be a vacuum and the converter is known as a *vacuum diode*. However, the electron gas stream produces a space charge similar to the effect discussed in Chapter 9 in connection with ion engines and the electrodes have to be placed extremely close together to obtain a reasonable current flow. Such spacing of the order of 10 μ or less is next to impossible to maintain under conditions of mechanical and thermal stress in flight and the alternative is to introduce some positive ions to neutralize the space charge. Again cesium is a good material to use for this, as it is readily ionized by the high energy electrons. The cesium is neutral overall and hence the converter in this form is known as a *plasma diode*.

The actual state and the detailed processes undergone by the particles between the electrodes in the plasma diode are difficult to diagnose exactly. However, the general state may be interpreted as shown in Fig. 11.6. Electrons from the emitter given sufficient energy equal to the work function ϕ_E, plus sufficient kinetic energy E_k to overcome the space-charge potential barrier, are accelerated into the main body of the plasma toward the collector by a potential drop. They collide with plasma particles and those electrons with sufficient energy ionize cesium atoms. Energy is lost equal to E_p, thus reducing the useful potential E. The latter is

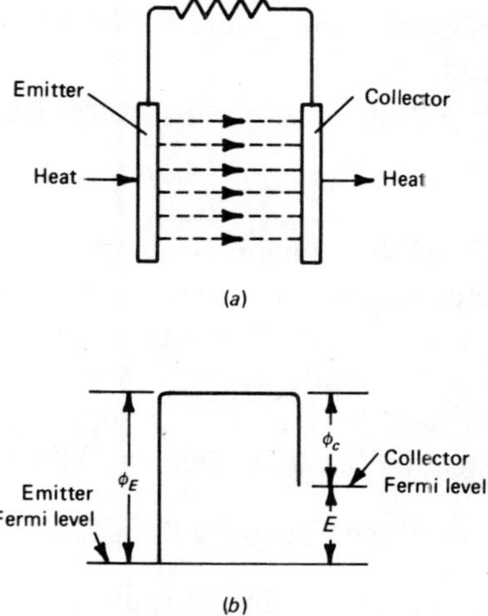

Fig. 11.5 Thermionic energy—ideal.

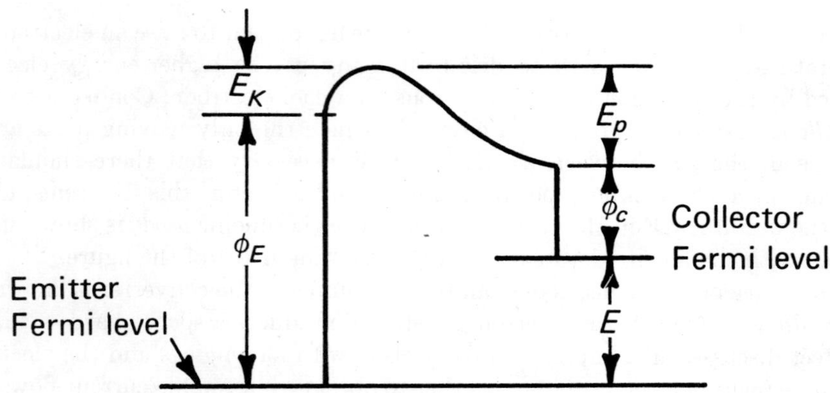

Fig. 11.6 Thermionic energy—actual.

then equal to $(\phi_E - \phi_c) - (E_p - E_k)$. The cesium, in addition to the essential space-charge reduction, has two other useful effects. Some cesium atoms are absorbed on the collector and this lowers its work function ϕ_c, thus leaving more potential available for useful work. Also, if the Cs pressure is high enough, Cs is absorbed in the emitter. This lowers its work function, but the important advantages are lowering of the emitter plate temperature for a given current density (or increasing current density at a given limiting temperature) and more stable electron emission. Its action may be likened to that of doping thermoelectric materials by altering the Fermi level.

Thermionic converters yield outputs 10–20 W/cm^2 of electrode surface at 0.7 to 1.3 V for emitter temperatures of the order of 3000°F. Collector temperatures are about 1300°F. Efficiency is again Carnot-limited as the device is really a heat engine with electrons as the working fluid and values of efficiency are obtained of 10–20%, with 15% being representative.

Current density, j, is a function of temperature. By Richardson's equation

$$j = AT_e^2 e^{-\phi_E/kT_e} \tag{11.5}$$

where A is a constant, k is the Boltzmann constant, ϕ_E is the emitter work function and T_e is the emitter temperature. This is shown plotted in Fig. 11.7 for arbitrary values and the strong dependence of j on T_e can be seen. High emitter temperatures do mean greater loss due to radiation but current density increases more than proportionately, hence the need for high T_e.

Efficiency is also dependent on T_e as follows. The efficiency is expressed as

$$\eta = \text{electrical output/heat input}$$

The output is the product of current and output voltage, $jE \times E$ or $j(\phi_E - \phi_c - E_p + E_k)$ from Fig. 11.6. The input may be taken as this output plus losses, which are mainly thermal. By far the greatest loss is by radiation. Denoting the radiation loss per unit area as q_r, then

$$\eta = \frac{jE}{jE + q_r} = \frac{E}{E + q_r/j} \tag{11.6}$$

From this the importance of high current density is seen, as although q_r increases with temperature, j increases more rapidly.

The emitter material is critical, as it must have resistance to high temperature and to any corrosion with cesium vapor. Rhenium has been satisfactory but is costly and not always available in the required shape. It has better properties when chemically vapor deposited than when used in crystalline form, notably ability to capture cesium atoms, hence yielding a much higher current density.[2] This is interesting as an indication of the importance of materials engineering and solid-state physics in these areas of direct energy conversion. Tungsten and molybdenum are also used, with the latter less sensitive to cesium corrosion at the higher temperatures.

Considerable efforts are being made in developing thermionic converters and not only for use in space, as they have possibilities for surface power generation. An attractive possibility is the use of such a converter directly with a nuclear reactor, avoiding the use of a heat transfer loop and secondary energy-conversion

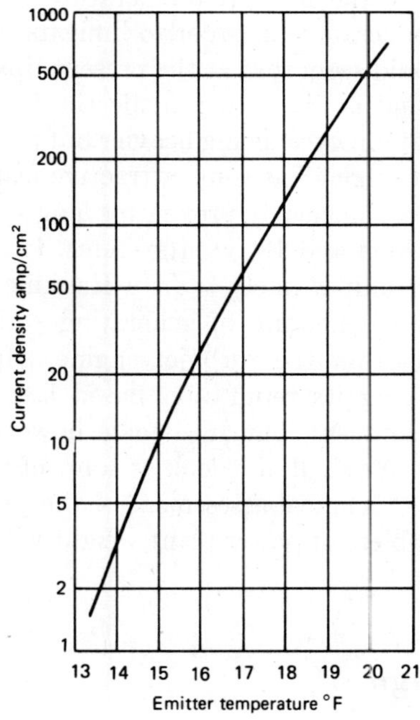

Fig. 11.7 Current density vs emitter temperature.

[2] *Sandia Science News,* Sandia Laboratories (June 1970).

equipment. It presents a formidable problem, as the difficulties of its own development are compounded with having to achieve compatibility with nuclear effects, but the overall result could be very worthwhile.

11.13 Thermodynamic Power

Three so-called "direct" energy-conversion devices have been discussed: thermoelectric, photovoltaic and thermionic. None of them appear well suited for large outputs, except possibly the last-named as an integral part of a reactor, mainly because their individual power units are each limited to voltage of about unity and the current density is low. The alternative is to use a conventional heat engine and generator, conventional at least in the sense that the cycle and apparatus are basically well-established and have a considerable history of development, but possibly having some unconventional features to optimize use in space. The heat engine cycles in mind are the Rankine, the Brayton and the Stirling. The last two are gas cycles entirely while the Rankine cycle is liquid-vapor or liquid-gas and herein lies the crux of the matter. The crucial component for space use turns out to be that for heat rejection, called the condenser in the Rankine engine, and simply a heat exchanger for a closed gas cycle. The only possible method of heat rejection to a sink in space is by radiation and the rate of heat transfer is proportional to the fourth power of the absolute temperature. The *radiator* must therefore be at a relatively high temperature in order to limit its size, and this also severely limits the efficiency. It would seem that at the present time, the Rankine cycle with a liquid-metal working fluid is likely to provide the best system with respect to specific mass, with the Brayton cycle being heavier but potentially simpler and more reliable. The Stirling cycle engine has some attractive features but it will be a long time before its reliability is sufficiently proven for long space missions and effort is better put on to the Rankine and Brayton systems. It is perhaps ironic that the first dynamic space power unit is likely to be a Rankine type which we generally associate with the heavy engineering of generating stations for our earthbound power. One might suspect that the turbojet engine, supreme in our era for atmospheric travel and having its reputation based largely in low specific mass, would be the proper development, but once again however space presents its own special design constraints. We shall now look at some of the characteristics of such plants in order to obtain a technical assessment of the performance and associated problems, but first the problem of power plant weight will be considered due to the radiator requirement.

11.14 Radiator Design

It is convenient to make an analysis in terms of the ideal or Carnot cycle efficiency η_i of the engine and the fraction η_E of this efficiency attained by the actual engine, i.e.,

$$\eta = \eta_i \eta_E$$

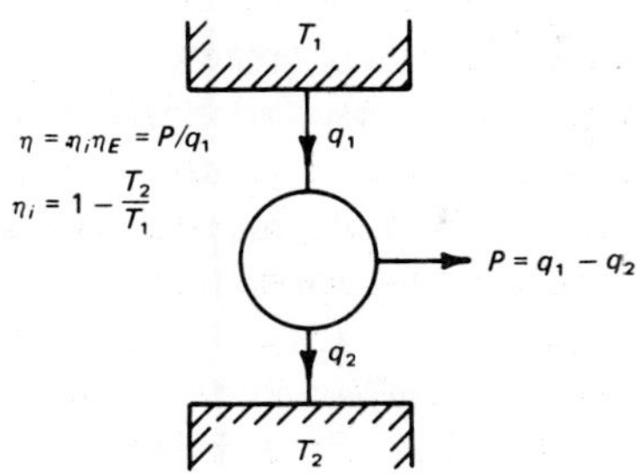

Fig. 11.8 Heat engine—schematic.

Using T_1 and T_2 as the maximum and minimum temperatures of the cycle, then $\eta_i = 1 - T_2/T_1$ and $\eta = \eta_E(1 - T_2/T_1)$. Figure 11.8 shows schematically the cycle symbols.

We wish to get an expression for the specific area of the radiator A/P, area per unit output, in terms of the cycle temperatures. Radiator area can be put in terms of T_2 and q_2, since $q_2 = \epsilon\sigma A T_2^4$ where ϵ = emissivity and σ = Stefan-Boltzmann constant. Thus $A = q_2/\epsilon\sigma T_2^4$. The output $P = \eta q_1 = \eta_E \eta_i q_1 = \eta_E q_1(1 - T_2/T_1)$. Hence

$$\text{specific area} = \frac{A}{P} = \frac{q_2}{\epsilon\sigma T_2^4} \cdot \frac{1}{\eta_E q_1(1 - T_2/T_1)}$$

q_2 can be put in terms of q_1, since $q_2 = q_1 - P = q_1 - \eta_E q_1(1 - T_2/T_1) = q_1[1 - \eta_E(1 - T_2/T_1)]$, hence

$$\frac{A}{P} = \frac{q_1[1 - \eta_E(1 - T_2/T_1)]}{\epsilon\sigma T_2^4 \eta_E q_1(1 - T_2/T_1)} = \frac{1 - \eta_E(1 - T_2/T_1)}{\epsilon\sigma T_2^4 \eta_E(1 - T_2/T_1)} \tag{11.7}$$

Differentiating A/P with η_E and T_1 constant and setting equal to zero yields a *minimum* value for T_2/T_1, thus

$$\frac{T_2}{T_1} = \frac{8\eta_E - 5 + (25 - 16\eta_E)^{1/2}}{8\eta_E} \tag{11.8}$$

This value of T_2/T_1 is almost independent of η_E as it has a limiting value of 0.8 as η_E goes to zero and a value of 0.75 when η_E is equal to unity. Hence we may take the optimum value of T_2/T_1 as 3/4, which means that $\eta_i \approx 0.25$ and hence with η_E probably being about 50%, then for minimum radiator area the overall efficiency can be only 10–12%. So this is quite different from our land-based efforts of making T_2 as low as possible. Radiator areas can be calculated from Eq. 11.7. For example, for $\epsilon = 1.0$, $\eta_E = 0.5$, and $T_1 = 1340°\text{F}$ (a temperature approximating the SNAP 8 system), then $A/P \approx 4.25\ \text{ft}^2/\text{kW}$ from the above analysis. The area would be halved if T_1 were increased to 1680°F.

However this is but part of the story as the expressions are valid only for minimum radiator area or mass M_R, and the power plant mass M_{pp} also changes as its thermal conditions are altered. Thus the criterion might well be the combined system mass, $M_c = M_{pp} + M_R$. A difficulty arises because there is more than one variable, e.g., the mass of the power plant (reactor) will vary with the temperature level and some function ought to be inserted to take care of this. We can either plot the overall specific mass for assumed values of power plant specific mass or, assuming the latter is a constant, find a minimum by differentiating in the usual way. The resulting expression for the optimum value is cumbersome but T_2/T_1 can be plotted in terms of a parameter $(M_{pp}/q_2)(\epsilon\sigma T_2^4/K_R)$,[3] where $K_R = M_R/A$, mass/area of the radiator. It can be concluded that for the probable range of values, the optimum value of T_2/T_1 is of the order of 0.5–0.6.

It should be noted that both radiator mass and area values are important, the mass for the obvious reason as increasing the M_w component of the initial mass M_0, and the area because the geometry has to be such that each part of the radiator surface is exposed fully to space and "sees" as little of the other part of the hot surface as possible. For small powers, a nuclear reactor has a given minimum size that does not increase greatly with increase of output. Thus at higher power levels, the radiator becomes increasingly large and the major part of the system mass. As it has to be developed for maximum exposure, the vehicle configuration may be largely dictated by radiator requirements.

In addition, the radiator must be able to withstand meteorite damage. This is a difficult problem because even the smallest puncture of the prime surface will allow the contained liquid to escape. Therefore, either the matrix must be constructed of material to withstand all but the most extremely unlikely damage or it must be made in a sufficient number of individual segments so that several may be damaged without lowering the plant performance below an acceptable limit. Another possibility exists in a novel radiator system proposed by Weatherston and Smith[4] as shown in Fig. 11.9. A long, wide, endless belt of very thin material passes continuously through the liquid material to be cooled, picking up heat from it and thence passing out in a loop of large radius to lose the transferred heat by radiation to space. Because of the absence of drag and gravitational forces, the belt can be of almost any desired length, subject to initial folding restrictions during launch. It can be made of very thin material and can radiate from both sides. It can be punctured many times before losing its effectiveness or strength. One of the major design problems would be to devise absolutely tight seals where the belt passed in and out of the radiator fluid if the transfer of heat to the belt were convective. If the transfer were conductive only, that is, the belt passed over the outside of a drum containing the fluid to be cooled, then the problem would be one of surface contact to ensure a satisfactory and constant heat transfer coefficient.

11.15 Rankine Cycle Engines

A simple Rankine cycle is shown in Fig. 11.10, (*a*) for *T–s* coordinates, (*b*) for *p–v* coordinates. The dome showing the liquid saturation line is typical of water and

[3] For example, E. Stuhlinger, "Ion Propulsion for Space Flight," pp. 282 ff. McGraw-Hill, New York, 1964.

[4] R. C. Weatherston and W. E. Smith, "A New Type of Thermal Radiator for Space Vehicles," Cornell Aeron. Lab. Report No. DK-1369-A-3, June 1960.

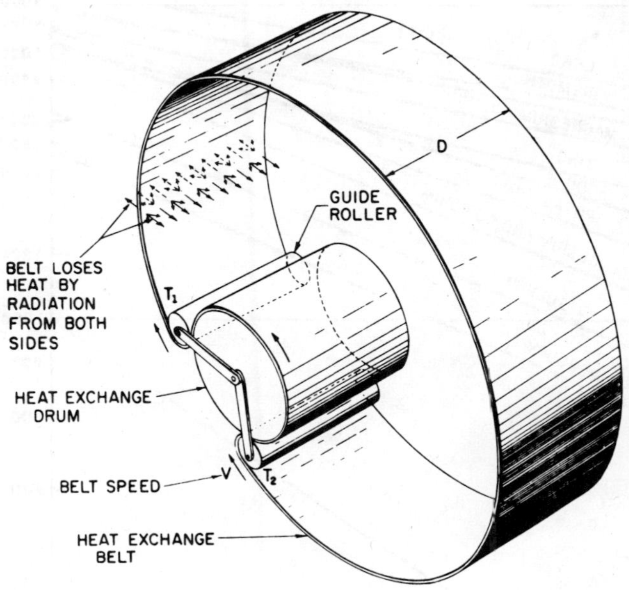

Fig. 11.9 Traveling belt radiator (Weatherston and Smith.)

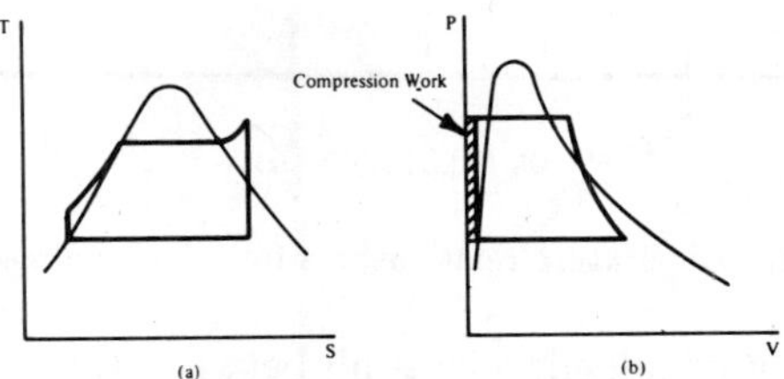

Fig. 11.10 Rankine cycle: (*a*) *T-s* diagram; (*b*) *p-v* diagram.

the liquid metals (some other fluids may have a retrograde saturated vapor line, one of positive slope). It is shown with a small degree of superheat although this is not always used for space applications. The *p–v* diagram shows more clearly the small amount of work needed for the compression process as the fluid is a liquid in this case. This contrasts strongly with the Brayton cycle in which the work of compression may be more than half the total expansion work.

The combination of a Rankine engine and a nuclear reactor can be made in several different ways. One is to use the engine working fluid directly as the reactor coolant. This avoids losses and allows the highest temperature but the problem of radioactivity and corrosion is considerable. It also rules out superheat, as the reactor is not designed for gas cooling. In another method a second loop can be used, using the reactor fluid in a "boiler" to transfer heat to the engine fluid. This allows cycle superheat to minimize turbine erosion by excessive working in the wet region during expansion. This allows flexibility in both fluids for optimum properties.

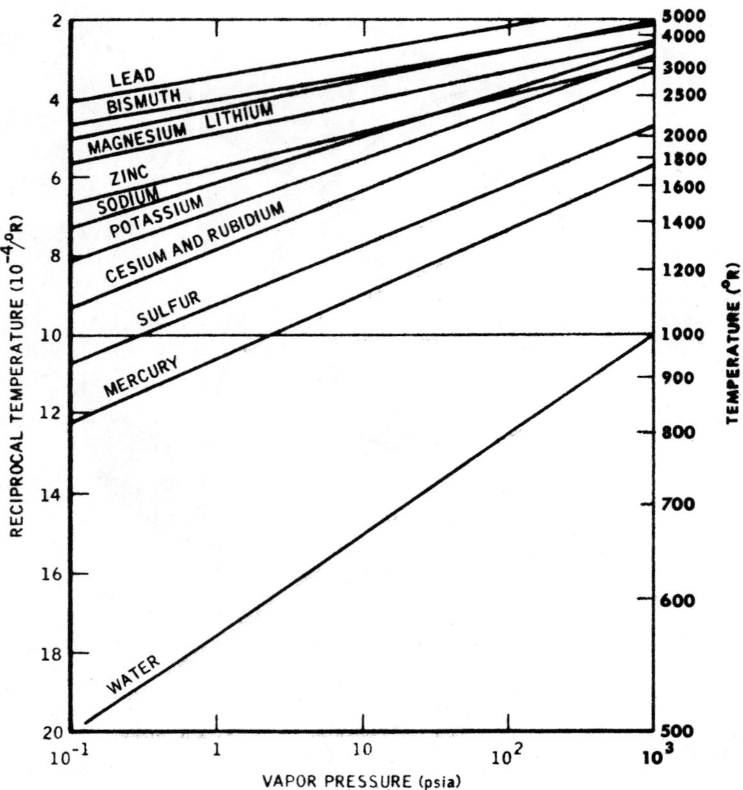

Fig. 11.11 Pressure-temperature relationships for alkali metals (from Dieckamp.)

Again, a third loop may be added, operating between the condensing phase of the engine cycle and the space radiator. Temperature drops in each loop are wasteful but allow the most flexibility of cycle arrangement and operational use.

With one, two or three loops, it is apparent that there are problems in choice of working or transfer fluids. Water is ruled out because of the saturation pressure concomitant with the temperatures required. Most attention has been given to liquid metals, notably mercury, lithium, sodium, potassium, rubidium and cesium. Figure 11.11 shows the temperature-pressure relationships[5] for a number of liquid metals and for water, illustrating the unsuitability of the latter. Some of the characteristics which must be considered are corrosion and materials compatibility, heat capacity, heat transfer coefficient and nuclear properties if used in a reactor. No one metal is suitable for all ranges and for the uses under consideration now, a Na–K eutectic is a likely candidate, with mercury and potassium being given attention. Lithium has attractive properties in its low density and low saturation pressures.

[5] H. N. Dieckamp, "Nuclear Space Power Systems," Space Power Systems, Pt. I, AGARDograph 123, Propulsion and Energetics Panel, AGARD , 1969, p. 190.

The SNAP 8 power plant may be used as an illustration of the probable form for the 1970's. It is shown schematically in Fig. 11.12. There is a Na–K loop from reactor to boiler, with a maximum fluid temperature of about 1300°F. The power fluid is mercury through boiler, turbine and condenser, with a turbine inlet temperature of about 1250°F and 250 psia. Finally there is another Na–K loop from condenser to radiator. As of 1970, over a year (>10,000 hr) of ground testing of the major rotating components had been successfully completed, that is the liquid-metal pumps and the turbogenerator. The output of 35 kWe is only about that of a small car, but SNAP 8 with its nuclear reactor will be a very important milestone when launched for a planetary mission.

There are many possibilities for variations of the simple Rankine cycle, but only a few can be mentioned here in outline. It was stated earlier that superheat was not probable with a nuclear reactor of the type likely to be used in the near future. However, the use of superheat can improve the efficiency considerably by raising the maximum temperature while still keeping the pressure at a reasonable level corresponding to the saturation temperature. The Feher cycle[6] is an extension of the Rankine cycle principle to a region where the minimum cycle pressure is above the critical pressure for the working fluid (CO_2 for example). Thus there is no liquid-vapor phase change and the compression work is still low as in the Rankine cycle.

The working fluid may also be an organic fluid, which can have the great advantage of not being corrosive like the alkali metals. Also, most of them have retrograde saturated vapor lines and expansion results in superheating, so that no problems of erosion with wet vapors arise. Efficiencies are low because such fluids cannot stand a very high temperature without decomposing, but regeneration can improve this. An ORACLE (Organic RAnkine cyCLE) electric generator is being developed with such an organic working fluid, developing about 6 kW for normal space missions. Its power source is radioisotopes in the first instance. It utilizes a jet condenser, which helps in overcoming cavitation and zero-gravity problems for the system pump.

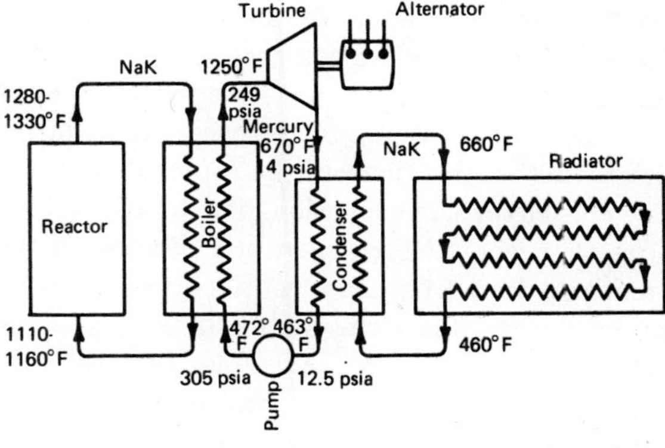

Fig. 11.12 SNAP 8 power system—schematic.

[6] E. C. Feher, "The Supercritical Thermodynamic Power Cycle," I.E.C.E.C. Conference, 1967.

Although the simple Rankine cycle engine with liquid-metal working fluid appears to be the leading contender for first use in space, there are difficulties. These seem to have been largely overcome but a space test will have to decide. The two major problems are corrosion, which of course is mainly one of materials, and of the boiling and condensation processes, as these must occur in a zero-gravity field. Our knowledge of heat transfer with phase change is based on the gravitational field of the earth and special designs will be necessary for space use. Also the absence of gravity can raise some peculiar fluid distribution problems in other ways which must be allowed for in design.

11.16 Brayton Cycle Engines

The closed-cycle gas turbine operating on the Brayton cycle offers many advantages for space use. There are no problems of liquid-vapor phase changes as the working fluid is always a gas, and no corrosion problems, as an inert gas can be chosen for that fluid. Also it has had a successful history in very lightweight construction as an aircraft engine, which is not true of the Rankine cycle engines. It has therefore every prospect of rapid development for long-term use.

The major disadvantage, and it is a considerable one, is the large radiator area required consequent on the nonisothermal heat rejection process. All heat transfer processes are carried out by forced convection in the gas phase, with coefficients much less than for the liquid phase and condensation or boiling phases of the Rankine cycle. These heat-transfer coefficients are improved by higher datum level of pressure in the closed-cycle system but there is a limit to this due to stress considerations and size of rotating components.

The general characteristics of the gas turbine have been dealt with in Chapter 4. The differences in this application to shaft power exclusively are (1) the effect of temperature on efficiency and (2) the use of regeneration or exhaust heat to preheat the gas before the heat addition proper. For the ordinary surface gas turbine, increased turbine inlet temperature not only increases power but improves efficiency, although the latter is less marked when component efficiencies are high. However for space application, the prime criterion is system mass, and optimizing procedures indicate that a wide range of turbine temperature has little effect on efficiency when mass is minimized.[7] The use of regeneration is shown on a cycle diagram in Fig. 11.13. Full lines *a–d* and *c–f* show complete regeneration in an ideal cycle with constant specific heat, that is, the air from the compressor at *f* is raised up to the limit of temperature *a* after turbine expansion. The broken lines *ae* and *bf* show the practical case of less than complete regeneration because of the need of a finite temperature difference to limit the heat transfer area required. The amount of regeneration is usually expressed in terms of *effectiveness*, which is the ratio of heat actually transferred to that theoretically possible. In terms of temperature, the effectiveness of the regeneration shown in broken line in Fig. 11.13 would be ef/af = actual temperature change/maximum possible change. Values of effective-

[7] W. G. Harrach and R. T. Caldwell, "System Optimization of Brayton-Cycle Space Power Plants," ASME Paper No. 63-WA-87, 1963.

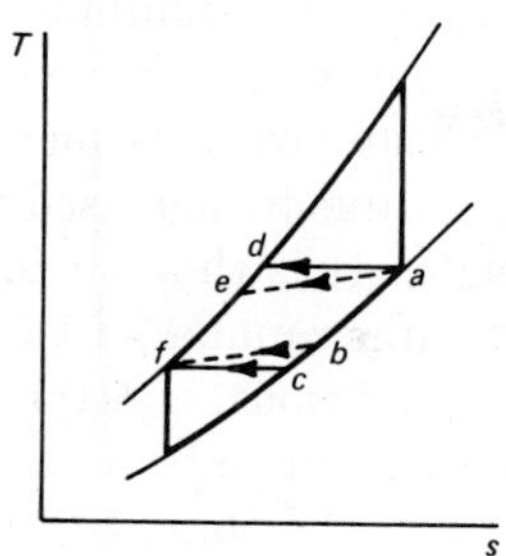

Fig. 11.13 Regenerator effectiveness.

ness of up to 0.9 are possible, although required area increases rapidly as effectiveness increases.

The advantages of regeneration are threefold. The major one is increased efficiency, as the heat transferred from the exhaust process to the cool air saves heat input from the source. Almost as useful is the reduction in optimum pressure ratio for maximum efficiency, as this allows more efficient, lighter and smaller turbo-components. A third advantage is that the performance is less sensitive to component efficiency, as some of the losses are recovered as "heat" in the regeneration process. The major disadvantage is increased size and weight, as considerable surface is needed for high effectiveness. Also some pressure loss is caused, which reduces the output.

The fact that there is a choice of working fluid complicates the problem of optimum design. The inert gases usually considered include helium, neon, argon, krypton and xenon, with molecular weights of about 4, 20, 60, 83 and 131, respectively. This is a considerable range and a corresponding range of gas constant R, adiabatic index c_p/c_v, specific heat c_p itself, and viscosity. The main effects on performance and design are the number of stages needed in the turbomachines, in heat transfer characteristics and in Reynolds number, and some of these are conflicting. Thus the number of required stages decreases with increased molecular weight, e.g. for a given condition, with argon and gases of higher molecular weight requiring only a single turbine stage, neon would require 2 and helium 10. Unfortunately heat transfer properties have the opposite characteristic. Reynolds number, which depends on the fluid properties of density and viscosity, shows increasing values for increasing molecular weight, again over a considerable range. As performance of turbomachines depends on Reynolds number and values are likely to be in a region where the effect can be marked, the exact value can be important. The physical size itself can also be important, again for Reynolds number and also for clearances, which increase relative to wheel diameter, and elsewhere as size decreases which can also lead to marked loss.

A total system weight analysis must include the reactor, shielding and meteorite protection for the radiator. Because the radiator size varies with the degree of regeneration effectiveness, the relative weights of recuperator and radiator are important, that is, the increased weight of the latter for meteorite protection must be realistically assessed. It is pointed out by Harrach and Caldwell that the specific weight of the whole system may have a minimum value at a particular system power output. The increase in specific weight above the minimum value results from the rapidly increasing radiator weight with its meteorite protection, which is not counteracted by the reduced weight due to increased cycle efficiency and a relatively constant reactor and shield weight. It is thus obvious that an optimum design is critical and that it depends on a large number of factors, some of which depend on the mission itself, assessment of conditions, and the state of the art at the time.

While Brayton cycle engines appear to be assessed as second to Rankine cycle engines at the present time, it could be that they will find a place for particular missions in the future in a certain size and mission range. Because of what seems to be very reasonable certainty of performance, the Brayton cycle engines may be required if the mixed-phase characteristic of Rankine types presents difficulties in the future.

11.17 Some Other Propulsion Power Units

It is worth discussing briefly two other proposals for propulsion, both being essentially electrothermal means involving reactors and turbomachinery. They are introduced, not particularly for their own sake since as far as is known they are not being developed, but as an indication of the great variety of power systems possible for use in space and the opportunity for the applications of thermodynamics. The first scheme, suggested by Ackeret,[8] is an attempt to eliminate the radiator, thus avoiding both heat loss and weight. The second scheme, due to Resler and Rott,[9] likewise has this objective and also to improve on the first method.

Ackeret proposed a unit in which liquid helium would be pumped to a high pressure, heated in a boiler whose heat supply comes from a reactor, thence expanding in a turbine whose load is a generator. Following expansion, the helium can be reheated in another reactor, followed by turbine expansion yielding electrical work. Another reheat by reactor is followed by electrical heating (from the generators) to a high temperature and then expansion in a nozzle to give thrust. It is shown schematically in Fig. 11.14. The helium would be stored at low pressure to avoid heavy containers, as the pump would take little power and not weigh a great deal. Radiators are avoided, as there is no heat engine cycle and a high final temperature could be used because it is attained in a component with no moving parts.

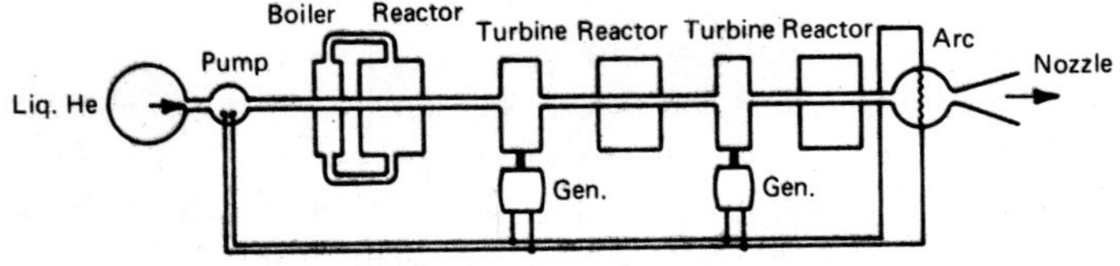

Fig. 11.14 Ackeret propulsion scheme.

[8] J. Ackeret, "A System of Rocket Propulsion Using Reactors and Gas Turbines," *Proc. IXth. Int. Astron. Cong.*, Amsterdam 1958, Vol. I (1959), 277.

[9] E. L. Resler, Jr. and N. Rott, "Rocket Propulsion with Nuclear Power," *ARS Journal*, 30 (Nov. 1960), 1099.

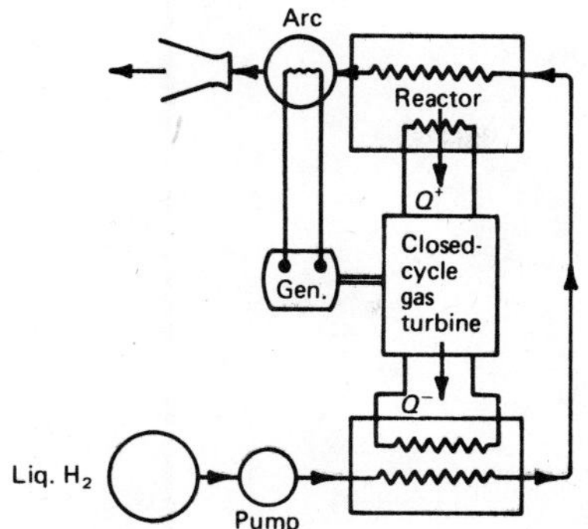

Fig. 11.15 Resler-Rott propulsion scheme.

The reactors need only be low temperature units. While this does eliminate radiator area and the need for protected surface exposed to meteorites, heat exchange area is large in the reactor reheat processes because of the low pressure.

The Resler–Rott proposal used a closed-cycle regenerative gas turbine working between a reactor as heat source and liquid hydrogen as heat sink. Again the turbine power could run a generator to provide electrical heating of the hydrogen, which is the propellant in this case. It is shown schematically in Fig. 11.15. Liquid hydrogen is pumped through a heat exchanger acting as heat sink for the closed-cycle gas turbine and then passes to a reactor, which also acts as heat source for the closed-cycle unit. Thence the hydrogen receives further heat electrically by virtue of the generator coupled to the turbine unit and expands through a nozzle for thrust.

The specific enthalpy of the hydrogen at nozzle entry is given by the enthalpy gained in the reactor plus that given by the engine power. The former is c_pT_R, where T_R is the reactor temperature. The latter is $Q^+ - Q^-$, heat into reactor minus heat rejected, and $Q^- = \dot{m}h_{fg}$, where $\dot{m}$ = mass flow of hydrogen and h_{fg} is the enthalpy of evaporation of hydrogen ("latent heat"). Thus with T_F the final temperature at nozzle entry,

$$c_pT_F = c_pT_R + [(Q^+ - Q^-)/\dot{m}]$$

and with $\dot{m} = Q^-/h_{fg}$, then

$$c_pT_F = c_pT_R + h_{fg}[(Q^+/Q^-) - 1]$$

For an ideal (Carnot) engine, $Q^+/Q^- = T^+/T^-$, and with $T^+ = 2300°R$ (reactor) and $T^- = 36°R$ (liquid hydrogen), then the gain is tremendous. Actual engines would not give these ideal figures but nevertheless the performance could be impressive. The key of course is the low sink temperature, that of liquid hydrogen. Note that the closed-cycle plant would have to use helium, since it is the only gas with a condensing temperature below that of hydrogen. The closed-cycle gas turbine has had considerable development and therefore the system appears capable of being brought to a usable stage relatively quickly. There remains the question of whether a specific impulse $\lesssim$ 1000 sec is useful for a large vehicle which would be required for a plant of this nature—that is, is there a suitable mission for its properties?

CHAPTER 12

Speculations and Evaluation

The airbreathing engines and rocket motors discussed so far have been either operational or in course of development, even if only in the laboratory. What of the future, looking perhaps to journeys outside the solar system? Here we have to think in terms of journeys at extremely high speed, approaching the speed of light, or of journeys of very long duration which can utilize an external power source. For journeys outside our planetary system, the vehicle speeds become relativistic, raising all kinds of interesting questions relating to time but which are outside our intended scope here. In fact, although what we shall shortly discuss, namely the use of radiation as a rocket drive, is a fascinating topic in itself, we shall touch only the very basic parameters involved in a performance evaluation. Eventually such means of propulsion may be needed but they are a long way off at present.

12.1 Radiation, The Photon Engine and the Solar Sail

The basis of these exotic thrust systems is then the use of radiation rather than propellant mass as heretofore. All radiation travels at the same velocity, the "speed of light," a term in common use but somewhat circumscribing like the term "speed of sound." This velocity is about 186,000 miles/sec, hence the specific impulse, which is then fixed, is about 3.05×10^7 sec. Under the Einstein principle, mass and energy are interchangeable via the well-known relation $E = mc^2$ and it is usually convenient to interpret emission and absorption as corpuscular in terms of particles with mass and to deal with transmission as wavelike in terms of radiation. The emission of radiation implies a change of momentum and hence a force which, if directed properly, can be used as useful thrust. Similarly the absorption or reflection of radiation again implies a force from change of momentum. Thus an electric flashlight experiences a rearward thrust and the objects on which its light falls experience thrust due to "radiation pressure." In principle, the *photon engine* is little more than a flashlight and the *solar sail* is a surface utilizing radiation from the sun as a gigantic flashlight.

The fundamental propulsion unit is the *photon* or quantum which has no intrinsic or rest mass as it travels at light velocity, but has momentum and energy. The momentum p of a photon is given by

$$p = h\nu/c$$

where h is Planck's constant, 6.623×10^{-34} joule/sec, ν is the radiation frequency and c is the velocity of light. The energy E of a photon is given by

$$E = h\nu$$

Using the Einstein relationship $E = mc^2$, where m is the photon mass, then we can say

$$p = mc$$

We thus have the usual relationship of $E \propto$ momentum $\times$ velocity but there is no factor of one-half, as in Newtonian mechanics. Since c is fixed, we note that the total momentum and energy are both proportional only to the total mass. Thrust is directly proportional to rate of charge of momentum which with constant c is then equal to $c\,dm/dt$. The power is $dE/dt = c^2 dm/dt$ and the power-thrust ratio is then c. With c having the very high value it does, then photon propulsion is the extreme case of the power-thrust dilemma discussed in Chapter 6.

The basic rocket equation $\Delta U = c \ln M_r$ shows that not only the specific impulse is important in attaining a given velocity increment but also the mass ratio and this is extremely small in photon propulsion. According to Stuhlinger,[1] a mass ratio of $1 + 10^{-8}$ may be obtainable on the basis of modern reactor and light source engineering. Using subscript p for a photon rocket and subscript c for a chemical rocket,

$$\Delta U_p / \Delta U_c = I_{sp} \ln (M_r)_p / I_{sc} \ln (M_r)_c$$

Using $c_c = 450$ and $(M_r)_c = 10$, then

$$\frac{\Delta U_p}{\Delta U_c} = \frac{3.05 \times 10^7 \ln (1 + 10^{-8})}{(450) \ln 10} \approx 3 \times 10^{-4}$$

that is, the capability of the photon-generating process must improve so that the mass ratio of the photon rocket is increased by more than 3000 times in order to achieve a comparable final velocity with a chemical rocket.

What form might a photon rocket take? The basic equations discussed above show that the wavelength of the radiation is immaterial in the analytical sense, so the whole electromagnetic spectrum is at our disposal. However the practical aspects weigh very heavily in limiting the possibilities. In the first place, radiation is usually emitted in all directions, i.e., isotropically, and has to be directed or *collimated* into one general direction to provide controllable thrust. So the whole propulsion unit consists of a power supply, a propellant supply (i.e. photon source) and a collimator or reflector, and it is the compromise between these that limits the present effectiveness of photon devices. The high-frequency, short-wavelength end of the spectrum includes cosmic rays, gamma-rays and particles in accelerators. There does not appear to be prospect of harnessing cosmic rays, and particle accelerators are many orders of magnitude greater in mass-thrust ratio than are feasible. Nuclear reactions are a source of gamma-rays, but the very nature of radiation of this wavelength precludes its effective collimation by its power of penetrating materials. It is possible that a gaseous-core reactor, either on the fission or the fusion principle, might eventually be able to provide a directed beam of high-energy radiation by virtue of electromagnetic effects analogous to those

[1] E. Stuhlinger, "Ion Propulsion for Space Flight," McGraw-Hill, New York, 1964.

described for particle containment in Chapter 10. At the other end of the radiation spectrum lie radio waves but here the conversion efficiency for long wavelength radiation is very low indeed. In the center of the spectrum we have visible light bounded on either side by ultraviolet and infrared radiation and this is a usable region. Such radiation is produced by materials at high temperature, with frequency increasing with temperature. Thrust per unit area increases with temperature and thus high values are indicated. The ultimate limit is in the capability of the system material to conserve its structural properties at high temperature but also materials lose their reflecting or collimating ability with decreasing wavelength, which goes hand-in-hand with increasing temperature. Thus it is the complete system which must be optimized and not only one component. Within possible limits of present-day technology in this region of near-visible radiation are heated filaments, arcs and plasma jets, which can produce photons in quantity that can be directed with reasonable efficiency. It should be noted that all the energy produced in a space vehicle must eventually be dissipated as radiation, even the energy ordinarily thought of as "lost" (i.e., via the second law of thermodynamics and process inefficiencies), but this low-grade energy requires a very large area and hence mass for its dissipation. Thus although it is not lost as thrust, the thrust-weight ratio is poor compared to that of the prime device, which is already extremely low compared to conventional engines. All the thrust obtained is proportional to the power and power generation implies mass.

The immediately preceding remarks apply to propulsion by emission of radiation from the vehicle, so now let us look at the opposite situation in which the vehicle is propelled by receiving radiation. The only possible source is the sun which emits radiation with a flux of $(3 \times 10^{25})/r^2$ joule/(sec)(m^2) or about 0.125 kW/ft^2 at the earth distance. Using the latter figure and the velocity of light at 186,000 miles/sec, the radiation pressure amounts to 9.4×10^{-8} lb/ft^2. If we assume complete reflection then the thrust will be double this value or 1.88×10^{-7} lb/ft^2.

With this value of pressure, the area needed for a thrust of one pound is over five million square feet, so it is obvious that solar radiation pressure is of little value for interplanetary missions, particularly as the most interesting journeys are at greater distances from the sun than is the earth. However fractional thrusts are sufficient for many orbital maneuvers (*cf.*, the resistojet, Sec. 8.3) and radiation pressure is a possibility for some of these. There is of course the question of direction and some control is possible with a moveable reflecting surface, very much like the sail of a boat. Hence the use of the expression, *solar sail.* While the power is free, a reflecting surface has to be provided. This is simple, as aluminized Mylar (plastic sheet) about one tenth of a mil thick is suitable, the corresponding mass being about 1×10^{-3} lb/ft^2. The thrust-mass ratio for the sail material is then 1.88×10^{-4}, which is much better than the photon engine and comparable with electric rockets. The solar sail then may have a place where its simplicity appeals, but only for certain missions where the limitations of a directional, external source of energy can be utilized, yielding only microthrust.

12.2 Comparison of Propulsive Means

It is apparent that in many cases, the most effective means of space propulsion is not clear-cut. This includes the type of power plant, even if the thrust device is apparent. If assumptions are made with respect to mass-thrust ratio or specific mass α values of the many types of apparatus available, then it is possible to map out areas of optimum design. Many diagrams have appeared showing the most likely areas of applications of thrusters and power supplies, and these can be very useful. They suffer however from changes in the state-of-the-art, and can therefore be misleading unless authoritative and up-to-date. Nevertheless it may be useful to give a generalized method of preparing one type of such diagram for the thrust device to which the appropriate figures can be given as information becomes available. The basic method follows that of Littman[2] who gave specific figures valid at the time of publication. Here no attempt will be made at quantitative results, the object being to show the general method of analysis.

We start with the criterion being to minimize the combined mass of power plant and propellant, $M_w + M_p$, denoted by M_{wp}. The propellant mass, assuming constant thrust, is $\dot{m}_p t$ and hence

$$M_{wp} = M_w + M_p = M_w + \dot{m}_p t$$

which with thrust $F = \dot{m}_p I_s$ becomes

$$M_{wp} = M_w + (Ft/I_s) \tag{12.1}$$

Taking thrust and mission time as constants, differentiating Eq. 12.1 with respect to I_s and setting equal to zero yields

$$I_s \big|_{\text{opt}} = \left(\frac{Ft}{\partial M_w/\partial I_s}\right)^{1/2} \tag{12.2}$$

Now if we have available known or estimated values of power plant mass M_w and power output P_w, a curve of M_w vs P_w can be plotted. We need $\partial M_w/\partial I_s$ and first put P_w in terms of I_s, i.e.

$$P_w = \frac{P_j}{\eta_j} = \frac{\dot{m}_p c^2}{2g_c\eta_j} = \frac{Fc}{2\eta_j} = \frac{Fg_c I_s}{2\eta_j} \tag{12.3}$$

Then the steps are as follows:

(1) Assume values of F and I_s
(2) Calculate P_w, using an appropriate value of η_j for the particular type of thrust device in question
(3) Find M_w from $M_w - P_w$ curve
(4) Plot M_w vs I_s
(5) Find the slope $\partial M_w/\partial I_s$ at various values of I_s
(6) Substitute back in Eq. 12.2 and solve for t
(7) Find M_{wp} from Eq. 12.1
(8) Plot M_{wp} vs t for constant values of F, with I_s as parameter

The type of diagram that results is shown in Fig. 12.1. The heaviest solid lines indicate boundaries for the particular types of thrust device analyzed. In the

[2] T. M. Littman, "A Critical Evaluation of the Ion Rocket," 2nd Symposium on Advanced Propulsion Concepts, Boston, Mass., 1959.

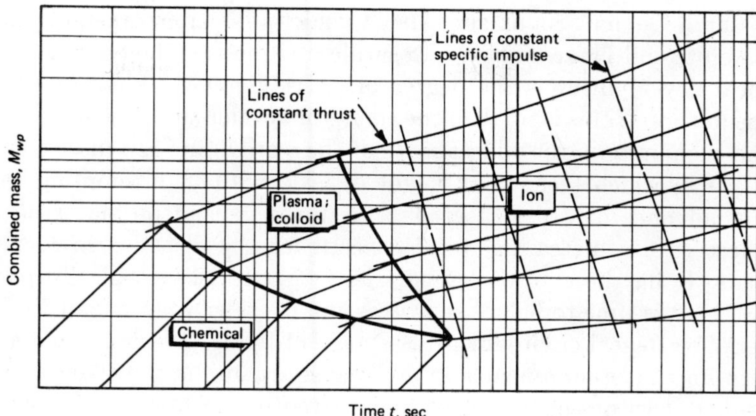

Fig. 12.1 Generalized evaluation diagram (after Littman).

illustrative figure following Littman, there are three such divisions—chemical, plasma and colloid, and ion—but this is arbitrary. The lighter solid lines are lines of constant thrust, while the broken lines are those of constant specific impulse. In this way, some initial concept is gained of the type of thrust device that is most likely to be best-suited to the mission at any given time of the existing or estimated future technology.

12.3 What of the Future?

Chapter 1 started with the premise that development in aerospace technology has been strikingly rapid, with quantum jumps of progression culminating at the end of the 1960's with a landing on the moon. What lies ahead? Steady progress or "giant steps for mankind"?

Perhaps first we should look at the stimuli which may guide us in estimating possibilities. The thirty years starting about 1940 have seen a continuous pressure to develop aerospace technology, starting with World War II as weapon technology, continuing thereafter largely by the political rivalry of the NATO countries and the U.S.S.R. The necessary ingredient of money was available, at least in the United States, with the amount expended on both airbreathing and rocket propulsion being stupendous compared with previous history. A great deal of the progress in commercial transport was made available by support for military requirements. The large passenger aircraft of the Boeing 707 and Douglas DC8 class descended from the refuelling tanker aircraft for bombers, while many turboengines in commercial transport are derated military engines or their technology derived directly from such previous experience. The key is not in the actual defense connotation but in the money made available for research, development and prototypes. In space, the effort has had to be almost entirely supported by governments, as any possible commercial return is entirely incommensurate with the cost and effort involved for any but minor projects. A mixture of defense awareness and national prestige led through the I.C.B.M's, Sputniks and Vanguards to the Mercury, Mariner, Venera,

Explorer, Luna, Gemini, Soyuz and Apollo vehicles for interplanetary probes and manned satellite and lunar missions. Regardless of motive, the essential feature of technological development is the degree of effort involved, which in our system means expenditure. If this is so, then the course of development becomes a matter of politics and economics in the widest sense of the words. One can discuss only some possible ways in which technology might develop without attempting to predict any rate of advance or order of growth, as these depend on the socioeconomic milieu. There is also increasing social concern for the direction of technological development. It has been said that anything that is technologically possible, is done, and that it is time to halt this process and consider what is best for mankind in the overall ecological environment. Shortage of funds, growing social awareness and concern for the environment may well cause a pause in the steep rate of technological progress witnessed in the last quarter-century. A brief summary of lines of development will be given with airbreathing engines and rockets considered separately.

12.4 Developments in Airbreathing Engines

The major development in complete engines would seem to be in new hybrid types for high supersonic and hypersonic flight. The hybrids may be combinations of turbojet, turbofan and ramjet, or combinations of airbreathing and rocket components. The *ram rocket* is a distinct possibility here, including a fan stage for lower speed use. The rocket acts mainly as the power source for an ejector, which essentially becomes a jet compressor, with fuel burned downstream to provide greater thrust. The fan would contribute pressure at low speeds but would not be required at high speeds when the ram effect would take over. Such a device or variations of it have possibility for use from takeoff to high supersonic speeds. Their development in the near future may depend on the acceptance or otherwise of the SST as a desirable means of transport in general use.

With respect to the various components of the turboengine, the tendency is to increased turbine temperature, higher pressure ratio and larger by-pass ratio. Higher temperatures are being made possible by blade cooling and improved materials. Values over 2000°F are already in use and will be increased steadily until stoichiometric conditions are approached. This will greatly increase specific thrust, making smaller and lighter engines possible, but at the same time the SFC will get worse on this account. Increased pressure ratio will help the SFC, while the specific thrust will be affected to only a small extent (see Figs. 4.12 to 4.15). Other factors helping to improve fuel consumption will be better compressor and turbine efficiencies and better propulsive efficiency. The component efficiencies are already high and are likely to improve only incrementally. If one takes 95% polytropic efficiency for both compressor and turbine as a reasonable aerodynamic limit, at a pressure ratio of 50/1 with $k = 1.4$ for air and $k = 1.3$ for hot gases, then the corresponding adiabatic efficiencies are about 92% for compression and 96.8% for expansion. There is more likely to be improvement in pressure ratio per stage, in greater air flow per unit area, and in stability range (range of flow between surge and choking at a given speed). The first two factors mean lighter engines and the third factor means greater flexibility and range of overall engine operation. Propulsive efficiency is improved by increase of by-pass ratio in a turbofan engine and hence the move toward higher values. Figure 4.21 shows that at $M = 0.9$ great improvement is possible up to a by-pass ratio of 6 or 7 to 1, but that thereafter the gain is marginal. This is particularly true remembering that the fan diameter must increase with by-pass ratio and hence increase weight and drag.

It is quite possible that the turboengine of the future may use hydrogen as fuel. The saving in weight is very large, as the heating value of hydrogen is about 2.7 times that of JP4. The drawback is the much lower density, about one-quarter of that of existing petroleum fuels. This is of course compensated to a large extent by the improved heating content and, if values of SFC and the aircraft aerodynamic design are improved, then the necessary tankage space may be no more than the present requirement. For operation on and near the ground, the use of hydrogen would be beneficial environmentally, as at least CO, CO_2 and intermediate hydrocarbons would be eliminated from the exhaust. Oxides of nitrogen would still be present and there is some question of the effect of large quantities of water vapor being distributed in the stratosphere where the normal convection and mixing processes are minimal.

12.5 Developments in Rockets and Rocket Vehicles

Rocket propulsion has reached the point where chemical rockets can be built in any size and operate at the design performance with almost complete reliability. Space missions, however, are all ballistic and the next era should be one of powered flight. Thus one of the most urgent needs would seem to be the perfection of a nuclear power plant and associated thrust device. Because such systems by their very nature must operate for long periods without maintenance, testing is going to be a lengthy business. Even when satisfactory ground performance is concluded, space testing will require a further period.

Another major objective would be a large orbiting laboratory, continuously manned, with capacity for testing of apparatus intended for space use. In addition of course, such a vehicle would be used for scientific and engineering purposes concerned with measurements of earth features and for space observation without the interference of the earth's atmosphere. Such a vehicle would be a mammoth affair and have to be made in sections that would be joined together in situ.

The provision of such a station or other similar continuously manned vehicles obviously requires a commuting service, i.e., reusable, shuttle vehicles for transporting men and materials on a routine basis. Such a shuttle is also required for all manner of other operations, such as regular lunar missions. At present, an Apollo lunar mission requires the main booster and two further stages to put the lunar vehicle into orbit and all these units are discarded in space immediately after the burning time is completed. The cost of the units is said to be of the order of $175 million and it is to reduce such colossal launching costs that a reusable vehicle system is required.

The shuttle scheme might consist of an Integral Launch and Reentry Vehicle (ILRV) of two stages. A piloted booster would carry a space vehicle piggyback to an earth orbit and then return to base. The spacecraft would continue its mission, say to dock at the orbiting laboratory, and then return to earth, landing in a reusable condition. Both vehicles might have wings or lifting surfaces to enable them to maneuver in the atmosphere.

An alternative is to design one vehicle that would take off, fly through the atmosphere, go into orbit, make its orbital rendevous and return. Rocketdyne has projected such a vehicle[3] with a "composite engine" (C/E), shown diagrammatically in Fig. 12.2 and comprising turbojet, ramjet, and rocket operations. Four phases of operation may be distinguished as follows:

1. *Launch phase*: *Air-Augmented Rocket Mode*. The rocket, in addition to its intrinsic thrust, induces air and acts as an ejector. The fan may supercharge the augmented air when this is required at certain operating conditions.

2. *Mid-speed phase*: *Ramjet Mode*. At low supersonic speed, ramjet operation takes over, initially with subsonic combustion but with transition to supersonic combustion at high Mach number (Scramjet mode).

3. *Final phase*: *Rocket Mode*. At a Mach number of possibly 15, the intake is shut off and the rocket restarted, with provision for more complete expansion by nozzle end addition. This phase would accelerate the vehicle to orbital speed.

4. *Flyback and landing phase*: *Fan Mode* (*Ramjet Optional*). With the mission completed, retrofiring would put the vehicle into reentry path when it would decelerate to low supersonic Mach number. Ramjet operation would be used at this stage to bring the vehicle rapidly back to near its base. Then speed would be reduced to a subsonic value and the fan would take over to give a loiter and landing capability.

The rocket and ramjet fuel would be liquid hydrogen, with rocket oxidizer being obtained from air liquefaction during the early part of the journey from launch to ramjet mode. Hydrogen at 36°R would not liquefy a stoichiometric amount of air necessary for maximum thrust, but if initially precooled to 25°R with part of the

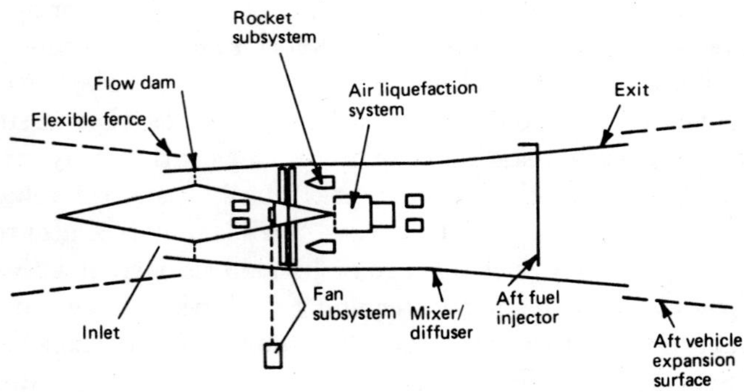

Fig. 12.2 Proposed composite engine for shuttle service—diagrammatic (after Escher).

[3] W. J. D. Escher, "Composite (Rocket/Airbreathing) Engines: Key to the Advanced (Non-Staged) Space Transport Vehicle," presented at Univ. of Tennessee Space Institute, August 1969.

hydrogen recycled back to the tank, then the process is possible. The mass ratio O_2/H_2 for combustion is 8, hence obtaining the oxidizer even as air rather than oxygen provides a significant relief of takeoff weight.

This review of one particular propulsion scheme is given as an example of the type of engine project that will become necessary for space work to develop. It combines the elements of many of the propulsion units discussed here and is a fascinating example of the wide aerothermodynamic thinking required in aerospace propulsion.

Bibliography

The following books have been useful to the author and are suggested for further reading or for an alternative presentation.

J. V. Foa, "Elements of Flight Propulsion," Wiley, New York, 1960.

P. G. Hill and C. R. Peterson, "Mechanics and Thermodynamics of Propulsion," Addison-Wesley, Reading, Mass., 1965.

R. G. Jahn, "Physics of Electric Propulsion," McGraw-Hill, New York, 1968.

H. S. Seifert and K. Brown, "Ballistic Missile and Space Vehicle Systems," Wiley, New York, 1961.

E. Stuhlinger, "Ion Propulsion for Space Flight," McGraw-Hill, New York, 1964.

G. P. Sutton, "Rocket Propulsion Elements," 3rd ed., Wiley, New York, 1963.

D. L. Turcotte, "Space Propulsion," Blaisdell, New York, 1965.

M. Vertregt, "Principles of Astronautics," 2nd ed., Elsevier, Amsterdam, 1965.

"Space Power Systems," Parts I and II, AGARDOgraph 123, reproduced by Clearinghouse for Federal Scientific and Technical Information, Springfield, Va., 1969.

INDEX

About the Author

DENNIS G. SHEPHERD has been a member of the faculty of the College of Engineering at Cornell University since 1948 and in 1965 was appointed Director of the Sibley School of Mechanical Engineering at Cornell. In addition to his many years of teaching, he has been Chief Experimental Engineer at A.V. Roe Canada Ltd. in Toronto, as well as an engineer with Power Jets Ltd. in England, engaged in the development of turbojet engines. He was awarded a Guggenheim Fellowship (1954–55) and an O.E.E.C. Senior Visiting Fellowship (1961–62) and, in addition to *Aerospace Propulsion*, has authored several other publications. Professor Shepherd is a member of A.S.M.E., the Institution of Mechanical Engineers, the Combustion Institute, and Sigma Xi, and holds a B.S. degree in Engineering (Physics) and Engineering (Mathematics) from the University of Michigan.